Short-Reach Optical Wireless Communication

IEEE Press
445 Hoes Lane
Piscataway, NJ 08854

Short-Reach Optical Wireless Communication

By Directed Narrow Beams

Ton Koonen
Full Professor Emeritus
Department of Electrical Engineering
Eindhoven University of Technology
Eindhoven, The Netherlands

IEEE Press
WILEY

Published by John Wiley & Sons, Inc., Hoboken, New Jersey.
Published simultaneously in Canada.

Library of Congress Cataloging-in-Publication Data Applied for

Hardback ISBN: 9781394300389

Cover Design: Wiley
Cover Image: © amgun/Shutterstock

Set in 9.5/12.5pt STIXTwoText by Straive, Pondicherry, India

"Numquam minus otiosus sum, quam cum otiosus."
(I am never as busy as during my leisure hours.)
after Marcus Tullius Cicero, 106–43 BC, in De Officiis III, 1, 1

Dedicated to Annemie,
Martijn, Robin, Laurens,
and my parents

Contents

About the Author *xi*
Preface *xiii*
List of Acronyms and Symbols *xv*

1 Introduction *1*
1.1 Motivation *1*
1.2 Scope of the Book *5*
1.3 Structure of the Book *7*
References *10*

2 Photonic Home Area Networks *13*
2.1 Introduction *13*
2.2 Economics of Home Networks *17*
2.3 Optical Fibers for In-Home Networks *23*
2.4 Residential Gateway *27*
2.5 High-Capacity Transmission for Wirebound Delivery of Services *27*
2.6 High-Capacity Transmission for Wireless Delivery of Services *31*
2.7 Converged Indoor Networks *36*
2.8 2D Radio Beam Steering in a Radio-Over-Fiber System *37*
2.8.1 Radio Beam Steering *38*
2.8.2 Wavelength-Controlled 2D TTD Radio Beam Steering *40*
2.8.3 Integrated 2D TTD Beam Steering Circuit *44*
2.8.4 Conclusions *45*
2.9 Dynamic Capacity Allocation *45*
2.10 Mode Group Division Multiplexing *47*
2.11 Optical Wireless Communication *48*
2.11.1 Eye Safety *48*
2.11.2 Optical Wireless Communication Using Visible Light *50*
2.11.3 Optical Wireless Communication Using Infrared Light *51*

2.12 Evolution Trends *53*
2.13 Concluding Remarks *55*
References *56*

3 Optical Wireless Communication Technologies *63*
3.1 Introduction *63*
3.2 OWC's Application Domains *65*
3.3 Indoor OWC Techniques *66*
3.4 Visible Light Communication Techniques *69*
3.5 Light Fidelity (LiFi) Techniques *71*
3.6 Beam-Steered IR Optical Communication Techniques *72*
3.7 Localization *73*
3.8 OWC Receiver *74*
3.9 Bidirectional Wireless Networks *75*
3.10 Beam-Steered OWC versus Wide-Beam OWC *77*
3.10.1 Electromagnetic Interference *77*
3.10.2 Data Capacity per Device *77*
3.10.3 Latency *79*
3.10.4 Energy Efficiency *80*
3.11 Photonic Integrated Circuits for Optical Wireless Communication *82*
3.11.1 Photonic Integrated OWC Transmitters *83*
3.11.2 Photonic Integrated OWC Receivers *86*
3.12 Concluding Remarks *89*
References *89*

4 Beam-Steered Indoor OWC System *95*
4.1 Introduction *95*
4.2 The BROWSE System Concept *97*
4.3 2D Beam-Steering Using Diffractive Means *100*
4.4 The Indoor Fiber Backbone Network *105*
4.5 Concluding Remarks *106*
References *107*

5 Beam-Steering Techniques for Indoor OWC *109*
5.1 Introduction *109*
5.2 Active Beam-Steering Devices *109*
5.3 Passive Beam-Steering Devices *111*
5.3.1 1D Steering by Diffraction Gratings *113*
5.4 2D Diffractive Beam Steerer Using Crossed Gratings *121*
5.4.1 Beam Steerer Design *122*
5.4.2 System Experiments *123*

5.4.3 Concluding Remarks *127*
5.5 2D Diffractive Beam Steerer Using an AWGR and 2D Fiber Array *128*
5.5.1 Beam Steerer Design *129*
5.5.2 System Experiments *135*
5.5.3 Concluding Remarks *141*
5.6 2D Beam Steering by Mechanical Translators *142*
5.6.1 Concluding Remarks *143*
5.7 Summarizing Conclusions *143*
References *145*

6 User Localization *149*
6.1 Introduction *149*
6.2 Localization in an Indoor Beam-Steered System *153*
6.3 Localization When Using Discrete Diffractive Beam Steering *154*
6.3.1 Corner Cube Retroreflectors *154*
6.3.2 Localizing the User Devices *163*
6.3.3 System Experiments *168*
6.3.4 Concluding Remarks *173*
6.4 Localization When Using Analog Beam Steering *173*
6.4.1 Center-of-Gravity Algorithm *174*
6.4.2 Impact of Lens Aberrations on Beam Alignment Accuracy *175*
6.4.3 System Experiments *181*
6.4.4 Concluding Remarks *183*
6.5 Summarizing Conclusions *183*
References *183*

7 Broadband OWC Receiver with Wide Field-of-View *187*
7.1 Introduction *187*
7.2 Receiver Based on a Matrix of Photodiodes *189*
7.2.1 Frequency Characteristics *192*
7.2.2 Bias Conditions in the Photodiode Matrix *193*
7.2.3 SNR Performance *198*
7.2.4 Comparison with Angular Diversity Receiver *199*
7.3 Capturing the Optical Beam by the Photodiode Matrix *201*
7.3.1 Theoretical Analysis of the Beam-to-Photodiode Matrix Coupling *202*
7.3.2 Numerical Analysis of the Beam-to-Photodiode Matrix Coupling *204*
7.3.3 Impact of the Fill Factor of the Photodiode Matrix *208*
7.3.4 Shifting the Receiver's Aperture over the Beam's Footprint *211*
7.3.4.1 Impact of Crosstalk from Neighboring Beams *211*
7.3.4.2 Impact of Cropping the Beam's Spot by the Lens' Aperture *218*
7.4 Implementing the Receiver *220*

7.5 Experimental Validation *223*
7.6 Scaling the PD Matrix *225*
7.7 Concluding Remarks *226*
References *227*

8 Bidirectional All-Optical OWC Indoor System *229*
8.1 Introduction *229*
8.2 Hybrid OWC Downstream/RF Upstream Systems *230*
8.3 All-Optical Bidirectional OWC Systems *233*
8.4 Bidirectional OWC System Architecture *234*
8.5 Upstream Optical Path Design *235*
8.6 System Demonstrator *238*
8.7 Conclusions *243*
References *243*

9 Conclusions and Outlook on Further Research *247*
9.1 Conclusions *247*
9.2 Outlook and Suggestions for Further Research *250*
9.2.1 Applying the OWC Key Functions in Other OWC Domains *251*
9.2.2 Photonic Integration *251*
9.2.3 Next 2D Beam-Steering Concepts *251*
9.2.4 Next Localization Schemes *253*
9.2.5 Next OWC Receiver Concepts *253*
9.2.6 New OWC Networking Aspects *254*
References *258*

Appendix A: 3D Ray Tracing *261*
Index *283*
Publications of the Author *291*

About the Author

Ton (Antonius Marcellus Jozef) Koonen was born in Oss on Oct. 20, 1954. He received an MSc cum laude in Electrical Engineering from the Eindhoven University of Technology (TU/e) in 1979. He subsequently worked in applied research at Philips' Telecommunicatie Industrie (1979–1984), AT&T Network Systems (1984–1994), and Lucent Technologies–Bell Laboratories (1994–2000), initially as a Member of Technical Staff and since 1987 as a Technical Manager. Ton was a part-time professor in the Telecommunication Systems group at the University of Twente (1991–2000). He was appointed as a full-time professor of "Breedbandige netwerken" (Broadband Networks) in the Department of Electrical Engineering (EE) at TU/e on Jan. 1, 2001. Ton chaired the Electro-Optical Communication Systems group (2004–2021) in the EE department. He also was vice-dean of the EE department (2012–2020) and Scientific Director of the Institute for Photonic Integration (formerly COBRA; 2016–2019). Ton officially retired on Feb. 19, 2021. After his retirement, he continued research and acquired a PhD in Dec. 2023 with a thesis on optical wireless communication.

Ton is a Bell Labs Fellow (1998, the first one in Europe), IEEE Fellow (2007), Fellow of OPTICA (formerly OSA, the Optical Society of America, nowadays OPTICA; 2013), and High-Level Visiting Scientist at Beijing University of Posts and Telecommunications (111 Program, 2018–2023). He was a principal investigator in the Dutch Gravitation programs "Center for Integrated Nanophotonics" and "Networks" (both running from 2013 to 2024). He is a co-initiator and co-leader of the Dutch NWO Perspective program "FREE – Optical Wireless Superhighways" (Sep. 2021–Aug. 2026). Ton received the Dutch ICT Regie Award in 2009. In 2012, he acquired the prestigious Advanced Investigator Grant of the European Research Council on optical wireless communication, followed by a

Proof-of-Concept Grant in 2018. With this research, Ton was one of the six finalists in the national Huibregtsen prize competition in 2018. At his valedictory lecture on Sep. 24, 2021, he received the honorable Dutch Royal distinction "Ridder in de Orde van de Nederlandse Leeuw" ("Knight in the Order of the Dutch Lion").

Ton Koonen initiated and managed a wide range of national and international projects in optical systems research and secured more than € 30M in funding. He has authored and co-authored more than 750 conference and journal papers, with 13,531 citations, and h-index of 54 (according to Google Scholar, Feb. 20, 2025), and he filed/holds 13 patents. His research interests include optical fiber communication networks, fiber-to-the-home, indoor networks, spatial division multiplexing, radio-over-fiber techniques, and most recently optical wireless communication.

Preface

Optical wireless communication (OWC) brings a number of distinct advantages beyond radio-based wireless communication (such as 4G and 5G mobile, Wi-Fi): the visible and near-infrared (near-IR) optical spectrum offer orders of magnitude more bandwidth than the radio frequency (RF) spectrum (even in the THz frequency bands), crosstalk between users is eliminated by using narrow beams and/or opaque walls, neither the connection is hampered by electromagnetic interference nor does it generate EMI itself, and by steering the narrow beams the energy can be brought only to those places where and when needed, thus yielding high energy efficiency. The small footprints of the beams also support a high user density. By deploying near-IR light such as commonly used in fiber-optic communication, relatively high beam powers (up to 10 mW at wavelengths beyond 1.4 μm) can be used without eye safety issues, and OWC systems can build on the vast set of mature optical modules already widely used in fiber-optic networks. In contrast to systems using widely diverging beams (such as LiFi systems), systems using narrow (nearly) collimated beams ensure a good link power budget that enables high-capacity connections with a longer reach at high user densities and with enhanced privacy.

This book surveys research efforts in high-capacity, beam-steered OWC systems, zooming in on those using narrow two-dimensionally (2D) steered infrared beams for indoor (/ short-to-medium range) service delivery. After discussing the indoor OWC network architecture and its design aspects, it summarizes the key functionalities required, gives a non-exhaustive concise review of the research efforts for these functionalities reported in the literature, and presents the novel solutions we propose for their realization:

- *2D steering of multiple narrow beams by means of a passive diffractive unit* (containing a pair of crossed gratings, or alternatively a high-port-count arrayed waveguide grating router) and remotely tuning the wavelength of each beam's signal.

- *Accurate and autonomous localization of the OWC receivers* in order to aid the beam steering for downstream and upstream communication, by means of passive optical retro-reflector techniques and single-sided beam scanning algorithms without the need for a pre-existing return channel to confirm adequate beam alignment.
- *Broadband wide aperture and wide Field-of View (FoV) energy-efficient optical receivers,* which are based on scalable two-dimensional matrices of photodiodes followed by a single preamplifier.

The solutions have been designed, analyzed both analytically and numerically, and implemented, and first implementations have been experimentally validated in laboratory demonstrator systems, culminating in *bidirectional optical wireless links* with an FoV of up to ±10° (full angle) that can bring data at Gigabit Ethernet speeds to up to 128 densely spaced users individually, e.g., for real-time streaming of high-definition video content.

Gratefully I would like to acknowledge the funding received from the European Commission, in particular in the Advanced Investigator and Proof of Concept personal grant programs of the European Research Council (ERC), and the funding from the Dutch Government, in particular in the Gravitation program of the Dutch Research Council (NWO). The funding enabled the research work described in this book.

The work could not have been done without the invaluable support of my colleagues at the Eindhoven University of Technology. In particular, I would like to thank Dr. Ketema Mekonnen for his great technical support and Dr. Eduward Tangdiongga for his valuable discussions, as well as the PhD students who I had the privilege to guide for their great work in the exciting domain of OWC: Dr. Zizheng Cao, Dr. Joanne (Chin Wan) Oh, Dr. Xuebing Zhang, Dr. Ngoc Quan Pham, Dr. Yu Lei. Furthermore, I thank the postdoctoral researchers Dr. Fausto Gomez Agis and Dr. Amir Khalid, and technician Mr. Frans Huijskens for their valuable work. I am also indebted for the careful reading and the scrutinizing of my text by Prof. Dominic O'Brien, Prof. Volker Jungnickel, and Prof. Sonia Heemstra de Groot. I owe very much to my father and mother, who supported me unconditionally. And last but certainly not least, my gratitude and love goes to my "homefront": my three sons and daughters-in-law Martijn and Annemarie, Robin and Eva, Laurens and Misha and our little grandson Quin; and above all my wife Annemie for all their support and endless patience.

Ton Koonen

List of Acronyms and Symbols

Acronyms

2D	two dimensional
ADR	angular diversity receiver
AoA	angle of arrival
ASE	amplified stimulated emission
AWGR	arrayed waveguide grating router
BER	bit error ratio
BFL	back focal length
BROWSE	Beam-steered Reconfigurable Optical-Wireless System for Energy-efficient communication (ERC Advanced Grant project)
BS-ILC	beam-steered infrared light communication
BW	bandwidth
CapEx	Capital Expenditures
CC	corner cube
CCC	central communication controller
CMOS	complementary metal oxide semiconductor
CoG	center of gravity
CPC	compound parabolic concentrator
CWDM	coarse wavelength division multiplexing
CW	continuous wave
DBR	distributed Bragg reflector
DMT	discrete multi-tone (modulation)
DS	downstream
DSB	double side-band
DWDM	dense wavelength division multiplexing
EDFA	Erbium-doped fiber amplifier
EFL	effective focal length
EMI	electro-magnetic interference

ERC	European Research Council
FoV	field of view
FP	Fabry Perot (cavity)
FSR	free spectral range
FttH	fiber-to-the-home
FttR	fiber-to-the-room
FWHM	full width at half maximum
GPON	gigabit passive optical network
GEPON	Gigabit Ethernet Passive Optical Network
HD	high definition (video)
IEEE	Institute of Electrical and Electronics Engineers
IM	intensity modulation
IoT	Internet of Things
IR	infrared
ISM	Industrial, Scientific and Medical (frequency bands)
ITU	International Telecommunication Union
LD	laser diode
LED	light-emitting diode
LiFi	light fidelity
LO	local oscillator
LoS	line of sight
MD	mobile device
MDU	multi-dwelling unit
MEMS	micro electro-mechanical system
MGDM	mode group division multiplexing
MIMO	multiple input multiple output
MMF	multi-mode fiber
MZI	Mach Zehnder Interferometer
MZM	Mach Zehnder Modulator
NA	numerical aperture
NLoS	non-line-of-sight
NPV	net present value
NRZ	non-return-to-zero (modulation)
NWO	Nederlandse Organisatie voor Wetenschappelijk Onderzoek (Netherlands Organization for Scientific Research)
O/E	optical-to-electrical (conversion)
OOK	on–off keying (modulation)
OPA	optical phased array
OpEx	operating expenses
OWC	optical wireless communication

OXC	optical cross-connect
PAA	phased array antenna
PAM	pulse amplitude modulation
PD	photodiode
PDL	polarization-dependent loss
PIC	photonic-integrated circuit
PMMA	polymethylmethacrylate
POF	plastic optical fiber
PON	passive optical network
PRA	pencil radiating antenna
PRBS	pseudo-random bit sequence
P2P	point-to-point
P2MP	point-to-multipoint
QAM	quadrature amplitude modulation
RAP	radio access point
REAM	reflective electro-absorption modulator
RF	radio frequency
RG	residential gateway
RoF	radio over fiber
RR	retro-reflector
RSS	received signal strength
Rx	receiver
SDM	spatial division multiplexing
SGC	surface grating coupler
SLM	spatial light modulator
SMF	single-mode fiber
SOA	semiconductor optical amplifier
SSB	single sideband
TDM	time division multiplexing
TDMA	time division multiple access
TIA	transimpedance amplifier
TRL	technology readiness level
TRx	transceiver
TTD	true time delay
Tx	transmitter
US	upstream
USP	unique selling point
UTC	uni-traveling-carrier (photodiode)
UWB	ultra-wideband (radio)
VCSEL	vertical cavity surface emitting laser diode
VIPA	virtually imaged phased array

VLC	visible light communication
WDM	wavelength division multiplexing
Wi-Fi	radio-based wireless connection standard
XG-PON	10-Gigabit-capable passive optical network

Symbols

a_{CC}	side length of corner cube
α	incidence angle (of optical beam)
$\nabla\lambda_{\mathrm{tun}}$	wavelength tuning range
$\nabla\lambda_{\mathrm{FSR}}$	wavelength Free Spectral Range
η_A	fill factor of area
η_{AP}	fraction of number of rays captured by an aperture
η_{antenna}	radiation efficiency of antenna
η_{CC}	fraction of incident power retroreflected by corner cube
η_{matrix}	fill factor of photodiode matrix
λ	wavelength (of light, or radio wave)
φ, ψ	beam steering angles (in two dimensions)
ω	radial frequency
c_0	speed of light (in vacuum)
C_d	capacitance of photodiode
D_{A}	directivity (of an antenna)
D_{beam}	beam diameter
D_{lens}	lens diameter
f	focal length (of a lens)
G_{A}	antenna gain
H	room height
K	numbers of rays in a beam (in ray tracing)
K, M	dimensions of matrix (of photodiodes)
L	length and width (of square area covered by optical beam steerer)
M, N	dimensions of matrix (of fibers in AWGR optical beam steerer)
m	order of interference
p	defocusing parameter
P_{beam}	power of a beam
$\mathcal{R}$	responsivity (of a photodiode)
R_d	resistance (parallel resistance in photodiode)
R_s	resistance (serial resistance in photodiode)
T_{eq}	equivalent beam-to-photodiode power coupling factor
w_0	mode field radius (of Gaussian beam at its minimum width)

1

Introduction

1.1 Motivation

Since its first introduction in field trial networks in the early 1970s, optical fiber has rapidly become the medium of choice for data transport in communication networks, starting with long-distance links, then in regional and metropolitan networks, and in recent years conquering the last kilometers to the home. Fiber widely outperforms copper-based cables by its very low losses and low dispersion and has pushed the transport capacity of communication networks to great heights. Moreover, its wavelength dimension provides a multiplicator for transport capacity, as well as extra traffic routing functionalities. Recently, exploiting the spatial dimension in few-moded fiber has raised the capacity and flexibility even further.

The powerful Internet as we know it today would not have been possible without fiber. Fiber has already arrived at the doorstep of many households. Extending its potential into the home, however, has not truly happened. The outdoor fiber highway is handed over to a variety of copper- and radio-based indoor networks, and only part of the fiber's high-bandwidth potential is experienced by indoor users.

On the other hand, there is a fast rise in capacity demand by the users and their devices. The variety of services offered by the Internet is booming, encompassing tele-shopping, tele-education, teleconferencing, entertainment, ultrahigh-definition video streaming, large file transfer, and many more; the traffic volume they generate is enormous (Fig. 1-1). The number of users of the Internet is increasing fast (Fig. 1-2), and most of them access the Internet by wireless devices (such as laptops and smartphones). In particular, video-based services require a large share of the capacity of the home access network. Moreover, there are also intra-home data transport needs, such as posed by the transport of broadband data streams and large files between in-home laptops, tablet computers, smartphones, video and audio

Short-Reach Optical Wireless Communication: By Directed Narrow Beams, First Edition. Ton Koonen.

Fig. 1-1 What happens on the Internet worldwide in 2023 in every minute (*source:* eDiscovery Today & LTMG)

sets, network-attached storage (NAS) units, scanners, printers, etc., as well as by transport of bursty, low-volume data between various home devices that are part of the Internet of Things (IoT; see Fig. 1-3). The volume of intra-home traffic may exceed the traffic volume going into/out of the home, making the load on the in-home network even heavier.

Hence, following their successes in the public network, photonic solutions should also be considered for the indoor network. Moreover, the use of multi-wavelength technology can enable the convergence of different service families within a single fiber infrastructure, which saves indoor network installation costs.

The users increasingly prefer wireless service delivery instead of wire-bound delivery, thus getting more freedom in their mobility and reducing installation efforts of the indoor physical cabling infrastructure. Radio-based wireless indoor networks are widely deployed, in particular Wi-Fi based networks that are

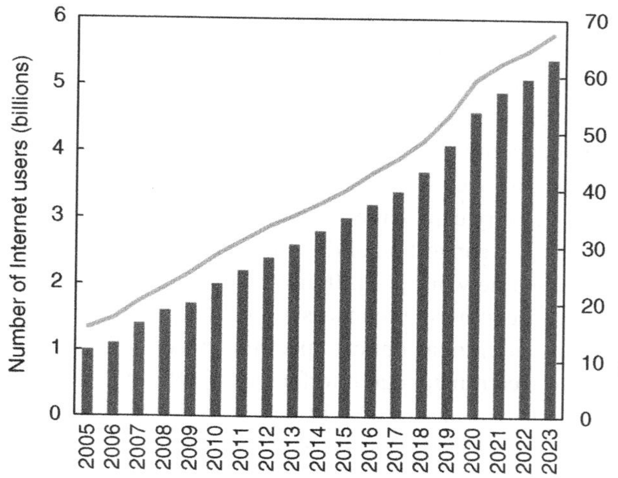

Fig. 1-2 Growth of the number of individuals using the Internet (*source:* ITU, Dec. 2022)

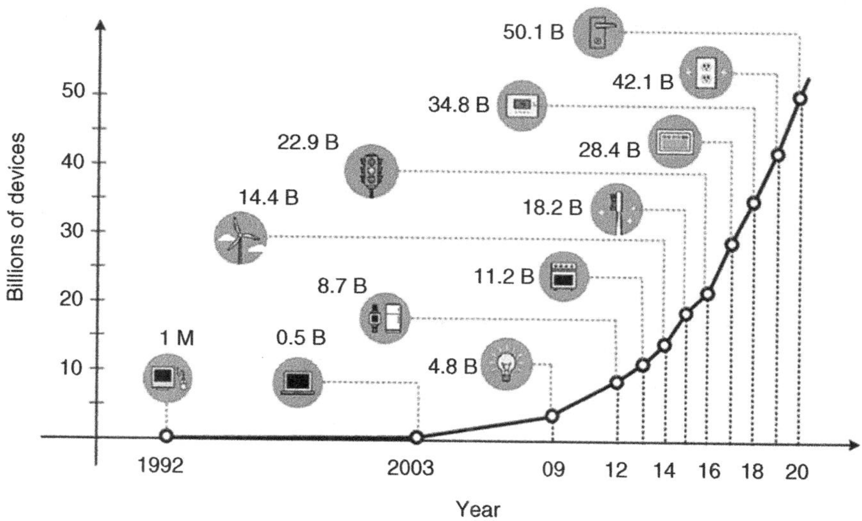

Fig. 1-3 Growth of the Internet of Things (*source:* NCTA)

operating in the 2.5 and 5 GHz bands. The use of mm-wave and microwave technologies (e.g., 60 GHz and sub-THz) is being considered for providing extra bandwidth and for creating pico-cell network architectures in order to increase the network's aggregate capacity by spatial multiplexing. Radio-over-fiber for such high frequencies can combine the best of two worlds, and photonic integrated circuitry enables powerful microwave signal processing capabilities [1-1, 1-2]. The load on radio-based networks is booming, fueled by the growth of the number of IoT devices, as well as the increasing bandwidth needs of broadband services. This growth is leading to radio spectrum congestion and incidentally, even resulting in service denials. Optical wireless communication (OWC) by means of dedicated free-space optical technologies can open a huge amount of additional spectrum and enhance wireless service delivery capabilities in many domains [1-3–1-5].

Indoors, OWC can counteract the congestion of the radio spectrum and offload much traffic from the radio-based network [1-6]. It also offers enhanced privacy and reduces crosstalk from/into neighboring rooms as it does not penetrate walls. Another advantage is that the additional optical spectrum needs no licenses. Moreover, OWC is immune to electromagnetic interference (EMI) and does not generate EMI itself. OWC, however, needs line of sight (LoS), which Wi-Fi technology does not. In OWC, even with nondirective receivers, LoS is needed, as the reception of LoS signals is much stronger than that of diffuse non-LoS signals. When using mm-waves (with frequencies in the range of 30–300 GHz), LoS is also needed, as these waves need directional antennas with high gain for establishing the transmission link power budget according to Friis' transmission equation [1-7]

$$\frac{P_R}{P_T} = G_T G_R \left(\frac{\lambda}{4\pi d}\right)^2 \tag{1-1}$$

where P_T is the power fed into the transmitter antenna terminal, P_R is the power out of the receiver antenna terminal, G_T and G_R are the gain of the transmitter and receiver antennas, respectively, d is the length of the mm-wave propagation path, and λ is the (short) wavelength of the mm-wave. Wi-Fi typically uses frequency bands in the 2.5 and 5 GHz region and thus has lower radio path losses than mm-waves; these Wi-Fi frequencies can even penetrate through walls (which compromises privacy). By offloading the broadband traffic, the OWC network creates adequate room for the radio wireless network to handle a large amount of lower volume (bursty) traffic flows, particularly the traffic for IoT devices, which may need to operate in non-line-of-sight (NLoS) conditions [1-8].

Outdoors, OWC can also bring a range of benefits and can offer relief for overloading of the radio spectrum and other issues. Unlike in indoor networks, atmospheric effects (such as fog, rain, snow, and air turbulence) can hamper the optical path, causing loss and phase fluctuations in it (scintillation [1-9]). For the particular case of underwater communication, radio transmission is not feasible and compared to sonar, OWC can provide transmission at higher data rates which, for example, is useful for video inspection and controlling unmanned underwater maintenance/repair robots [1-10]. While the short reach of OWC links for indoor applications typically ranges from a few meters to some tens of meters, the range can be considerably longer for outdoor applications. Medium reach free-space optics (FSO) applications include fixed optical wireless access to remote buildings, backhaul links for mobile telephony base stations, mobile communication on the road between vehicles and vehicle-to-roadside communication, street-level communication, inter-drones communication, communication in large outdoor areas such as sports stadiums and outdoor festivals. Examples of long to ultra-long FSO applications are ground-to-airplane communication, ground-to-satellite communication, and inter-satellite communication [1-11]. Medium- to ultra-long optical wireless links can be established with OWC technologies deploying narrow (nearly) collimated optical beams that are carefully steered to their receiver destinations. The reach of wide, diverging beams is strongly power limited. The steering of narrow beams needs to be highly accurate and their divergence must be very small, especially for establishing long-reach links. This requires sophisticated beam-steering and beam-shaping techniques. On the other hand, the Field-of-View (FoV) of the receiver typically does not need to be as large as for short-reach indoor communication, as long-reach links typically use only a single beam (or very few beams) and the incident angle at the receiver varies only over a tiny range.

1.2 Scope of the Book

The research described in this book focused on OWC system architectures and techniques to be applied in the domain of short-reach indoor communication. The goal is to enable the delivery of broadband services wirelessly to indoor users and devices at high user densities while complementing radio-based wireless techniques such as the IEEE 802.11 Wi-Fi suite. Such an OWC system should enable the wireless extension of the data highway capabilities of (fiber) access networks to users and devices individually, as well as support intra-building high-capacity data communication.

This research particularly focused on the architecture and enabling techniques for indoor short-reach OWC by means of narrow two-dimensionally (2D)-steered beams, which can bring the powerful functionalities of optical fiber wirelessly. Next to the general advantages of OWC (such as no need for spectrum licensing and the adding of the ample bandwidth available in the optical spectrum), beam-steered infrared light communication (BS-ILC) can bring a number of particularly powerful advantages with respect to radio-based wireless communication and wide-beam OWC:

- It can offer very high (fiber-like) capacity to users individually. The transport performance of a narrow infrared beam may theoretically even exceed that of an optical fiber, as, unlike a fiber, an optical beam is not affected by waveguide dispersion. As the beams are highly directive and each follow a single path, no multipath effects and their associated fading occur. The larger bandwidth of the beam link may strongly reduce the signal processing efforts required for compressing the signal's bandwidth, thereby reducing the latency time incurred during that processing. Moreover, the latency of the beam link is about 33% smaller than that of an equally long silica fiber link due to the refractive index of the fiber core glass (which is around 1.5). The latency reduction increases the link capacity in bidirectional operation, such as in machine-to-machine communication.

- High user densities. Due to the small footprint of a beam, very small picocells can be created, which enables dense spatial division multiplexing and thus allows narrow spacings between the users.
- A high privacy and security level. Light does not penetrate walls, and the small footprint ensures negligible crosstalk with nearby users.
- High-energy efficiency. The beams are directed on demand, only to those places where and when they are needed.
- Immunity for externally generated EMI, and it also does not generate EMI itself.
- It can build on mature optical device technologies already readily available and widely used in optical fiber networks.

BS-ILC also has a few disadvantages:

- LoS is needed for a beam to reach a user. Workarounds can be devised, such as applying multiple spatially separated beam steerers, which can create alternative paths and may be arranged in a multiple-input multiple-output (MIMO) setting, or by adaptively relaying the signal by reflective surfaces, at the cost of comprehensive control measures and signal processing.
- Accurate beam steering and user tracking are needed, which takes time and compromises the mobility of the user. Therefore, the BS-ILC is mainly intended to serve nomadic and static user devices that do not require connectivity while moving. This operation mode is often applicable in the delivery of broadband services such as high-definition video streaming and the exchange of large data files.
- The beam steering and tracking also make the higher layer protocols more complex; extra room (either in-band in the time domain or out-of-band in the frequency domain) is needed to coordinate the network control plane with the data plane.
- When using multiple beam steerers, the beam steering needs to be coordinated among themselves to avoid interference; this requires additional protocols.

1.3 Structure of the Book

The research described in this book focuses on the studies done to bring the benefits of optical communication into the indoor domain for individual users, both in their professional work environment and their residential environment. These findings have largely been published in a range of journal and conference papers and have also been laid down in a number of patents. The adjective "indoor" when used in combination with communication networks is used to distinguish it from "outdoor" networks, that is, in an environment not governed by strict regulations of standards and largely set up by the user(s) according to their personal wishes, not hampered by atmospheric effects, and typically involving short reach links (up to a few tens of meters).

Chapter 2 introduces the potential of optical technologies to enhance the performance of indoor communication networks, and among these technologies, the research in OWC described in this book. It discusses how photonic technologies can be introduced in indoor networks in order to bridge the gap between the capacity offered by the optical fiber-to-the-home (FttH) access network and the capacity accessible for the users, as well as to improve the intra-home network performance. Indoor networks have to carry a large variety of services, with widely differing needs regarding for instance bandwidth, quality of service, and reliability. They have to support both wired and wireless connectivity, at low costs and low energy consumption levels. The chapter reviews the architectures, the economics (both from an installation and an operational viewpoint), and the techniques for converged optical fiber indoor networks that are cost- and energy-efficient, and compares them with current copper-based solutions. Particular attention is given to high-capacity multimode (plastic) optical fiber techniques, radio-over-fiber techniques including adaptive radio beam steering, techniques for providing capacity on demand, and OWC techniques. An evolution roadmap is outlined that describes how the growing indoor communication demands can be met by gradually introducing these powerful techniques and network architectures.

Chapter 3 zooms in on OWC as the main topic of this book. OWC offers powerful solutions for resolving the imminent capacity crunch of radio-based wireless networks. OWC is not intended to fully replace radio wireless techniques such as Wi-Fi, but to rather complement these and off-load their high traffic loads. After discussing OWC's strengths in various application domains, a non-exhaustive tutorial overview of two major directions in OWC is provided: firstly, wide-coverage visible light communication (VLC) and LiFi, which are built on LED illumination techniques and share capacity among multiple devices, and secondly, communication with narrow two-dimensionally steered infrared beams that offer the unshared high capacity to devices individually. The latter constitutes the main area of research addressed in this book. A non-exhaustive review of the research on key techniques supporting beam-steered OWC is given: active and passive beam-steering techniques, device localization, and wide Field-of-View (FoV) receivers. Their application in bidirectional hybrid optical/radio networks and bidirectional all-optical wireless networks is discussed. The performance characteristics of beam-steered OWC are compared with those of wide-beam OWC (encompassing VLC and LiFi) and of the radio-based IEEE 802.11 systems (the Wi-Fi suite), regarding sensitivity for EMI, data capacity available per device, latency, and energy efficiency aspects. Photonic integration will become a key enabler for the introduction of indoor OWC; hence, a non-exhaustive review of photonic integrated circuit (PIC) solutions for OWC beam-steered transmitters and receivers is also given.

Chapter 4 outlines the architecture we propose for indoor broadband beam-steered OWC systems, termed the "BROWSE – Beam-steered Reconfigurable Optical-Wireless System for Energy-efficient communication" architecture. The architecture comprises a broadband fiber backbone network that feeds optical data streams to the rooms, in which they are conveyed by optical free-space beams to the individual users/devices. The BROWSE network design guidelines are derived, which serve as the basis for the design of the OWC key functionalities (such as beam steering, localization, and wide Field-of-View reception). The architecture is intended to provide broadband services wirelessly on demand where and when needed to individual users, tailored to their wishes. By using passive modules for the beam steering at the ceiling sites (the network access points), it intends to centralize the active service delivery functions and control functions as much as possible and to locate these remotely from the ceiling sites, in order to ease upgrading and maintenance and reduce energy consumption.

Chapter 5 focuses on techniques for adaptive indoor 2D beam steering to multiple users individually, even when they are not stationary. For downstream communication, beams directed at the users are emitted from a common ceiling unit in a point-to-multipoint scheme. Diffractive techniques are investigated where the wavelength of each data signal fed to the beam steerer determines in which 2D direction the associated beam is steered. Two options are analyzed in depth: using a pair of crossed diffraction gratings that enables semi-continuous steering and using an arrayed waveguide grating wavelength demultiplexer of which the output ports are two-dimensionally arranged in front of a lens shaping the beams enabling discrete steering. Both options include only passive elements that minimize maintenance efforts and failure risks and are readily scalable to serve more users individually by just feeding more wavelengths. Moreover, the signals can be generated remotely by tunable laser diode transmitters, which greatly facilitate system upgrades. For the upstream OWC link, continuous beam steering deploying a 2D electromechanical translator stage is investigated. Laser diodes with an arbitrary wavelength can be used, which is attractive for the cost-sensitive user terminal; there, diffractive steering is less attractive as wavelength tunable sources and their control entail higher costs.

Chapter 6 deals with techniques to localize the user, which is a key requirement for the automatic operation of indoor OWC networks. Accurate device localization for steering the beams is required. Self-calibrated localization has been investigated by beam scanning and deploying optical retro-reflector technologies at the target device, particularly foils containing matrices of miniature corner cube reflectors, which enable monitoring of the reflected power of the incident scanning beam. This approach avoids the need for a pre-existing return path, which eases the set-up of an OWC link per user. Also, parallax errors are avoided that can occur when the localization aperture and the data signal reception aperture are physically separated; the compensation of such errors would need a pre-established calibration process. For downstream, localizing each user individually is done by scanning with a relatively wide downstream beam in order to reduce the time needed for scanning the full user area in conjunction with monitoring the reflected power from the retro-reflecting foil at the user's device. For upstream, in order to save battery power and safegard eye safety on the user side, each user device launches a narrow beam at low power that fits within the aperture of the ceiling's receiver. By means of beam scanning and a center-of-gravity algorithm, accurate upstream beam alignment is achieved. The design considerations, performance analysis, implementation, and experimental validation of both localization methods are discussed.

Chapter 7 discusses the research efforts to create a broadband OWC receiver with large aperture as well as large Field-of-View. Such characteristics are attractive for avoiding tedious alignment efforts of the receiver with respect to the beam steerer. The approach proposed involves a novel concept of arranging identical high-speed photodiodes in a 2D matrix and interconnecting them in such a way that the bandwidth of the photodiode matrix receiver equals the bandwidth of an individual photodiode, whereas the optical reception area is increased. When applying the matrix and carefully designing the receiver's lens collecting the beam's power, the attainable receiver's performance enhancement in terms of Field-of-View and captured optical power has been assessed both analytically and numerically using extensive ray tracing. The studies include the tolerance against moving the receiver across the footprint of a beam, the signal-to-noise performance in comparison with other OWC receiver concepts, and the impact of upscaling the photodiode matrix. Implementation of the receiver's concept and its validation in a laboratory testbed are discussed.

Chapter 8 discusses the design and implementation of a bidirectional all-OWC system. Hybrid bidirectional systems have been reported that deploy downstream optical links and radio frequency (RF) upstream links, in which part of OWC's key advantages such as EMI immunity may be lost. Alternatively, all-optical bidirectional systems have been reported that duplicate the downstream technology and reuse it for the upstream link. This approach does not align well with the more cost- and space-effective point-to-multipoint (P2MP) physical architecture according to BROWSE, where a single ceiling unit serves multiple users. We propose a P2MP architecture for a bidirectional all-OWC system that builds on the key functions we designed, analyzed, and realized in Chapters 5–7. We validated its performance in our laboratory OWC system testbed and demonstrated real-time streaming of high-definition video signals at Gigabit Ethernet (GbE) speed to closely spaced users individually that deploy laptop computers.

Chapter 9 summarizes the main findings of this book and provides an outlook on how further steps can be made toward elevating the bidirectional broadband beam-steered OWC system to higher technology-readiness levels (TRLs).

Appendix A describes the methods for three-dimensional ray tracing that the author has developed for modeling the propagation of rays in homogeneous media, the refraction of a ray at a plane and at a spherical surface boundary between two media with different refractive index, and the reflection of a ray at a surface boundary. Thus, ray propagation through (multi-element) spherical lenses can be analyzed, including the occurrence of various lens aberrations. Also, a ray tracing model has been developed for a Fresnel lens with large numerical aperture that fully corrects for spherical aberration. Moreover, a ray tracing model for a matrix of corner cube retro-reflectors has been built. These methods have been implemented for numerical processing in MATLAB programs and have been extensively used for analyzing the imaging of the received beams onto the photodetector and thus analyzing the performance of the OWC receiver concepts presented in Chapter 7. They also have been used for analyzing the localization concepts using a matrix of miniature corner cube retro-reflectors presented in Chapter 6. The ray tracing programs actually enable us to do "virtual lab experiments" in which the parameters of the optical modules can be changed easily and thus provide valuable support for optimizing the design of an OWC system.

References

[1-1] J. Capmany, and D. Novak, "Microwave photonics combines two worlds," *Nat. Photonics*, vol. 1, June 2007, http://www.nature.com/naturephotonics

[1-2] J. Capmany, B. Ortega, and D. Pastor, "A tutorial on microwave photonic filters," *J. Lightwave Technol.*, vol. 24, no. 1, pp. 201–229, Jan. 2006, https://doi.org/10.1109/JLT.2005.860478.

[1-3] H. Haas, J. Elmirghani, and I. White, "Optical wireless communication," *Philos. Trans. R. Soc. A*, vol. 378, no. 2169, 11pp, Mar. 2020, https://doi.org/10.1098/rsta.2020.0051.

[1-4] D.C. O'Brien, S. Rajbhandari, and H. Chun, "Transmitter and receiver technologies for optical wireless," *Philos. Trans. R. Soc. A*, 21pp, Mar. 2020, https://doi.org/10.1098/rsta.2019.0182.

[1-5] A.M.J. Koonen, "Optical wireless systems: technology, trends and applications," *Webinar IEEE Phot. Soc.*, Feb. 21, 2018, https://www.photonicssociety.org/education-careers/webinars/optical-wireless-communication-webinar

[1-6] A.M.J. Koonen, K.A. Mekonnen, Z. Cao, et al., "Ultra-high capacity wireless communication by means of steered narrow optical beams," *Philos. Trans. R. Soc. A*, 19 pp, Mar. 2020, https://doi.org/10.1098/rsta.2019.0192.

[1-7] Friis' transmission equation for wireless links [on-line]. Available: https://en.wikipedia.org/wiki/Friis_transmission_equation

[1-8] J.P. Linnartz, C.R.B. Corrêa, T.E.B. Cunha, et al., "ELIoT: New features in LiFi for next-generation IoT," *Proc. ECOC2021*, Bordeaux, Sep. 2021, Paper WO5.

[1-9] A. Ahmed, S. Gupta, Y. Luthra, et al., "Analysing the effect of scintillation on free space optics using different scintillation models," *Proc. 6th Int. Conf. Signal Processing and Integrated Networks (SPIN)*, Noida, Mar. 7–8, 2019, pp. 799–804, https://doi.org/10.1109/SPIN.2019.8711574.

[1-10] T. Ke Wang, Y. Song, C. Wang, et al., "Evolution of short-range optical wireless communications," *J. Lightwave Technol.*, vol. 41, no. 4, pp. 1019–1040, Feb. 2023, https://doi.org/10.1109/JLT.2022.3215590.

[1-11] Laser satellite communication, cf. [online]. Available: https://www.tno.nl/en/sustainable/space-scientific-instrumentation/satellite-communication/laser-satellite-communication

2

Photonic Home Area Networks[1]

2.1 Introduction

Home area networks, unlike many other communication networks, are a domain where optical fiber has not made significant inroads yet. Fiber has arrived at the doorsteps of our homes in fiber-to-the-home (FttH) access networks, but its huge capabilities are yet to be extended up to the users' devices inside the home [2-1]. Current home networks are predominantly a mixture of different network technologies, each originally optimized to carry a particular kind of communication service. Coaxial cable networks were installed to connect television and radio receiver sets and video recorders, twisted-pair copper lines to connect telephone sets and facsimile machines, and Cat-5 unshielded twisted pair cables to connect desktop PCs, printers, scanners, data storage units, and so on. Power line communication (PLC) technologies have also emerged to offer data connectivity without requiring new cabling. In the last decades, wireless connectivity by means of wireless LAN IEEE 802.11 technology (Wi-Fi) has become more and more popular, replacing partly wired networks in serving laptop computers, tablet computers, smartphones, and many more wireless gadgets appearing in the indoor Internet-of-Things environment. The booming number of wireless devices is causing congestion in the Industrial, Scientific, and Medical (ISM) radio spectrum band, and their mutual interference is hampering reliable communication. Trends are emerging to transition to pico-cells, which cover smaller areas in the building (such as a single room, or even only parts of a room) and thus reduce the interference and congestion issues within a cell. The pico-cells approach implies the

---

1 This chapter is largely based on A.M.J. Koonen, E. Tangdiongga, "Photonic home area networks," *IEEE/OSA J. Lightwave Technol.*, vol. 32, no. 4, pp. 591–604, Feb. 2014, https://doi.org/10.1109/JLT.2013.2283145.

Short-Reach Optical Wireless Communication: By Directed Narrow Beams, First Edition. Ton Koonen.

requirement for more radio antenna stations, which need to be simplified for cost reasons and need connectivity to a wired indoor backbone network.

The coexistence of so many different indoor networks entails considerable installation and maintenance costs. It also complicates the upgrade to new service types, and the interoperation of services running on multiple platforms (e.g., between video sets and computers). Moreover, these networks show bandwidth limitations that prohibit the extension of the broadband capabilities of the (FttH) access network to the users. It is noteworthy that the bandwidth demand from in-home networks may even exceed the demand on the FttH access line, due to, for instance, the home-internal fast data streaming needs to and from servers. For example, transferring a DVD movie of 4.7 GB within 10 seconds requires a rate of 3.8 Gbit/s; a BluRay movie of 50 GB even 40 Gbit/s. All these issues could be alleviated significantly by a converged indoor network architecture with truly broadband capabilities, that is, a single broadband network architecture capable of carrying many types of services simultaneously without mutual interference nor design compromises. Research has started to look into optical fiber for such converged novel indoor network solutions, as illustrated in Fig. 2-1. By its virtues of high bandwidth, low losses, small dimensions, and insensitivity to electromagnetic interference (EMI), optical fiber is an eminent candidate to integrate a wealth of services within a single network.

In designing in-home communication networks, one may note a range of differences with access networks. The latter are owned by a network operator, are installed by professionals with high-tech tools, need to conform to standards, and run standardized protocols (e.g., in FttH, there are GEPON (Gigabit Ethernet:

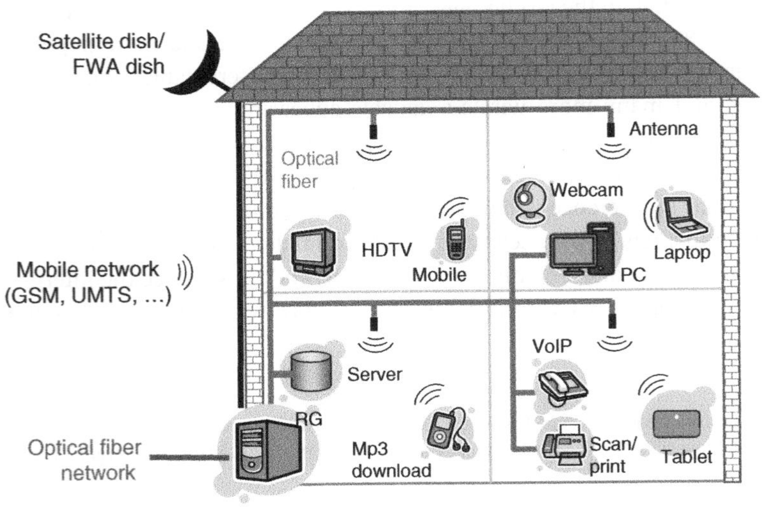

Fig. 2-1 Converged indoor network

Passive Optical Network), GPON (Gigabit PON), XGPON (10-Gigabit PON), etc.). They are installed with a return-on-investment perspective, and their costs are to be shared by many households. An in-home network, however, is (predominantly) privately owned. It should be easy to use ("plug-and-play") without requiring professionals, the user independently decides which equipment to install, and the entire cost is borne by a single household. Obviously, the design of home networks and their technologies need a quite different approach from that of operator-owned outdoor networks.

The interface rates with which consumer electronic devices want to communicate are rising fast. As illustrated in Figs. 2-2 and 2-3, the most demanding ones already require more than 10 Gbit/s, and the bandwidth demands keep growing in the near future [2-2]–[2-4]. The dominant connected consumer device in the home is the smartphone, with an average of six phones per home. Video traffic is driving the bandwidth needs, and virtual reality applications need even larger bandwidths. With respect to consumer demands, copper and WLAN network technologies (wireless local area network, notably Wi-Fi), however, are seriously lagging. Clearly, new network technologies are needed to bridge this gap. As already proven in outdoor network domains, optical fiber is the best-suited technology for wired connectivity, and optical wireless communication (OWC) will become an important complement to radio-based wireless communication.

In the remainder of this chapter, in Section 2.2, first, the economic viability of photonic network solutions is assessed in comparison with established

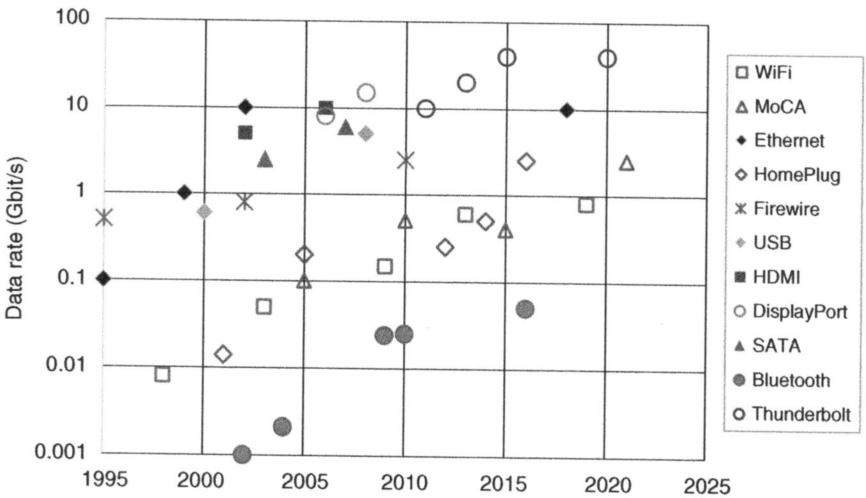

Fig. 2-2 Evolution of interface rates of consumer devices and of data capacity of wireless and copper-based indoor network technologies (from [2-2]/Anthony Ng'oma)

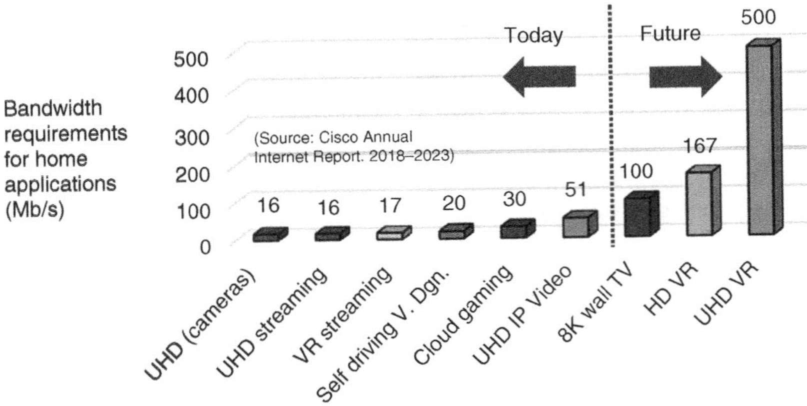

Fig. 2-3 Consumer devices are requiring ever more bandwidth [2-3, 2-4]

copper-based network solutions (notably Cat-5e), in order to identify which network topology is to be preferred for which application area. The chapter proceeds in Section 2.3 by discussing the pros and cons of different optical fiber types, identifying which type of fiber suits which application field, and paying special attention to multimode large-core polymer optical fiber (POF) as a fiber type particularly suited for cost-efficient ("do-it-yourself") installation in in-home networks. However, such POF has technical performance limitations, notably a severely restricted bandwidth. Section 2.4 briefly describes the functionalities of the residential gateway (RG) (the bridge between the access and the in-home network), and Sections 2.5 and 2.6 discuss techniques that help to overcome the bandwidth restrictions and enable the wirebound delivery of high-capacity data streams in particular via POF and with advanced modulation methods, including the wireless delivery by radio signals (notably by radio-over-fiber (RoF) techniques) of high-capacity services. Section 2.7 subsequently discusses how services for wirebound and wireless devices can be delivered together by means of a converged network approach. Advanced techniques for further-enhanced network performance are described in the following sections. In order to better meet fluctuating capacity demands and avoid congestion, two-dimensional (2D) radio beam steering by optical techniques in RoF systems is described in Section 2.8, and dynamic routing of signals in the photonic in-home network in Section 2.9. Mode group division multiplexing (MGDM) is discussed as an interesting option to open an additional dimension for increasing the capacity and routing flexibility of multimode fiber (MMF) in-home networks in Section 2.10. To solve radio spectrum congestion issues, Section 2.11 discusses OWC techniques and their strong potential to solve this congestion as an advanced alternative for radio wireless communication (indoor OWC will be the main topic in the remainder of this book). To conclude, in

Section 2.12, an evolution roadmap is outlined along which the overviewed communication techniques and network solutions can be introduced in order to meet the evolving in-home network capacity requirements. Concluding remarks about the potential of optical techniques for indoor communication are made in Section 2.13.

2.2 Economics of Home Networks[2]

The residential home is a very cost-sensitive environment. The lifetime costs of a home network infrastructure can be split into its installation costs (CapEx – capital expenditure) and operational costs (OpEx – operational expenditure) which include powering and maintenance [2-5].

The CapEx is largely dominated by the choices made for the network topology and for the network medium. In order to gain a basic insight into the impact of these choices, one may consider four fundamental network topologies, as illustrated in Fig. 2-4: point-to-point (P2P), bus, tree, and star. In practice, hybrid versions of these topologies can be used, depending on the design of the building. A residential gateway (RG) is interfacing the in-home network with the access network. In the P2P topology, individual cables run from the RG to every room in the building. The bus and tree are point-to-multipoint (P2MP) topologies, in which a riser cable connects the RG to a floor. In the bus topology, the rooms on that floor are served by add/drop nodes positioned along a horizontal cable, whereas in the tree topology, the rooms are served by individual horizontal cables connected to a star splitter node. Up to now, LAN-structured cabling solutions typically use a star topology with an Ethernet switch at the star splitting point (close to the RG); its physical topology corresponds largely to the P2P topology.

The P2P topology obviously requires the highest amount of cabling, which may lead to high installation costs and congestion issues in sharing the ducts, particularly in larger buildings. On the other hand, it eases the maintenance and upgrading per room. For smaller buildings, such as single-family homes, the amount of cabling may still be acceptable. In the P2MP bus and tree topologies, there is cable sharing and multiple access techniques are needed to give each room an appropriate share of the cable's capacity. When using fiber as the network medium, the network splitting nodes internally may use active optical-electronic-optical (O/E/O) signal conversion for signal routing (such as in Ethernet hubs and switches), or passive optical signal splitting functions such as optical power

2 This section is largely based on A.M.J. Koonen, H.P.A. van den Boom, E. Ortego Martinez, et al., "Cost optimization of optical in-building networks," *Opt. Exp.*, vol. 19, no. 26, pp. B399–B405, Nov. 2011, https://doi.org/10.1364/OE.19.00B399.

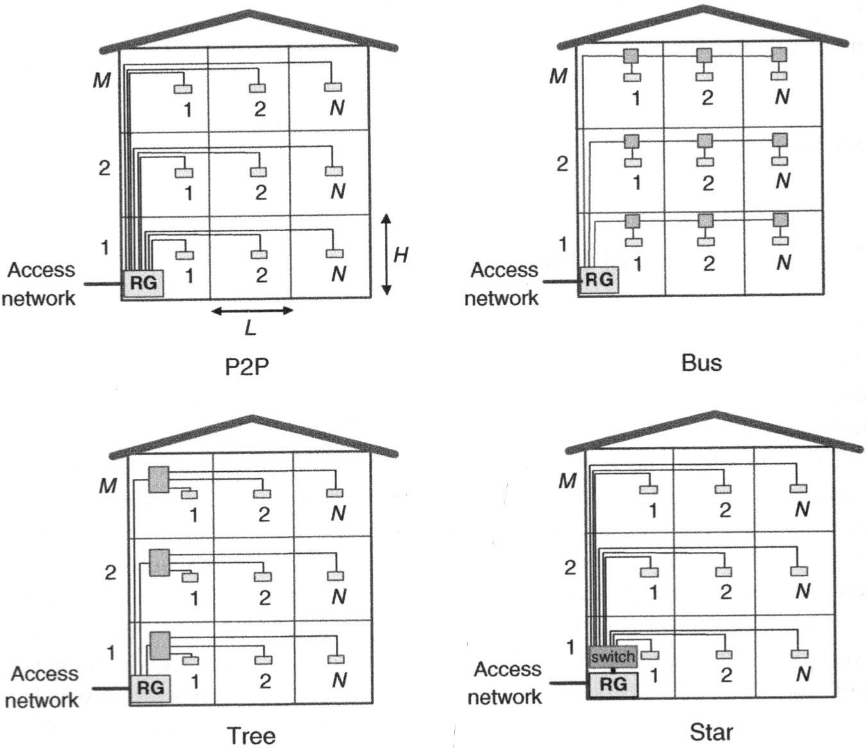

Fig. 2-4 In-building network architectures

splitting or wavelength routing. The bus topology is the most cable-lean and needs the least duct space; however, it also requires many add/drop nodes (and thus consumes more power in the case of active nodes). The tree topology needs only a single splitter node but requires more cabling and duct space. In general, P2P topologies may be preferred for smaller buildings (such as single-family homes), whereas tree or bus topologies are preferred for larger buildings (apartment buildings, office buildings, etc.). In the P2P physical topology, when an optical star coupler is deployed at the RG, a passive multipoint-to-multipoint (MP2MP) architecture can be established in which the rooms can communicate transparently with each other through this coupler.

For the network medium, optical fiber is the most powerful choice, as outlined in the introduction of this chapter. Moreover, as it is EMI-insensitive and not electrically conducting, it can be co-installed with electrical power cabling in the same ducts, hence avoiding the costly installation of new ducts. For safety reasons, this is not allowed for coaxial cables nor for Cat-x cables; they require separate (new) ducts. Cat-5e is presently the most popular cabling for connecting Ethernet-based devices and therefore a good benchmark for assessing the economics of fiber-based home networks. Several optical fiber types may be considered for in-home deployment (see Section 2.3): silica single-mode fiber (SMF), silica multimode fiber (MMF), and multimode plastic optical fiber (POF). Large-core POF is particularly interesting due to its ease of connectorizing and splicing [2-6]; 1 mm core diameter POF requires only low-precision connectors or even does not need a connector as the bare ends may be simply cut and inserted into a module or butt-faced together. Silica SMF and MMF require dedicated tools, delicate handling, and skilled personnel for installation, and thus incur higher installation costs. On the other hand, for silica SMF (and to a lesser extent for MMF), a wide range of modules is readily available for upgrading the network by power splitting and wavelength multiplexing for hosting more users and services. Large-core POF is more lossy and has more dispersion than silica SMF and MMF. The main benefit of large-core POF is therefore its suitability for cost-effective "do-it-yourself" installation in indoor links with limited length and limited bandwidth needs.

The OpEx is considered over the lifetime of the home's network and is largely attributable to the power consumption costs. When SMF is used, the add/drop nodes and the splitter nodes in the P2MP topologies can be implemented by passive optical power splitting or wavelength routing modules, which minimizes the power consumption to only the power consumption of the O/E (opto-electronic conversion) modules at the edges of the network. These passive optical modules are fully transparent: they can operate with any type of signal format. An SMF-based in-home network with passive optical nodes is therefore not only the most power consumption-lean solution but also the most future-proof solution as it is capable to handle upgrading to any new signal format and to multiple service bundles carried on separate wavelengths. For SMF networks, optical power splitters and wavelength routers are commercially widely available and enable the implementation of physically P2MP network topologies which can offer advantages in installation (in contrast to today's P2P topologies for Ethernet-based home networks). Optical power splitters and wavelength routers are not readily available (yet) for MMF networks. Hence, the add/drop nodes and splitter nodes for MMF and POF networks need O/E conversion at their I/O ports plus internal electrical signal splitting/routing functions. These nodes may therefore not be able to transport any type of signal format (they are "opaque"), and they perform active functions which are power consuming. For Ethernet signals, such nodes are commonly known as Ethernet hubs and switches.

We have made CapEx and OpEx analyses for the various basic topologies such as P2P, bus, and tree, for the cases of SMF, MMF, and POF cabling, with Cat-5e cabling as a benchmark [2-5]. The analyses were made for three building scenarios: single-family residential home (3 floors, 4 rooms/floor; floor spacing 3.3

meters, duct length between rooms 8 meters), multi-dwelling units building (MDU, also known as apartment building; 10 floors, 16 living units per floor; floor spacing 4.0 meters, duct length between rooms 14 meters), and office building (10 floors, 50 office units/floor; floor spacing 3.8 meters, duct length between rooms 10 meters). The analysis results are critically dependent on the assumptions regarding the costs and the power consumption of the network elements. These assumptions have been based on Fast Ethernet (100 Mbit/s) equipment and on market surveys done in 2010; the survey outcomes are represented by the typical values listed in Table 2-1.

The fiber solutions allow the sharing of the ducts with the electrical power cabling, whereas the Cat-5e solutions need new ducts. Duct costs are dependent on the duct mounting methods (e.g., buried inside a wall or mounted onto a wall) and the duct diameter, so on the number of cables in the duct and the cable diameter. Building on the topology and network element assumptions, CapEx and OpEx for the various scenarios have been assessed [2-5].

From a CapEx perspective, it has been found that the installation costs for the residential home scenario are lowest when adopting a P2P topology. The optical fiber cables may be installed in the same (existing) ducts as the electrical power cables, so no new ducts are required; but this duct sharing is not allowed for Cat-5e cables due to electrical shock hazards. Taking into account this duct-sharing advantage of the fiber solutions, Fig. 2-5a shows the average installation costs per room derived for the various cabling choices. It should be noted that these

Table 2-1 Costs and power consumption of the network items

	Cat-5E	POF[a]	SMF	MMF
Installed cable costs	€1.8/m	€1.7/m	€1.74/m	€1.95/m
Max. link length	100 m	70 m	1000 m	550 m
Mounted connector costs	€13	€3	€15[b]	€14[b]
Media converter costs; power consumption	(Negligible); 0.65 W	€30; 0.85 W	€70; 1.15 W	€40; 1.15 W
Hub/tap costs; power consumption	€20; 0.2 W	€20; 0.2 W	€20; 0.2 W	€20; 0.2 W
Switch costs, power consumption	€10/port; 0.3 W/port	€10/port; 0.3 W/port	€10/port; 0.3 W/port	€10/port; 0.3 W/port

Prices are based on market surveys done in 2010.

[a] For Ø1 mm large-core PMMA POF.

[b] These prices vary considerably for the various connector types and their mounting methods; we assumed SC connectors, and about 10 minutes in-field mounting time per connector (labor costs about €10).

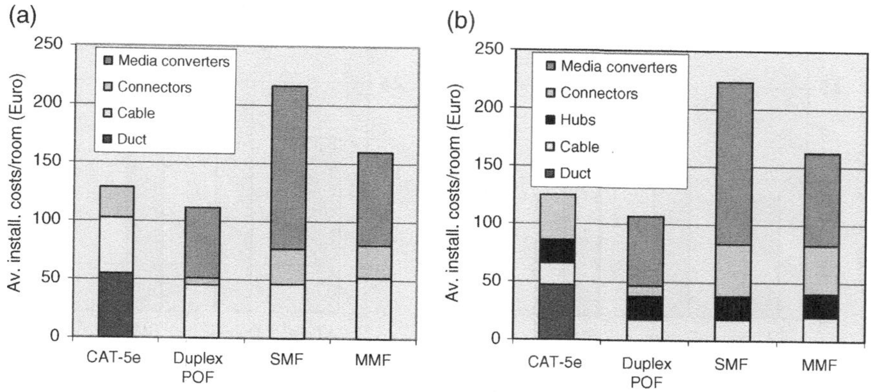

Fig. 2-5 Network installation costs per room for various cabling media. (a) Residential home, P2P topology. (b) Office building, bus topology

costs are calculated with the 2010 price levels from Table 2-1, which have changed since then; however, the relative price comparison between the cabling choices should still hold in reasonably good approximation. Fig. 2-5a illustrates that duplex POF cabling (i.e., two POFs jointly being a bidirectional link) is already cost-competitive with the mature Cat-5e solutions, notwithstanding the fact that POF technologies are just at the beginning of a learning curve. POF technologies have not reached large market volumes yet, and hence have not experienced yet the price reductions accompanying economy-of-scale. For the office building scenario, it is found that a bus topology is the most cost-effective. Again, as Fig. 2-5b shows, duplex POF cabling is already cost-competitive with Cat-5e cabling. In both scenarios, the ease of connecting and the lower costs of the O/E media converters are the main CapEx-saving factors offered by the POF solutions and put them in a better position than the silica SMF and MMF solutions.

From an OpEx perspective, the power consumption in the active network elements has been assessed using the data from Table 2-1, which includes the O/E media converters at the terminals of these elements and the active hubs and switches in the P2MP bus and tree topologies. For the residential home scenario, obviously, the P2P topology is the most power-lean. As shown in Fig. 2-6a, the POF solution consumes a bit more power than the Cat-5e one, but less than the SMF and MMF solutions. For large buildings, a tree topology needs fewer active elements than a bus topology. The advantage of reduced power consumption of the tree topology may, however, be offset by the larger amount of horizontal cabling needed, which can lead to duct congestion. Hence, we consider the bus topology as the preferred one for large buildings. Due to the active network nodes, the power consumption per room is slightly higher than for the residential home

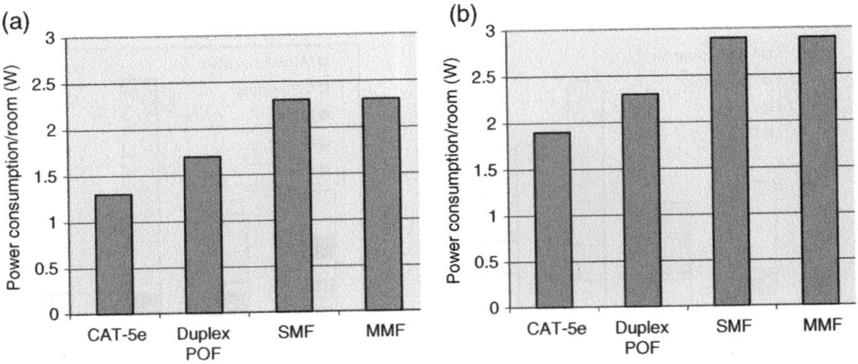

Fig. 2-6 Power consumption per room for various cabling media. (a) Residential home, P2P topology. (b) Office building, bus topology

scenario. Fig. 2-6b confirms that also for the office scenario, the POF solution performs slightly worse than the Cat-5e solutions but better than the SMF and MMF ones.

To assess the total costs incurred during the lifetime of the in-home network, both the CapEx (the installation costs, occurring once at the beginning of the life cycle) and the OpEx (the operational costs, accumulating during the life cycle) have to be taken into account [2-5]. These costs may be partitioned into labor costs, material costs, and energy costs. Assumptions have been made on the expected evolution of these costs in the future, on the breakdown in labor costs and material costs per installed network item, for a typical lifetime of 25 years of the in-home network infrastructure. For example, it is assumed that labor costs will rise with a money inflation rate of 2%/year, energy costs will rise by 5%/year, POF product prices will drop by 10%/year due to a growing market volume, SMF and MMF prices rise by 2%/year due to inflation, and Cat-5e cable prices will rise with 5% per year due to copper becoming more scarce. For the residential home scenario with P2P topology, the impact of these trends on the Net Present Value (i.e., eliminating the impact of money inflation) of the total network costs per room has been assessed.

Fig. 2-7 shows how the NPV evolves when installing the network in year n (where $n = 1$ means today) assuming a network lifetime of 25 years. The graphs basically underline that POF can be more cost-effective than SMF and MMF installations. Moreover, they show that if one has to install an in-home network today and duct sharing is possible (and the reach and bandwidth limits are not prohibitive for the POF solution), POF is already a better choice than Cat-5e from the viewpoint of minimization of total costs. If duct sharing is not possible, Cat-5e is the better choice, but in the near future POF may already offer an economically more attractive solution.

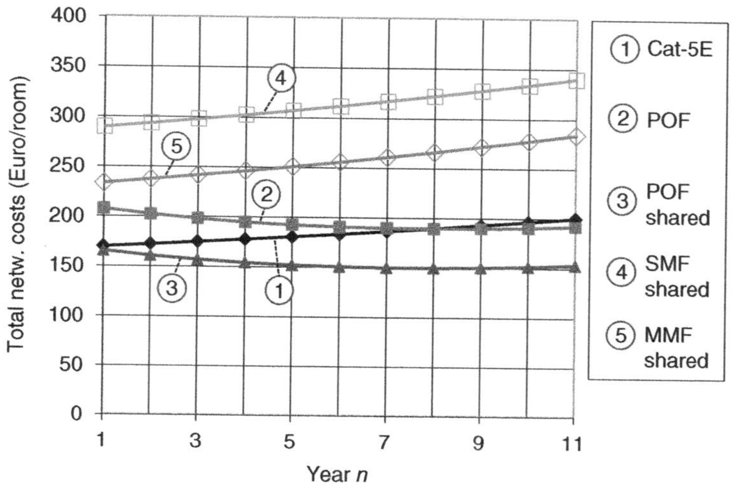

Fig. 2-7 Comparison of total lifetime costs of the network

In summary, it may be concluded that optical fiber solutions are attractive for in-home networks (and other in-building scenarios, such as office buildings) not only in a technical performance sense but also in an economic sense. In particular, when ease-of-installation is a first priority, solutions based on POF are attractive. Depending on adoption in the mass market and the associated decline of component prices, POF may gain wide deployment, notably in (do-it-yourself) home networks. When in the (near) future, upgrading to higher demands on bandwidth and reach is expected, silica SMF or MMF may be the better option. As a disclaimer, it should be noted that the cost analysis presented is based on 2010 market survey studies, and the results should be interpreted as an approximate comparison of costs to provide guidance among the media and network topology options.

2.3 Optical Fibers for In-Home Networks

The demands put on optical fiber for in-home applications differ in several respects from those for outdoor applications. As in-home link lengths are relatively short, the performances regarding bandwidth × length product and attenuation per unit length are less important. On the other hand, ease of handling (splicing and connectorization), bend-insensitivity, small cable diameter, flexibility, and small form-factor connectors are more relevant factors [2-6]. A larger core diameter and a higher numerical aperture make alignment errors in fiber connectors and

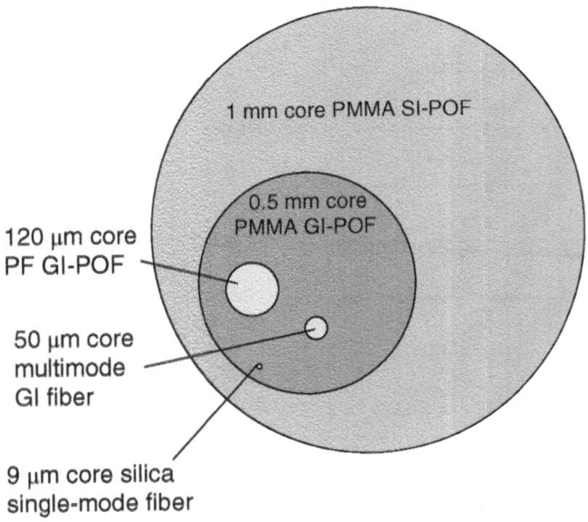

Fig. 2-8 Core diameters of various fiber types

splices less serious. Hence, the installation of multimode silica and plastic optical fiber is easier than that of single-mode fiber and thus entails lower labor costs. A larger core also implies that small dust particles and scratches on the fiber end-face are less harmful in connectors.

Fig. 2-8 shows a comparison of the core sizes of some well-known fiber types (all drawn with the same scaling factor). In silica fibers, a trench of glass with a lower refractive index around the core can enforce the modes to stay inside the core and thus makes the fiber less sensitive to bends. Bend-insensitive silica graded-index MMF (GI-MMF) has become available commercially, for example, with an 80 μm core diameter and large numerical aperture NA = 0.29; for a 3 mm bend diameter, it can have >7 dB less loss than OM1 and OM2 GI-MMF [2-2]. OM4 silica MMF with 50 μm core diameter and ≥4.7 GHz × km bandwidth × length product with 7.5 mm minimum bending radius is commercially available and particularly attractive for data center networks and LANs [2-7]. Also, bend-insensitive SMF conforming to standard ITU G.652.A2 with a minimum bending radius of 7.5 mm is on the market [2-8].

Plastic (/polymer) optical fiber (POF) is commercially available and particularly easy to handle due to its ductility and large core diameters of up to 1 mm [2-9]. The POF cables are easy to install together with electrical power wires by pulling them through the same ducts. The large core diameters allow manual butt-joint splicing by unskilled people; precision tools are not needed for this, nor a microscope. Connector ferrules may not even be needed; the POF can be cut by for instance a razor blade and the POF's bare end may be directly inserted in a module to make a fiber connection.

The optical properties of POF are, however, clearly less good than those of silica fiber. The production processes of silica fiber have since the 1970s evolved to very sophisticated levels, and silica fiber losses have attained extremely low levels. Polymethylmethacrylate (PMMA) is the most popular material for making POF; it is widely known as Perspex or plexiglass. The attenuation performance of PMMA-POF versus wavelength is shown in Fig. 2-9.

The losses are lowest around $\lambda = 550$ nm (green light), but still around 70–100 dB/km; the C—H bonds cause a quickly rising loss at longer wavelengths. PMMA-POF is therefore only applicable in the visible wavelength range, from 450 to 700 nm. Replacing these C—H bonds partly with C—F bonds gives an extended

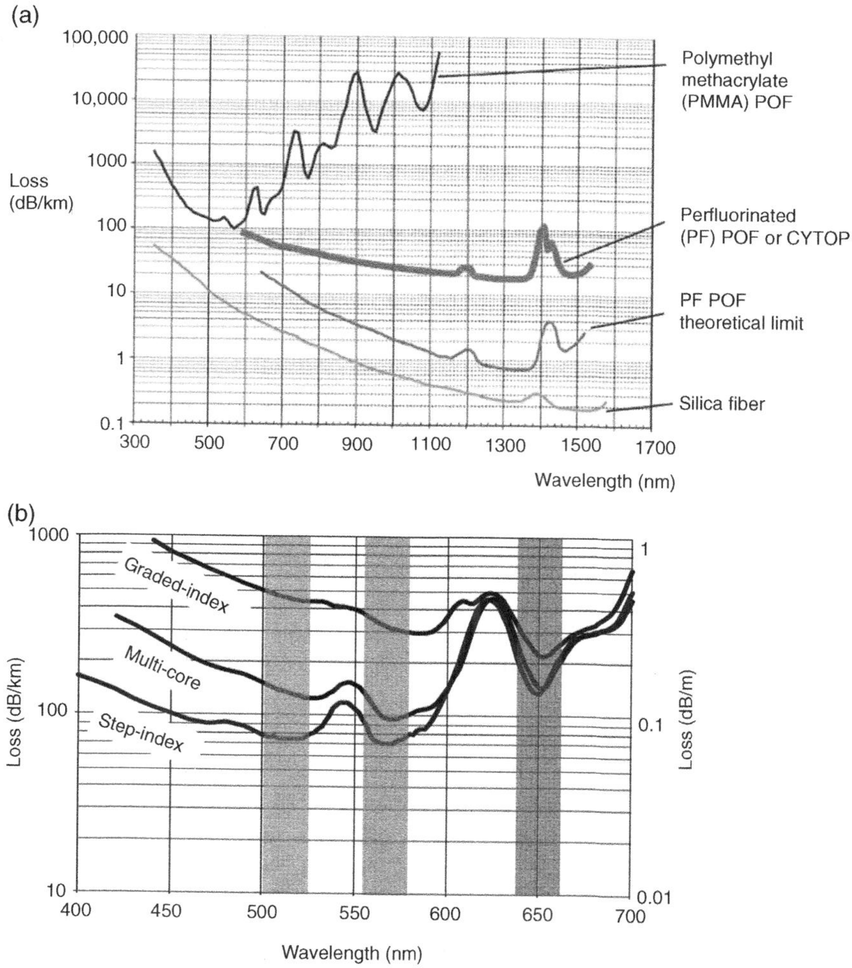

Fig. 2-9 Losses of POF. (a) Loss versus wavelength of silica fiber and of various types of POF. (b) Loss in the visible wavelength range of various types of PMMA-POF

low-loss characteristic up to more than 1350 nm; this so-called perfluorinated POF (PF-POF) thus allows the use of similar laser diodes and photodiodes as used for silica fiber systems [2-10]. Although the losses of PF-POF can go down to around 20 dB/km and thus are considerably lower than for PMMA-POF, they are still much higher than for silica fiber (ca. 0.3 dB/km at $\lambda = 1.3$ μm). The fluoride content of POF may bring some issues in installation with respect to low smoke zero halogen (LSZH) constraints. PF-POF is commercially available up to limited core diameters, that is, up to 250 μm, typically 50–62 μm. As outlined before, because a large core diameter is desired and in view of the relatively short links in in-home networks the fiber loss per unit length is less important, PMMA has become the most popular POF material. The limited wavelength operation range of PMMA-POF, $\lambda = 450$–700 nm, implies that visible light must be used for communication. As operational optical power levels are typically low this can be perceived safely by the naked human eye, which allows visual inspection for network diagnostics and thus eases installation, in particular for do-it-yourself purposes.

Several options for the refractive index profile of the core are available for large 1 mm-core diameter PMMA-POF. The most simple and easy-to-produce profile is the step-index profile (SI-POF). The 1 mm core diameter SI-POF with an NA = 0.5 has become the most popular, mass-produced POF and has been standardized according to IEC 60793-2-40 class A4a.2. Such a profile suffers from heavy multimodal dispersion and hence has a poor bandwidth × length product, in particular due to such a large NA. Up to link lengths of around 50 meters, simple on-off keying (OOK) modulated data streams of up to 1 Gbit/s can be carried. Fiber bends down to a radius of about 5 mm are allowable. The large core can also be composed of many small cores, which each guide a part of the total optical signal. Such PMMA multicore POF (MC-POF) offers a slightly better bandwidth, and most importantly a much better bending performance, down to 3 mm bending radius is allowable [2-11]. For data rates beyond 1 Gbit/s, a graded-index refractive index profile is needed, which offers much less modal dispersion. Large 1 mm-core PMMA graded-index POF (GI-POF) is, however, more sensitive to bending than SI-POF. Modified GI-POF designs have been proposed using a more complex core index profile [2-12] or a non-PMMA core material to yield a higher numerical aperture [2-13]; these designs have shown a bending sensitivity comparable with SI-POF. With smaller core diameters, so at the expense of more difficult handling, perfluorinated GI-POF can enable much higher data rates in the order of tens of Gbit/s [2-14]. Micro-structured POF designs have been explored in research laboratories, which can combine low bending losses with large bandwidths, but again at the expense of a smaller core size [2-15].

Sections 2.5 and 2.6 will expand on the suitability of various PMMA-POF types in wirebound and wireless system applications, respectively.

2.4 Residential Gateway

As illustrated in Fig. 2-1, the RG provides the interfacing between the (FttH) access network and the in-home network. It should translate IP addresses and modulation formats to match those of the in-home devices. It may provide the opportunity to external third parties to do maintenance and upgrading of services in the in-home network and to do remote diagnostics and performance measurements. On the other hand, it should also ensure the security and privacy of the residential users. In addition, the RG may host several home-internal functions, such as the local storage of data (file server) and the intermediation to allow interoperation of devices in the home. In view of the evolving user demands, the data rates of these home-internal data streams may surpass the data rate of the access network line (and are already doing that today; cf. 1 Gbit Ethernet home networks versus 100 Mbit/s VDSL or FttH access lines), which once more underlines the importance of a truly broadband services-converged home network.

2.5 High-Capacity Transmission for Wirebound Delivery of Services

Large-core POF is attractive due to its ease of installation, but on the other hand, it hampers high data rate transmission by its poor bandwidth × length performance due to modal dispersion. Three roads may be pursued to surmount the bandwidth barrier. Baseband amplitude modulation formats may be used in combination with strong equalization techniques at the receiver, for example, binary on-off signaling, or multilevel pulse amplitude modulation schemes which reduce the symbol rate (e.g., 4-level PAM, 8-PAM). Second, a single-carrier technique may be used with a quadrature amplitude modulation scheme, which combines more bits into a single symbol (e.g., 16-QAM carrying 4 bits/symbol, or quadrature phase shift keying, QPSK, carrying 2 bits/symbol). Third, multiple carriers may be used in parallel where each carries a fraction of the total data rate. Such orthogonal frequency division multiplex (OFDM) or discrete multitone transmission (DMT – the baseband variant of OFDM – techniques) effectively smear out the time of a symbol period and thus make the transmission less vulnerable to dispersion in the transmission path. As shown in Fig. 2-10, these techniques involve considerable signal processing at the transmit and the receive site. OFDM and DMT techniques are already widely used in wireless LAN, xDSL modems, DVB-C, and DOCSIS cable modems and thus can benefit from economy-of-scale. Moreover, the spectrum spreading of the multiple carriers in OFDM and DMT allows us to adaptively tune the modulation format and the power per carrier, in order to optimally match the

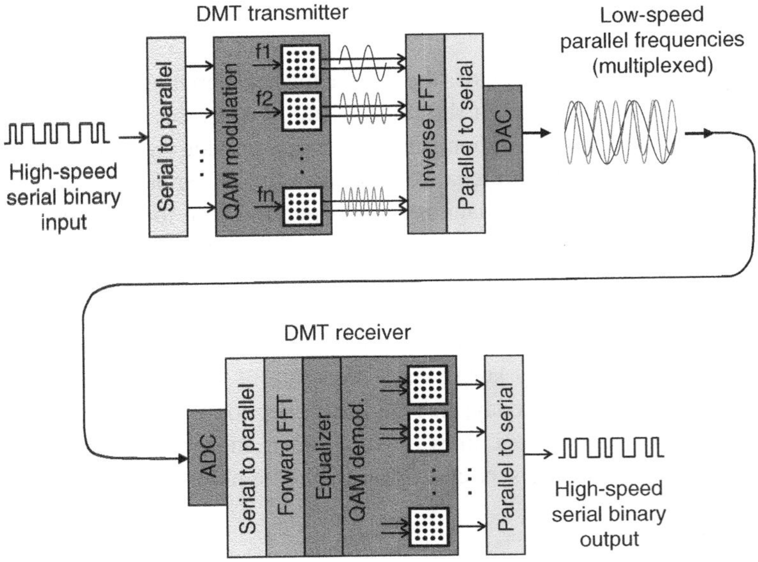

Fig. 2-10 Discrete multitone modulation

characteristics of the transmission path. By deploying this so-called bit- and power-loading, OFDM and DMT have proven to be very robust against changing dispersion conditions in the transmission path.

When the number of levels in multilevel PAM and QAM signaling increases, the symbol rate decreases and hence the transmission performance degradation due to bandwidth limitations becomes less. On the other hand, the noise margins between the signal constellation points become less too, which increases the transmission performance degradation due to noise. With OFDM and DMT, at particular time instants, the many subcarriers may add up leading to a high instantaneous peak level causing clipping in the transmitter and/or in the receiver. The performance of OFDM or DMT schemes thus is mainly limited by clipping noise, and this limitation becomes more severe with more subcarriers. Comprehensive studies which compare the suitability of the various modulation schemes for bandwidth-limited channels have been made. Fig. 2-11 [2-16] shows the optical power margin of a Gigabit Ethernet link (1.25 Gbit/s) versus the available bandwidth of the link normalized on the bitrate. The margin was defined at a BER = 10^{-3}, which is below the forward error correction (FEC) level. Using zero-forcing decision feedback equalization (DFE), 2PAM (equivalent to on–off signaling) performs the best at bandwidths exceeding 20% of the bitrate. For lower bandwidths, but still above 5% of the bitrate, 4PAM + DFE performs better,

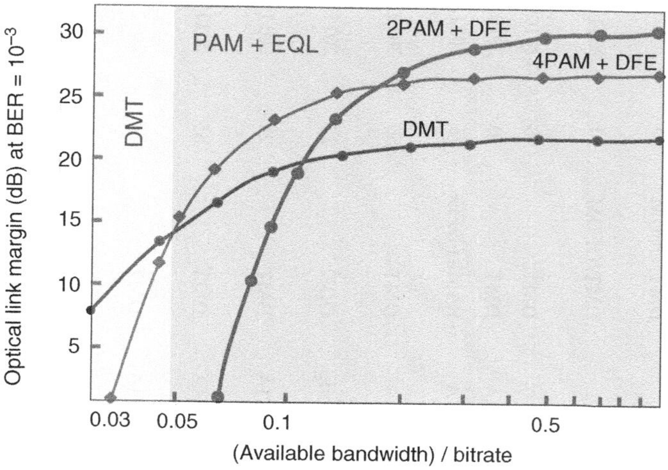

Fig. 2-11 Optical link margin versus bandwidth of Gaussian low-pass channel for a Gigabit Ethernet link (Adapted from [2-16])

whereas, for low bandwidths less than 5%, DMT is preferred. For DMT, symmetrical clipping was assumed. The margin was calculated for a Gaussian low-pass channel; SI-POF typically has a slower response decay which may make 2PAM the best choice for bandwidths even lower than 20% of the bitrate.

Research on high-capacity POF links was the topic of the European project POF-PLUS [2-17]. Real-time 2PAM experiments have been reported at 1.25 Gbit/s over 75 meters, using 1 mm core SI-POF, DFE, and a new Gigabit Ethernet compliant physical layer for POF [2-18]. Similar experiments were reported at 10.7 Gbit/s over shorter POF lengths (15 meters 1 mm core SI-POF using FFE (feed-forward equalization) and DFE, and over 35 meters 1 mm core GI-POF using maximum-likelihood sequence estimation (MLSE)) [2-19]. In another experiment, 10.7 Gbit/s in PAM-4 format was carried over 25 meters SI-POF and 60 meters GI-POF using DFE [2-19]. Using DMT with adaptive bit- and power-loading over 256 subcarriers, 5.6 Gbit/s was transported over 50 meters 1 mm core GI-POF using an eye-safe laser [2-20], and 10.7 Gbit/s over the same length using a high-power laser diode [2-19]. An overview of multi-Gbit/s system experiments using 1 mm core POF is given in Table 2-2 [2-21]. Even 51.8 Gbit/s gross (47.4 Gbit/s net) has been transported at $\lambda = 1302$ nm over a link of 100 meters PF GI-POF with a core diameter of only 50 μm, using DMT with power-loading and bit-loading up to 64-QAM [2-14].

Standardization for POF transmission systems is progressing. A 1 Gbit/s standard for POF based on OOK PAM was published in 2017 [2-22]. The G.hn standard contains medium-specific support for the PHY layer for POF using OFDM with

Table 2-2 Multi-Gbit/s system experiments (using 1 mm core PMMA-POF, unless indicated otherwise)

Data rate	POF type	Core Ø	Tx	Wavelength (nm)	Rx	Format	Length (m)	Year
1.25 Gb/s (real-time)	SI-POF	1 mm	Eye-safe RCLED	650	Large area receiver	OOK	50	2010
2.2 Gb/s (wired) +480 Mb/s (wireless)	GIPOF- PON (1 × 4)	1 mm	High-power laser	650	APD	DMT/OFDM	50	2012
3 Gb/s (wired) +480 Mb/s (wireless)	GI-POF	1 mm	VCSEL	665	APD	DMT/OFDM	50	2011
4.7 Gb/s	MC-POF	1 mm	Eye-safe VCSEL	665	APD	DMT	50	2010
5.3 Gb/s	GI-POF	1 mm	Eye-safe VCSEL	665	PIN+TIA	DMT	50	2010
5.8 Gb/s	GI-POF	1 mm	High-power laser	650	PIN+TIA	PAM2+DFE	75	2011
7.3 Gb/s	mPOF	140 µm	High-power laser	650	APD	DMT	50	2012
4 × 2.5 Gb/s	Ribbon		VCSEL array	665	PIN Diode array	OOK	25	2011
10 Gb/s	MC-POF	1 mm	High-power laser	650	PIN + TIA	DMT	25	2011
10 Gb/s	GI-POF	1 mm	High-power laser	650	PIN + TIA	DMT	35	2011
10.7 Gb/s	SI-POF	1 mm	WDM high-power lasers	405, 515, and 650	PIN + TIA	DMT	50	2012
10.7 Gb/s	GI-POF	1 mm	VCSEL	665	MSM PD + TIA	NRZ + MLSE + PLL	35	2012
47.4 Gb/s	PF GI-POF	50 µm	DFB	1302	Ø25 µm PD	DMT	100	2010

adaptive bit loading [2-23]; however, there have been no implementations reported so far. From 2022, studies are ongoing toward a 25 Gbit/s point-to-point over 15 m GI-POF in the IEEE 802.3dh project for automotive applications [2-24].

2.6 High-Capacity Transmission for Wireless Delivery of Services

Within the home, there is a booming need for wireless connectivity. Users often carry multiple wireless communication devices such as smartphones, tablet computers, and laptops. We are entering the era of "the Internet of Things" where not only human–human and human–machine communication but also machine–machine communication is becoming ever more intense. In order to counteract ISM radio spectrum congestion and mutual interference which limit the network's throughput, indoor radio networks consisting of many pico-cells are being created. To make this economically feasible, the antenna sites should be simple and the radio signal generation and processing functions be centralized. A fixed wired backbone network is needed to feed all the antenna sites from a central location that hosts the radio functions. Due to its virtues of high bandwidth and low losses, optical fiber is the preferred medium to build such a backbone.

In the RoF (radio-over-fiber) approach illustrated in Fig. 2-12, when transporting the RF signals modulated in their native analog format onto an optical carrier (i.e., ARoF, analog radio over fiber), the antenna sites only need to convert the radio signals from an optical to an electrical format, thus yielding the most simple antenna stations. However, the optical bandwidth available in a fiber link may be more effectively used when the RF signals are transposed to a lower intermediate frequency and put back onto their original position at the antenna site (i.e., intermediate frequency over fiber – IFoF). Alternatively, the radio signal may be digitized and transported on an optical carrier by digital modulation techniques (i.e., digital radio-over-fiber – DRoF), which brings the advantages of digital transport well-known for extended reach such as regeneration without SNR degradation and more network routing capabilities (such as add-drop and cross-connect). Both IFoF and DRoF require more active equipment at the antenna site than ARoF, thus need more locally supplied power and maintenance.

The centralized provisioning of the radio functions eases maintenance and upgrading of the wireless network. The many non-overlapping pico-cells enable radio frequency reuse and thus provide a higher total network capacity. The RoF pico-cell approach also brings energy savings in various ways. Less radio power needs to be emitted by the antenna because in a pico-cell the distance from the wireless device to the antenna is reduced. Reversely, the wireless device needs

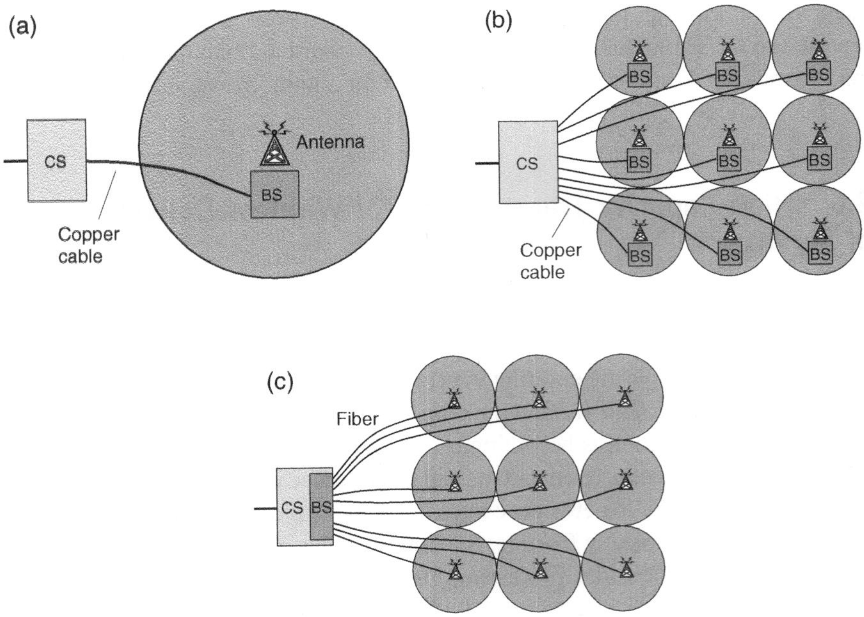

Fig. 2-12 Increasing the capacity of wireless networks by introducing pico-cells and radio-over-fiber (RoF) techniques. (a) Single large radio-cell. (b) Introducing pico-cells. (c) RoF-fed pico-cells

to emit less radio power to communicate upstream, thus prolonging its battery life. The pico-cells may be individually activated depending on the wireless traffic load in that cell, which allows several of them to be operated in "sleep mode." The central processing of the radio signals facilitates many advanced wireless network functions, such as smart antenna functions (multiple-input multiple-output – MIMO), adaptive radio beam-shaping and steering, and spectrum sensing for cognitive radio techniques, which also can contribute to improved delivery of wireless services and energy savings.

For indoor wireless communication, a wide range of wireless standards has been developed (and is still expanding). As listed in Table 2-3, a wide range of frequency bands is used, with a large variety of modulation formats and data rates. For higher data rates, Wi-Fi and 5G indoor are the dominant indoor standards. The higher frequency bands are mostly intended for short reach, that is, in-room usage, where 60 GHz even requires line-of-sight (LoS). Optical fiber can carry multiple signals independently on different wavelengths, and indoor fiber links are typically so short that they do not add delays large enough to have a noticeable impact on the radio medium access control protocols [2-25], hence RoF is well suited for indoor operation with multiple radio standards.

SMF has plenty of bandwidth to carry any standard. When intensity modulation is used, however, double sidebands around the optical carrier frequency occur,

Table 2-3 Indoor wireless technology standards

Technology	Standards	Coverage	Frequency bands	Modulation	Data rates (peak downlink)	Bandwidth
Wi-Fi	802.11g	Up to 38 m (indoor)	2.4 GHz	OFDM, DSSS	54 Mbit/s	20 MHz
	802.11n (Wi-Fi 4)	Up to 70 m (indoor)	2.4, 5 GHz	OFDM	150 Mbit/s (4 × 4 MIMO, in 40 MHz channel)	20, 40 MHz
	802.11ac (Wi-Fi 5)	Up to 70 m (indoor)	5 GHz	OFDM	867 Mbit/s (8 × 8 MIMO, in 160 MHz channel)	20, 40, 80, 160 MHz
	802.11ax (Wi-Fi 6)	Up to 70 m (indoor)	2.4, 5, 6 GHz	OFDMA, 1024 QAM	1200 Mbit/s (8 MU-MIMO)	20, 40, 80, 160 MHz
	802.11be (Wi-Fi 7)	Up to 70 m (indoor)	2.4, 5, 6 GHz	OFDMA, 4096 QAM	2400 Mbit/s (16 MU-MIMO)	20 ~320 MHz
5G indoor	5G NR	—	<1, 1–6, 6–60 GHz	OFDMA, 16~256 QAM	10 ~20 Gbit/s (8 × 8 MIMO and higher)	Up to 100 MHz
Bluetooth	802.15.1	10 m (class 1) 30 m (class 2, 3) 60 m (class 4) 240 m (class 5)	2.4 GHz	GFSK, π/4-DQPSK and 8DPSK with FHSS	1 Mbit/s class 1 ~50 Mbit/s EDR class 5)	1 MHz (79 bands in total)
Zigbee	802.15.4	70 m	868, 915 MHz and 2.4 GHz	OQPSK with DSSS	250 kbit/s	5 MHz (16 bands in total)
UWB	802.15.3/ ECMA368	10 m	3.1–10.6 GHz	QPSK/OFDM (MB-OFDM)	480 Mbit/s (MB-OFDM)	528 MHz for each sub-band (MB-OFDM)
				BPAM(DS-UWB)	1 Gbit/s(DS-UWB)	<7.5 GHz (DS-UWB)
WiGig	802.11ad	In-room, >10 m (with beam forming)	2.4, 5, 60 GHz	Single carrier/OFDM	7 Gbit/s	4 × 2.16 GHz

which due to fiber dispersion may suffer from fading; that is, for certain radio frequencies and at periodic length positions along the fiber, the two sidebands may show a π radians phase difference and extinguish each other. The length period is inversely proportional to the square of the radio carrier frequency; the fading effect is therefore in particular a problem at high radio carrier frequencies. Single-sideband (SSB) modulation can effectively avoid this fading but requires more complicated modulation circuitry. Alternatively, one may use a modulation scheme that suppresses the optical carrier, thus mitigating the fading and making the system insensitive to fiber dispersion. By driving a Mach Zehnder modulator (MZM) symmetrically around a bias point at V_π (the MZM's inflexion point), the carrier is suppressed and the resulting microwave frequency is twice the frequency of the drive signal [2-26]. Another dispersion-robust technique is optical frequency multiplying (OFM) [2-27], in which the optical frequency of the light source is periodically swept at a relatively low frequency, and the resulting frequency-modulated (FM) signal is converted by an interferometric filter (e.g., a Mach Zehnder Interferometer – MZI) into an intensity-modulated (IM) signal containing many harmonics of the sweep frequency. The data signal is intensity-modulated on the envelope of this signal. After photodetection, a simple electrical bandpass filter can select the desired higher harmonic to become the microwave radio signal to be emitted. It can be shown that this OFM technique is very robust against fiber dispersion and suppresses laser phase noise; for example, transmission of 120 Mbit/s in 64-QAM format at 17.2 GHz over 4.4 km of silica MMF has been demonstrated [2-27].

The very limited bandwidth of large-core POF links seriously limits the feasibility of transporting radio signals in their original spectral position. A relatively simple solution can be offered by spectrum translation, where the radio signal at the headend site is down-converted by mixing it with the help of an electrical local oscillator (LO) to a frequency band within the POF's bandwidth. At the antenna site, after photodetection, a similar LO is used to upconvert the signal to its original spectral position and subsequently be radiated as a radio wave. As illustrated in Fig. 2-13, with this approach, it has been demonstrated that a high-definition video signal up to 480 Mbit/s can be carried in an ultra-wideband (UWB) OFDM channel (TFC6, 3.696–4.224 GHz) over 50 meters 1 mm core PMMA GI-POF and 3 meters wirelessly, by down/up-converting it to/from the 0.836–1.364 GHz band for transport in the POF link [2-28].

Also, impulse-radio UWB techniques may be deployed in RoF links. Such techniques may aid the indoor localization of wireless devices [2-29]. They use specifically shaped narrow optical pulses covering the UWB spectrum up to 10 GHz. Typically, they require a small fiber dispersion, and are not compatible with large-core POF; data rates of 2 Gbit/s have been shown over 100 meters of 50 μm core diameter PF GI-POF [2-30].

The central radio signal processing enabled by RoF techniques offers nice opportunities for advanced radio signal processing. For example, while

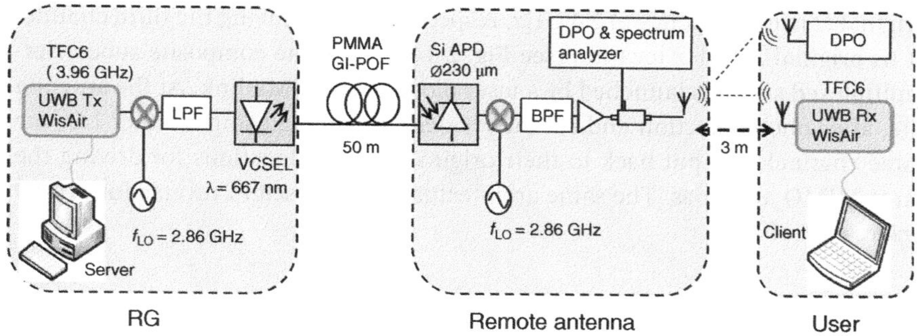

Fig. 2-13 UWB radio over 1 mm core PMMA GI-POF using spectrum translation

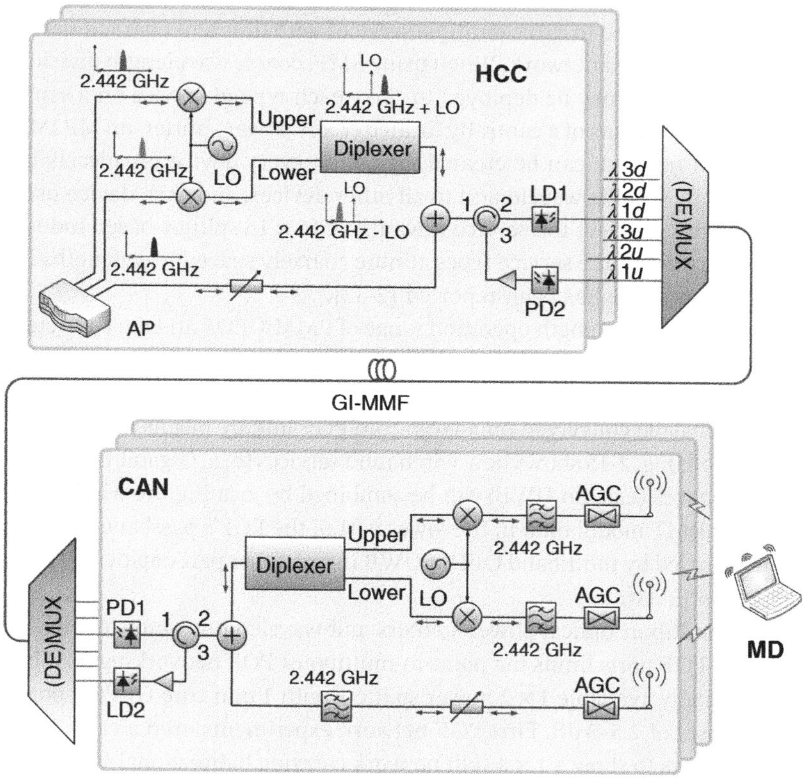

Fig. 2-14 3 × 3 MIMO radio over fiber using subcarrier multiplexing

performing the MIMO signal processing in the headend station, the separate MIMO channels may be carried over a single fiber by subcarrier multiplexing. Using an electrical LO in the headend station, a 3×3 MIMO system can be created by shifting two channels at the LO frequency distance below and above the original carrier frequency of 2.4 GHz, respectively, and leaving the third channel at its original 2.4 GHz location (see Fig. 2-14) [2-31]; the composite subcarrier-multiplexed signal is launched by a laser diode in the MMF link. At the antenna site, after photodetection and subcarrier-demultiplexing using another LO, the three channels are put back to their original 2.4 GHz locations for driving the three MIMO antennas. The same architecture can be used in reverse for MIMO upstream traffic.

2.7 Converged Indoor Networks

Optical fiber is well suited to carry multiple services with different characteristics within a single converged network. When using SMF, coarse wavelength division multiplexing (CWDM) may be deployed to carry each type of service on a separate wavelength. By means of a centrally located $N \times N$ power splitter, an MP2MP multiwavelength network can be created, by which every device broadcasts its output signal at its specific wavelength to all other devices, and each device uses a wavelength filter to select the service it wants. A 16×16 splitter-based indoor SMF network carrying nine service types at nine coarsely spaced wavelengths in the 1305–1610 nm range has been reported [2-32].

Due to the limited wavelength operation range of PMMA-POF and the restricted availability of sources operating at different visible wavelengths, wavelength multiplexing offers only limited possibilities for the convergence of services. Alternatively, services may be converged on a large-core POF link by means of electrical spectrum division. Fig. 2-15 shows how wirebound services (e.g., Gigabit Ethernet) and wireless services (e.g., on UWB) can be combined by locating the wirebound ones carried by DMT modulation in the lower part of the POF's passband and the wireless ones carried by multiband OFDM UWB in the upper part deploying spectrum translation [2-33].

The lack of multiport optical power splitters and wavelength routers equipped with large-core POF ports limits the point-to-multipoint POF networking possibilities. Commercially available 1×2 power splitters with 1 mm core GI-POF ports show excess losses of 2.5–3 dB. First POF network experiments used a cascade of 1×2 power splitters to show a 1×4 split network carrying bidirectional UWB signals over 30–50 meters of 1 mm core GI-POF [2-34, 2-35]. Using bulk-optic ball lenses and an interference filter, a two-wavelength red/green demultiplexer with PMMA-POF has been created [2-36]. Considerable advances in POF power splitter and WDM multiplexer/demultiplexer techniques are needed to enable more comprehensive POF networking.

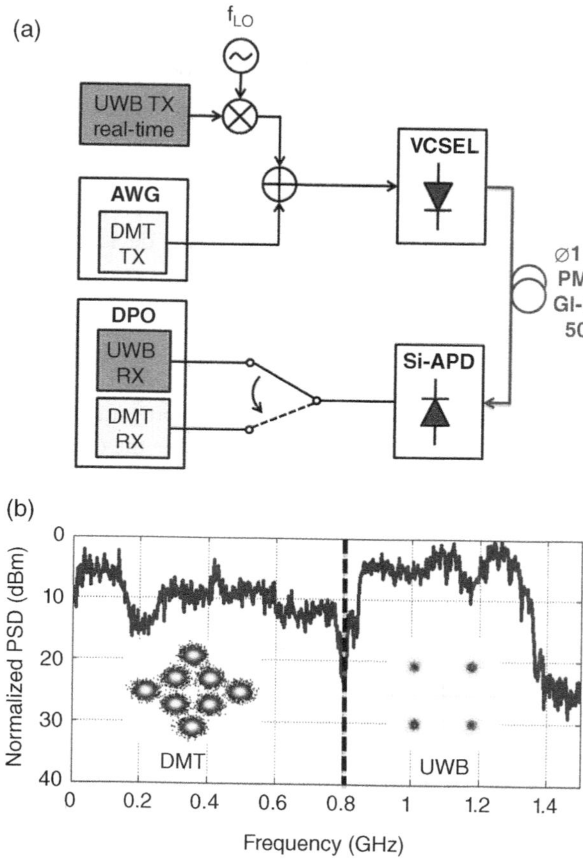

Fig. 2-15 Converged transport of wirebound and wireless services by spectrum translation over a 1 mm core PMMA GI-POF link. (a) System setup. (b) Spectrum measured after reception

2.8 2D Radio Beam Steering in a Radio-Over-Fiber System[3]

The need for wireless broadband connectivity is growing rapidly, not only because the number of devices to be connected is increasing but also because each device requires more capacity. The radio spectrum is getting exhausted, and devices start to interfere with each other. This is seriously affecting the data throughput per

3 This section is largely based on A.M.J. Koonen, and Z. Cao, "Optically controlled 2D radio beam steering system," and in A.M.J. Koonen, "Optical techniques for Gbit/s wireless indoor access" (keynote paper), in *Proc. Microwave Photonics Conference (MWP 2014)*, Sapporo, Oct. 20–23, 2014.

device. The data collisions due to interference make retransmissions of these data necessary and thus reduce the data throughput rate. The increased data rates also require higher received radio powers. To overcome the rising bottleneck in wireless connectivity, solutions to reduce the interference and improve the power efficiency of high-capacity radio systems are needed.

This section introduces an optical system concept for shaping and steering of radio beams in two angular dimensions, by which interference between the wireless devices is substantially reduced, radio power is directed to the devices more efficiently, and thus considerably less power is needed.

2.8.1 Radio Beam Steering

Phased array antennas (PAAs) are well known for radio beam-shaping and -steering. The phase differences between the antenna elements determine the direction in which the radio waves emerging from these elements interfere constructively, and thus the direction in which the resulting radio beam is pointing. The shape of the beam is determined by the number of active antenna elements. Typically, a PAA with N antenna elements gives a power gain of a factor N. When the PAA is composed of a 2D grid of antenna elements, the beam can be steered in two angular dimensions; this is illustrated in Figs. 2-16 and 2-17. Steering the beam direction is done by tuning the phase differences between the antenna elements, and typically involves comprehensive electronic signal processing to create the required phase differences between the radio-frequency (RF) signals fed to each antenna element. Moreover, the RF signal generation and modulation preceding this comprehensive processing also need to take place close to the antenna elements, which makes the whole antenna site complex to build and maintain.

RoF techniques have been developed which enable the simplification of the antenna sites considerably (see e.g., [2-27]). They allow to generate and modulate

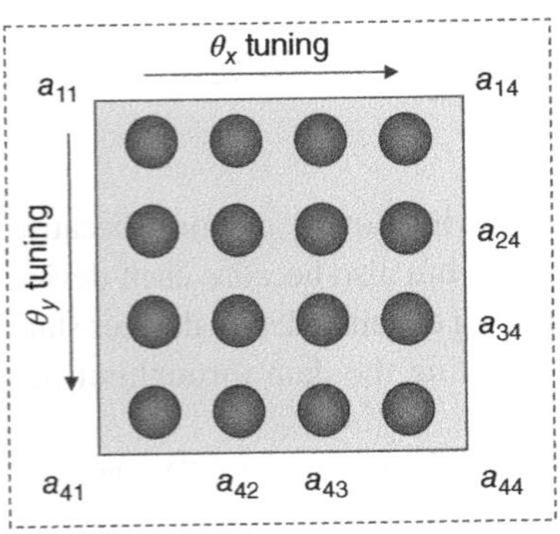

Fig. 2-16 4 × 4 phased array antenna for 2D radio beam steering

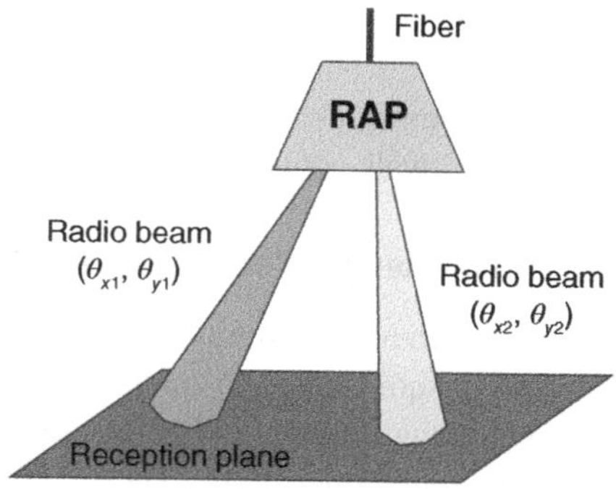

Fig. 2-17 2D steering of multiple radio beams from a radio access point (RAP) containing the phased array antenna

the RF signals remotely in a central site, and feed them from there via fiber to simplified antenna stations, where for the downstream RF transmission only optical-to-electrical signal conversion by means of a photodiode is needed followed by a simple RF power amplifier and the antenna. For the uplink, the possibly large power variations of the upstream RF signals from the users need to be accommodated into a single optical upstream channel which brings additional transport challenges.

Next to the remote generation and modulation of the RF signals, the system setup proposed in this section provides an additional advantage of RoF techniques, namely that the tuning of the phase differences between the antenna elements in a PAA can be done by relatively simple optical means (e.g., true time delay (TTD) lines by means of optical waveguides of different lengths, cf. Fig. 2-19), instead of by comprehensive electronic processing. In literature, the use of optical ring resonators with thermal heater elements to tune the phase shift per ring has been reported [2-37]. However, due to the strong frequency dependence of a ring, this phase shift is highly dependent on the frequency of the radio signal. When the radio signal is modulated and hence its bandwidth is extended, the phase shift is no longer the same across the spectrum of the signal, and thus the beam direction is blurred; this beam distortion is known as "beam squint." Typically, the ring resonator's bandwidth is below 2 GHz, which limits its suitability for broadband communication (e.g., for deploying a 7 GHz band at 60 GHz). The beam squint can be avoided by using TTD techniques which create a frequency-independent time delay between the antenna elements, thus allowing steering of a radio beam with a broad spectrum. An optical 1D TTD steering solution using bulky discrete components was reported in [2-38]. TTD techniques can be approximated by comprehensive electronic signal processing. Using optical micro-ring resonators, they can also be approximated by a complex photonic integrated circuit (PIC) with multiple rings [2-39]. Moreover, optical TTD solutions using comprehensive micro-electro-mechanical system (MEMS) switches have been proposed, which require complex electronic control circuitry [2-40]. However, the complexity of these one-dimensional (1D) TTD solutions makes a 2D TTD beam steering solution prohibitively complex.

2.8.2 Wavelength-Controlled 2D TTD Radio Beam Steering

We have introduced a simple yet powerful solution to provide remotely controlled broadband squint-free 2D radio beam steering by means of RoF techniques and wavelength-tunable optical signal processing which is accomplished by an integrated optical 2D TTD circuit [2-41] and based on earlier 1D TTD work in [2-42, 2-43]. The circuit enables discrete steering of the radio beam in two angular directions by the stepwise tuning of only a single parameter, namely, the wavelength λ of the input light carrying the radio signal. This light signal can be fed from a remote location via optical fiber to the antenna site. By feeding multiple light signals each at a different tunable wavelength and each carrying a radio signal at a different radio frequency, multiple radio beams can be steered individually from a single antenna in two dimensions by tuning their wavelengths separately.

As an application example, Fig. 2-18 shows an indoor network where in each room of a building a radio access point (RAP) is located. This RAP contains a phased array antenna that directs radio beams toward the mobile devices in the room.

The key optical element in the beam steering system is a stepwise tunable optical delay line, composed of an arrayed waveguide grating router (AWGR) with feedback loops. This concept has been reported before in various forms [2-42–2-44]. As

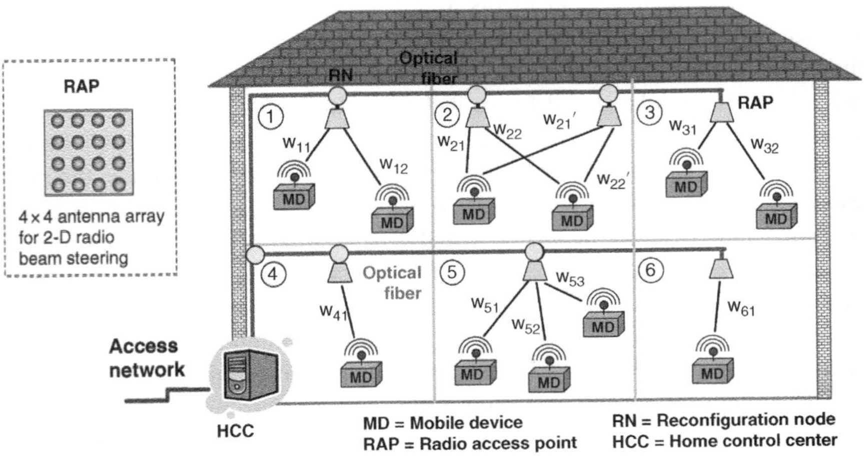

Fig. 2-18 Fiber-wireless in-home network with smart radio beam steering

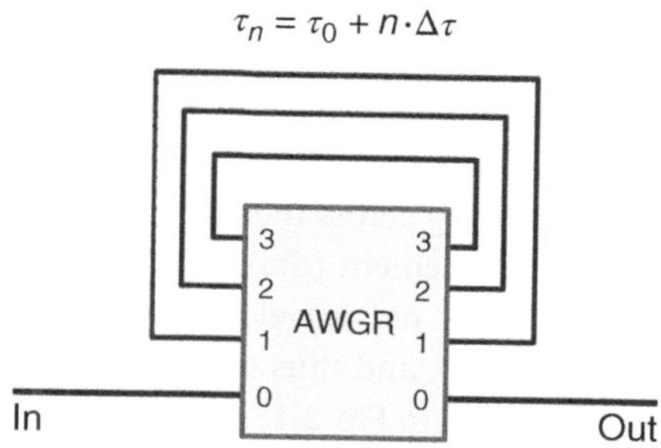

Fig. 2-19 Stepwise λ-tunable TTD element, with delay step size $\Delta\tau$

illustrated in Fig. 2-19, depending on the value of its wavelength λ, the light entering the AWGR at input port 0 exits the AWGR at a specific output port n ($n = 0, 1, 2, \ldots$). From this output port, the light with wavelength λ_n enters a feedback loop with a length such that it is delayed over a time $\tau_n = \tau_0 + n\Delta\tau$, and then re-entering the AWGR. Due to the internal symmetry of the AWGR, this light is subsequently exiting at output port 0. The light exiting this TTD element has thus experienced a delay τ with respect to its entry time, where τ is a stepwise function of the wavelength λ of the light (see Fig. 2-20). The cyclic behavior of the AWGR causes not only the light with wavelength λ_n to exit from output port n and thus to experience a time delay τ_n, but also light with a wavelength that is larger by a multiple of the free spectral range (FSR) of the AWGR. Hence, the stepwise delay characteristic is periodic with a wavelength period equal to the FSR.

The AWGR can be designed to have a finite bandpass characteristic per port. This means light is directed from an AWGR input to an AWGR output not only for a particular wavelength but also for a small range of wavelengths around that

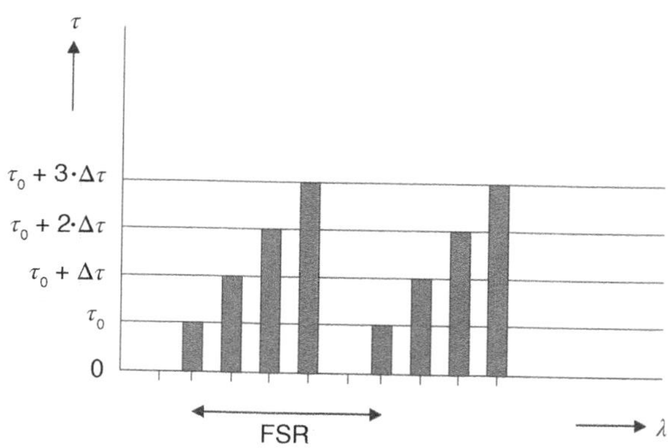

Fig. 2-20 Stepwise tunable time delay τ

wavelength. Thus, when an RF signal with a wide bandwidth is modulated on an optical carrier with wavelength λ_n, where the modulation process implies a spectral broadening of that carrier (but within that small range), this modulated carrier still experiences a constant delay time value τ_n (and not a delay time which is gradually changing across the signal's bandwidth). Hence, the setup in Fig. 2-21 acts as a real TTD element (under the assumption that the delay in the feedback loops themselves is not wavelength-dependent, which is a very good approximation of the reality), and thus the RF beam steering is squint-free.

As depicted in Fig. 2-18, 2D radio beam steering is needed to cover a certain area. This beam steering can be achieved by using a 2D array of antenna elements; Fig. 2-16 gives an example of a 4×4 array that steers both in the θ_x angular dimension and in the θ_y angular dimension, by tuning the corresponding time delays between the columns and rows of the 4×4 matrix of antenna elements, respectively. Multiple beams may be directed simultaneously by a single 2D-phased antenna array, as illustrated in Fig. 2-17, by deploying a different radio frequency for each of them.

We propose a concept with which the 2D beam steering can be realized by tuning just one parameter, namely, the wavelength λ of the input light. Fig. 2-21 illustrates how this single-parameter tuning can be implemented by means of a photonic-integrated circuit (PIC) which jointly uses TTD elements with a large

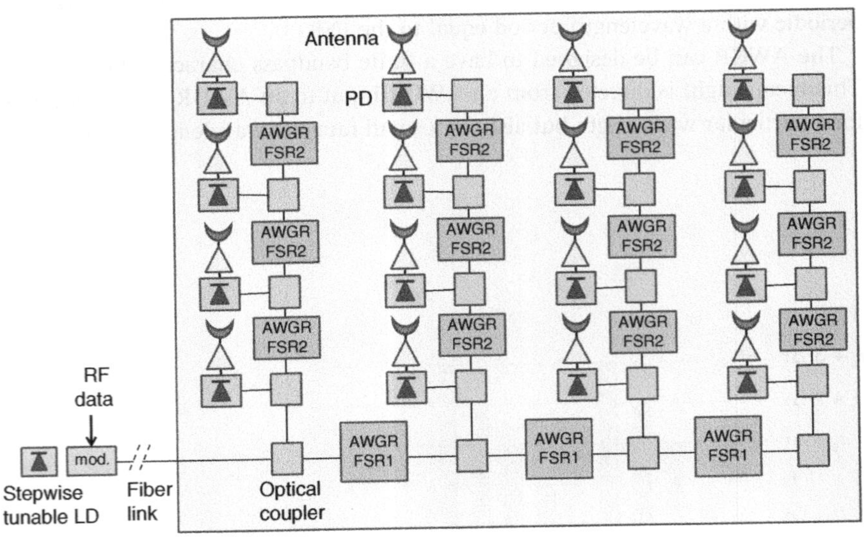

Fig. 2-21 Wavelength-tunable 2D beam steering PIC including a combination of TTD stages with a large FSR and TTD stages with a small FSR, co-integrated with O/E RF drivers for the antenna units

FSR and TTD elements with a small FSR, co-integrated with O/E RF conversion and driver stages for the antenna elements. An AWGR with a smaller channel spacing and a smaller FSR requires a larger structure than an AWGR with a larger channel spacing and a larger FSR; for a smaller channel spacing, the length difference between its internal waveguiding structures needs to be larger. Hence, to minimize the total required area of the integrated optical chip, it is advantageous to design the TTD elements in the lower row in Fig. 2-21 such that they have a small FSR, and the TTD elements in the columns to have a large FSR as they are more numerous.

As shown in Fig. 2-22, the respective FSRs must be designed such that a full wavelength passband of the AWGR with a large FSR_2 comprises the full width of the small FSR_1. Thus, while tuning λ through that full wavelength band the angular direction θ_y remains fixed, while the angular direction θ_x is stepwise tuned over its full range. While extending the λ-tuning over the full FSR_2 range, due to the periodicity of the TTDs with the small FSR_1 the angular direction θ_x is periodically tuned over its full range, while the angular direction θ_y is slowly and stepwise tuned over its full range. With this tuning scheme, both θ_x and θ_y are step-wise and consecutively tuned over their full ranges, and a 2D area is covered by the beam steering while tuning only the parameter λ of the input light signal.

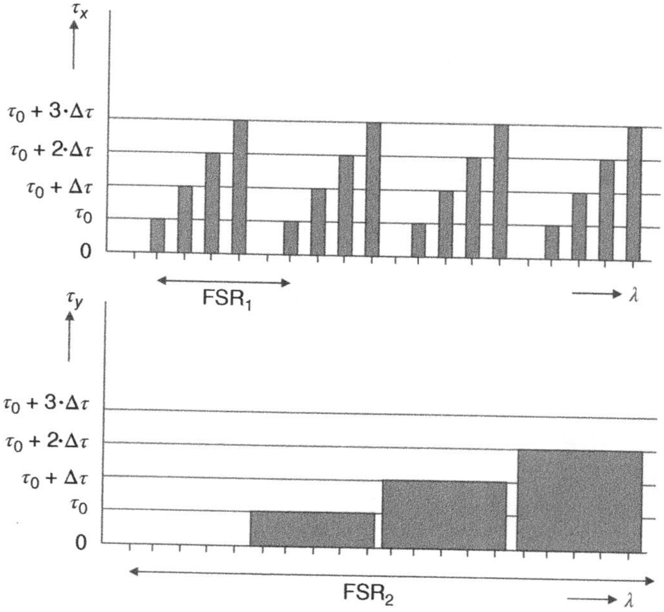

Fig. 2-22 2D beam steering by tuning only the wavelength λ of the input light

2.8.3 Integrated 2D TTD Beam Steering Circuit

The first integrated optical circuit has been made for a 2 × 2 PAA, in silicon-on-insulator optical integration technology [2-45]. Fig. 2-23 shows a photograph and the performance characteristics measured on a 1D TTD circuit with seven feedback loops. One of the loops was defective, hence, the graphs show the performance of the phase delays versus the frequency of the RF signal for six different wavelengths of the optical input signal, in which each wavelength activates one of the loops. As shown in Fig. 2-23, the phase delays increase linearly when the frequency is increased, with a stepwise increasing slope when the wavelength increases. This behavior confirms the stepwise TTD tuning, as a constant time delay corresponds to a linear phase delay slope with increasing frequency.

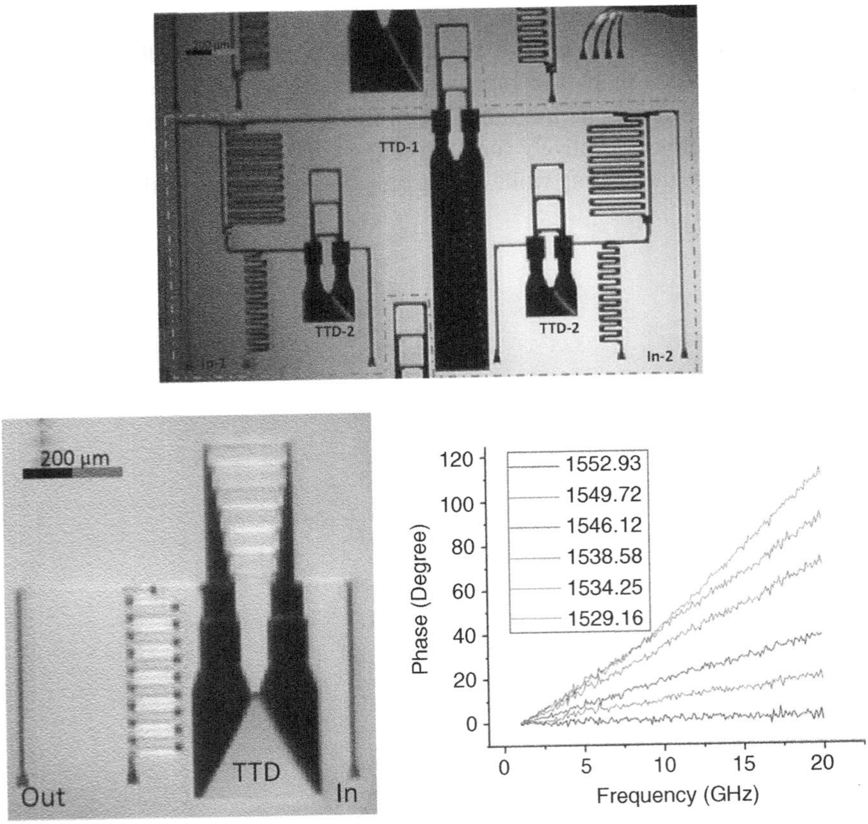

Fig. 2-23 First realization of a 2D TTD optical-integrated circuit for a 2 × 2-phased antenna array, and measured performance of a basic 1D TTD optical integrated circuit with 6 working feedback loops. (Adapted from [2-45])

2.8.4 Conclusions

We propose an optically controlled 2D radio beam steering system, based on a novel integrated optical circuit with TTD elements arranged in two groups which each take care of the beam steering in one dimension and which have interlaced periodic characteristics in order to jointly provide beam steering in two dimensions. The feasibility of a basic integrated 1D TTD stage has been shown.

The integrated optical circuit is located at the antenna site and can be remotely driven via a fiber by a stepwise wavelength-tunable laser diode. The antenna site thus can be compact and robust, requiring minimum installation effort. Moreover, the beam steering can be done by varying only the wavelength of the input light signal, which can be fed from the remote location containing a tunable laser diode transmitter by means of a fiber link; therefore, this solution eases the control algorithms for fast beam steering and avoids the burden of comprehensive electronic signal processing at the antenna site, with the associated maintenance and power consumption issues.

Because a fiber link can carry multiple wavelength signals simultaneously, each coming from a separate tunable laser diode at the central site, the proposed system also allows a multitude of radio beams, each individually steerable in two angular dimensions, can be generated. The beam steering requires constructive interference between the radio waves emitted from the antenna elements, and therefore the steering of multiple beams individually requires that each beam carries a different radio frequency.

2.9 Dynamic Capacity Allocation

In order to improve the operational efficiency of the installed network resources, dynamic routing of capacity inside the indoor network may be introduced such that local traffic needs are met on demand. For instance, users carrying wireless broadband devices may roam throughout the building, and wireless capacity may be intensely needed in rooms where many of them are meeting whereas other rooms are not or only sparsely populated and therefore need less capacity; this situation may alter frequently due to the roaming of users. To accommodate these locally changing traffic demands, dynamic wavelength routing in SMF-based indoor networks may be used to allocate wavelength channels to rooms or to clusters of rooms, as illustrated in Fig. 2-24a and b. Studies have shown that by using time-division-multiplexing inside wavelength channels, and moving time slots to other wavelength channels when the original wavelength channel gets congested (this idea is also being taken up as multi-link operation (MLO) in IEEE 802.be (also known as Wi-Fi 7), to be adopted in 2024 [2-46]), the network congestion

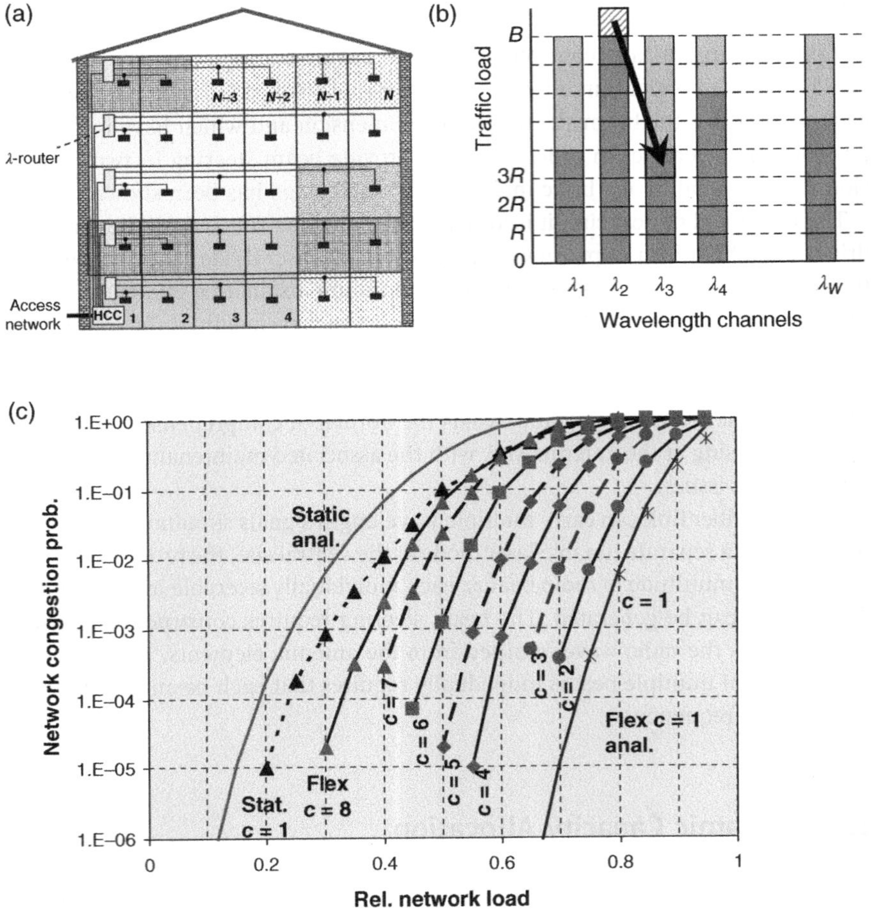

Fig. 2-24 Dynamically or statically assigning capacity to clusters of rooms in a building (with c rooms per cluster) (Reprinted with permission from [2-47] © Optica Publishing Group). (a) Assigning wavelength channels to rooms (with a cluster size $c = 2$). (b) Moving a cluster to another wavelength channel. (c) Network congestion probability versus relative network load for static and dynamic assignment of wavelength to a cluster of c rooms

probability can be reduced significantly without installing extra network resources [2-47]. This reduction will be larger when the clusters are smaller, as the capacity can then more efficiently be rearranged with a finer granularity. Thus, dynamic capacity allocation on demand enables the network to carry more traffic while maintaining the same network resources installed; as Fig. 2-24c shows, by refining the allocation granularity, the traffic load may be more than tripled while not affecting the network congestion probability.

The flexible assignment of wavelength channels to rooms can be done by tunable wavelength conversion of the RoF signals followed by a static wavelength router in the network splitting node. This approach has been demonstrated in an SMF network by means of cross-gain-based wavelength conversion in a semiconductor optical amplifier (SOA) applied to an OFM radio signal; tunable routing has been shown of 150 Mbit/s 64-QAM WLAN signals in the 1.5 μm band [2-48].

2.10 Mode Group Division Multiplexing

An alternative option to create multiple independent transmission channels within a single MMF is given by mode group division multiplexing (MGDM) [2-49, 2-50]: many guided modes inside an MMF may be partitioned in sets of modes, and each set (mode group) can constitute a transmission channel. As illustrated in Fig. 2-25, N lasers may be employed each launching a specific mode group in the fiber, and at the receiving end, each element of a set of M photodiodes receives a mixture of the transmitted signals. Due to mode mixing inside the fiber, the mapping of the transmitted signals to the received signals can be described by a transfer matrix **H** (which changes when the mode mixing conditions change). Provided that these conditions change slowly, the signal processing at the receive side (possibly aided by pre-coding at the transmit side) can unravel the mingling of the channels and yield the originally transmitted channels again. To some extent, this MGDM approach ressembles MIMO techniques that deploy multiple transmit and

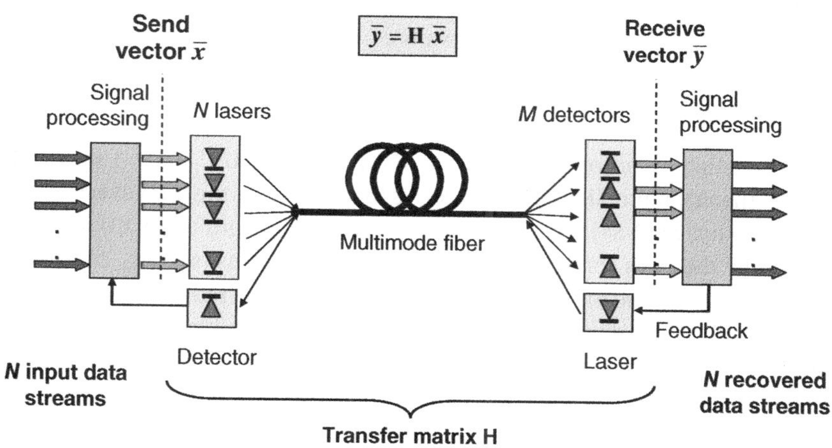

Fig. 2-25 Mode group division multiplexing

receive antennas in radio-based communication systems. Using the MGDM approach, the transmission of 3 × 10 Gbit/s has been demonstrated by means of three mode groups over a link of 20 meters silica GI-MMF with 185 μm core diameter [2-51]. The MGDM concept is also feasible for GI-POF [2-50]; transmission of 2 × 10.7 Gbit/s data rate has been shown over a link of 10 meters GI-POF with 62.5 μm core diameter [2-52].

A similar mode-multiplexing approach for few-moded fiber has been introduced to increase the capacity of long-haul fiber links. In this so-called spatial division multiplexing (SDM) approach individual modes are deployed, not sets of modes (in order to avoid multi-modal dispersion which would be detrimental in long-haul links) [2-53]. Within each mode group, additional multiplexing dimensions may be employed, such as WDM, which can further expand the network's capacity and routing flexibility. Using SDM in combination with dense wavelength division multiplexing (DWDM) and comprehensive signal modulation (such as QAM-64), data transport capacities in few-moded fibers beyond 10 Pbit/s have been achieved.

2.11 Optical Wireless Communication

The abundant proliferation of wireless devices in conjunction with their growing capacity needs is causing radio spectrum congestion and mutual interference among them, thus hampering wireless connectivity. The free-space optical spectrum can offer an alternative with plenty of unlicensed bandwidth. Optical wireless communication (OWC) may use the infrared, visible, or ultraviolet part of the spectrum. OWC does not cause electromagnetic interference (EMI) nor is it affected by it, and it can easily be contained in order to enhance privacy (e.g., within a room as it does not penetrate walls).

2.11.1 Eye Safety

Eye safety regulations as formulated in the ANZI Z-136 and IEC 60825-1 standard series state the maximum light intensity levels that may enter the human eye without causing damage [2-54]. Ultraviolet light with a wavelength up to 300 nm will be blocked by the eye's cornea, and up to 400 nm by the lens (see Fig. 2-26).

Between 400 and 1400 nm, most of the light will reach the retina. From 1400 to 3000 nm, it will be absorbed in the vitreous body, the inner transparent part of the eye. Beyond 3000 nm, it is blocked by the cornea. The retina runs the highest risk of being damaged. With a fully open pupil (7 mm diameter, typically), the maximum optical power that may enter the eye (the Accessible Emission Limit, AEL, for eye-safe Class 1 laser products) varies from 39 to 390 μW in the visible range (400–700 nm). From 1400 to 4000 nm, up to 10 mW may be allowed. Fig. 2-27

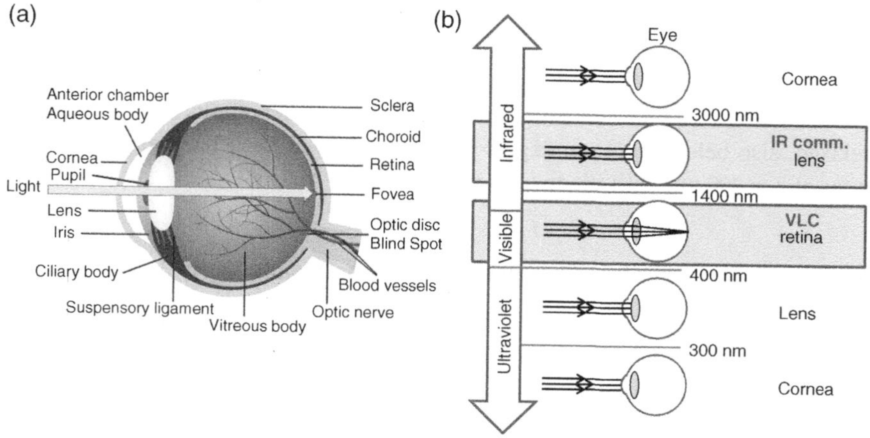

Fig. 2-26 Penetration of light in the human eye. (a) The human eye. (b) Beam of light entering the human eye

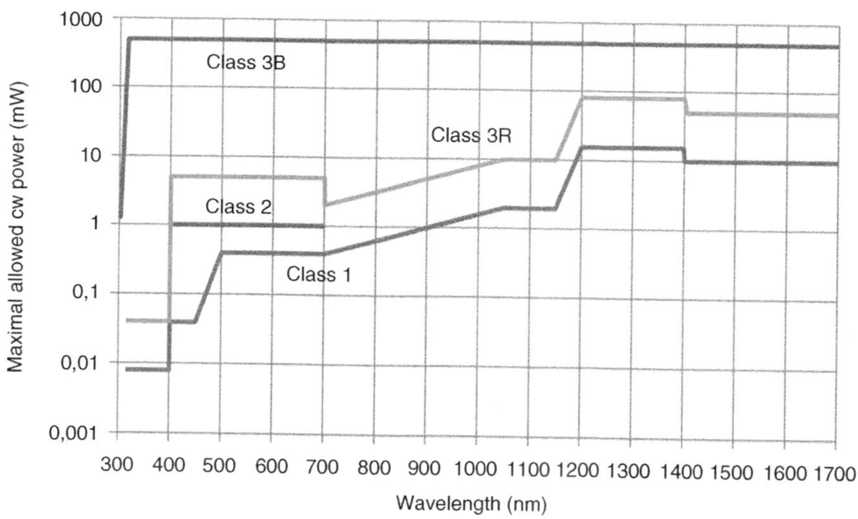

Fig. 2-27 Laser safety according to norm EN 60825-1:2007

shows the maximum allowed CW power from a laser versus its wavelength for various laser Classes [2-55]. The graphs only hold for static, point-like laser sources such as collimated or weakly diverging laser beams. A Class 1 laser is safe for normal use, that is, when a laser is viewed with the naked eye or with the support of typical magnifying optics such as a telescope or microscope (except when they have a large aperture). A Class 3B laser brings hazards if the eye is exposed directly, but not when the light is diffusely reflected from paper or other matte surfaces. Lasers of Class 2 and higher have to carry a warning label. For distributed Lambertian sources with a wider source diameter, the eye safety limits are significantly relaxed, also below $\lambda = 1\ \mu m$ (e.g., >50 mW is allowed for a Lambertian Class 1 laser at $\lambda = 900$ nm with diameter 3 mm [2-56]).

2.11.2 Optical Wireless Communication Using Visible Light

LEDs are increasingly being used for lighting purposes. As they can be modulated at moderate data speeds, visible light communication (VLC) can be "piggy-backed" on illumination for applications at home, in offices, industrial or medical environments, public transport, and other settings, as shown in Fig. 2-28. White-light LEDs typically are based on a blue-light LED plus a phosphorescent material. For office lighting, light intensities of 200–800 lux are desired; the blue part of the spectrum represents around 400–1700 mW/m^2. The phosphorescent material introduces slow decay effects and limits the VLC bandwidth to about 3 MHz. By detecting the LED's blue part of the spectrum only, the bandwidth can be extended to about 20 MHz. By means of adaptive DMT techniques, high data rates may be carried within this limited bandwidth [2-57]. Using DMT with 127 bit- and power-loaded subcarriers plus clipping to limit the aggregate DMT peak signal excursions, a gross bitrate of 513 Mbit/s has been achieved with a single LED over a distance of 30 cm to a silicon avalanche photodiode with Ø3 mm diameter preceded by a blue filter [2-58]. By applying wavelength multiplexing techniques,

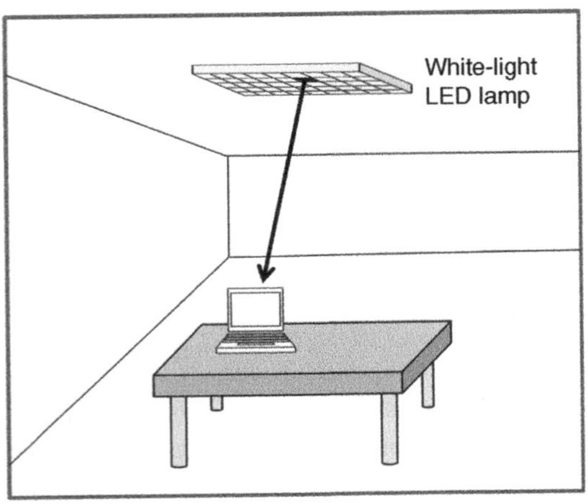

Fig. 2-28 Visible light communication (Lee et al. [2-57]. © [2008] IEEE)

the VLC data rate may be further increased. With a single white LED module composed of a red, a green, and a blue LED, individually driving the LEDs with DMT signals and performing RGB demultiplexing at the photodetector, an aggregate gross transferred data rate of 3.4 Gbit/s over 10 cm reach has been demonstrated [2-59]. The distance can be increased by using an array of LEDs. With an array of 40 white-light LEDs and DMT bit-loading up to 128-QAM, a distance of 0.75 m was achieved [2-57]. The optical communication channel may partly consist of line of sight (LoS) paths from the LEDs to the photodetector, and partly of a diffuse path resulting from reflections from objects and walls [2-60]. The diffuse component may yield a cut-off frequency of about 10 MHz for a medium-sized room but is much smaller than the LoS components (typically 20–30 dB electrical below the LoS signal). The limited field-of-view (FoV) of the photodetector restricts the delay between the received LoS paths. For a typical FoV of 120° and a medium-sized room, optical channel bandwidths well above 90 MHz are found, indicating that the achievable VLC data rate is mainly limited by the bandwidth of the LEDs.

2.11.3 Optical Wireless Communication Using Infrared Light

As an alternative to VLC, OWC may use infrared wavelengths, with particular attention to carefully obeying eye safety levels. Design compromises need to be made regarding the FoV: on the one hand, the FoV needs to be large enough to provide adequate coverage; on the other hand, the FoV of the transmitter must be small enough to deliver enough power density at the receiver. By using multiple transmitter and receiver units, each with a smaller FoV, the total FoV can be increased while preserving enough received power density [2-61]. Imaging diversity may be deployed, where an array of light sources is mapped to an array of detectors. By means of a seven-channel diversity receiver equipped with a hexagonal MSM InGaAs photodetector array, when using MIMO techniques a data throughput of 5 Gbit/s can be achieved [2-62].

Alternatively, angular diversity may be deployed, where the sources emit at different angles and the receiver has multiple detectors looking into different angles [2-63]; this angular diversity approach is illustrated in Fig. 2-29 [2-64]. A three-channel angular diversity system in which each transmitter and receiver had an FoV of about 8°, achieved a data rate of 1.25 Gbit/s at $\lambda = 830$ nm over a range of 3 meters. Increasing the FoV to 30°, and using larger APD detectors, an increased area in a range of 5–6 meters could be covered with a data rate of 280 Mbit/s. Higher capacities can be envisaged by reducing the FoV and adopting wavelengths beyond 1400 nm where the eye safety standards allow higher light flux levels. With narrow light beams, appropriate area coverage at data rates

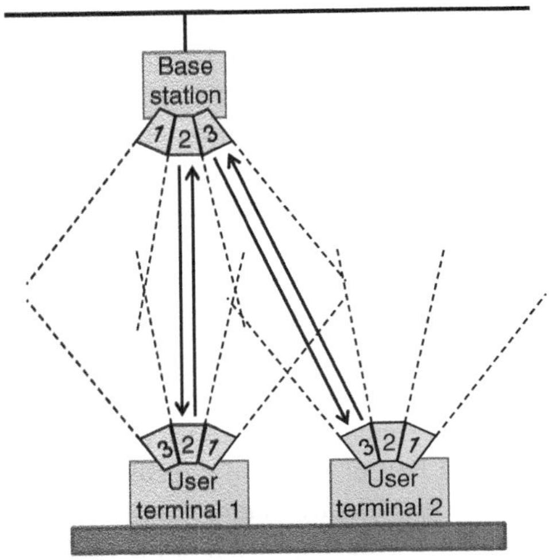

Fig. 2-29 Angular diversity system for optical wireless communication (Le Minh et al. [2-64]/© [2010] IEEE)

beyond 1 Gbit/s may be achieved by adaptive light beam steering techniques, for example, using spatial light modulators, micro-mechanical mirrors based on MEMS technology, or diffractive modules [2-65, 2-66]. Beam steering techniques will be discussed in detail in Chapter 5.

An indoor optical wireless link may also be the extension of an FttH access link. By deploying center launching into the fiber in order to increase the bandwidth of an MMF link and DMT modulation techniques, bidirectional transmission of 30 Gbit/s at $\lambda = 1.55$ μm over a link of 4.4 km MMF in conjunction with a 1.5 meters directional optical wireless link has been demonstrated [2-67].

OWC by means of 2D-steered infrared beams can achieve significantly higher aggregate data rates than the wide-beam OWC approaches discussed above, by building on the spatial multiplexing which is enabled by the small footprints of each beam. This concept aligns with the beam-steered system proposed as the BROWSE concept in Chapter 4 (see Fig. 4-1). Moreover, each beam can be pointed at an individual user, which guarantees high privacy and security, and also can be very energy-efficient as the beams only go where and when needed. A beam actually can provide the capacity of a fiber (and even more, as it does not suffer from wave-guide dispersion). A narrow beam may be regarded as a "virtual fiber" without needing a physical connection. It avoids the mobility problems of a wirebound link, although the beam steering processes require time and thus only support limited mobility, such as nomadic mobility, and static links.

Beam-steered OWC is the main topic of this book, including its system architecture and key enabling techniques, and will be discussed in detail in Chapters 3–8.

2.12 Evolution Trends

Optical fiber can play a major role in establishing an indoor network that is capable of transporting a wide variety of services and meeting the ever-growing capacity needs of the users. Taking account of the techniques discussed in the previous sections of this chapter, for the introduction of fiber-based indoor networks an evolution roadmap may be as depicted in Fig. 2-30. The present state of indoor copper-based networks is expected to fade out in the near future, and they are to be replaced by fiber-based networks. Each fiber network evolution phase is likely to overlap the next one, as the installation of indoor networks needs sizable capital investments which need to be depreciated over a longer period.

From the present status of separate networks each laid out for a particular set of services, fiber may be introduced first in converged P2P topologies, using POF in the home area and bend-insensitive silica MMF or SMF in the larger (semi-)professional buildings (or in the home when POF solutions cannot offer the reach and bandwidth needed). The various types of services may be integrated into this single infrastructure by carrying them on an IP basis. Wi-Fi base stations can be connected on an IP basis too.

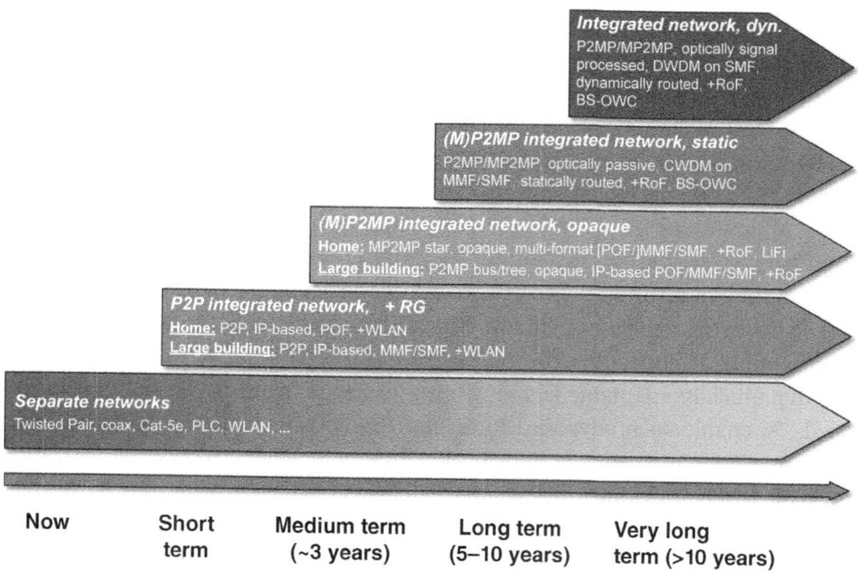

Fig. 2-30 Evolution roadmap for fiber-based indoor networks

In the medium term (around 3 years from now), active P2MP topologies may be introduced, using opaque (O/E/O converting) network nodes; the nodes can do the routing electronically and do signal restoration which permits the use of (lossy) POF cables as well as silica SMF and MMF cables. In larger buildings, bus or tree topologies may be adopted, which save duct space and cabling costs. Within the home area, as cable lengths are much shorter an MP2MP star topology with a multi-format electrical switch located in the RG may be adopted, which can handle not only IP signals but also other signal formats (video and audio streams, control signals for domotica, etc.). To serve the booming needs for wireless connectivity, radio pico-cells may be introduced which can be fed by means of (analog or digital) RoF techniques, as well as created by the first LiFi OWC systems.

In the longer term (some 5–10 years from now), passive optical P2MP topologies may be introduced, in which the passive transparent network nodes do just optical power splitting or static wavelength routing. Such functions are readily available with SMF, but not (yet) with MMF or POF; hence, the network is likely to be based on bend-insensitive SMF. The optical signals can pass these nodes transparently, without signal conversion, which makes the network compatible with any signal format and readily enables service upgrades. In larger buildings, a bus or tree topology may be preferred; as the dispersed network nodes are fully passive, attractive savings on power consumption and maintenance may be obtained with respect to the active nodes of the medium-term scenario. In the home area, a passive star coupler or multiport wavelength router (e.g., arrayed waveguide grating router, AWGR) may be deployed. By means of coarse wavelength multiplexing (CWDM) techniques, separate sets of services may be distributed and routed in separate wavelength channels, for instance, to handle the RoF signals. Beam-steered OWC systems may be introduced for providing high-capacity wireless services with low latency and EMI-immunity (such as in intra-data center networks, in industry 4.0 settings for machine-to-machine communication, etc.). In residential/office indoor settings, they can also release spectrum to RF-based wireless networks for serving the plethora of low-bandwidth devices in the Internet of Things.

In the more distant future (beyond some 10 years from now), P2MP topologies with dynamically adaptable signal routing in the network nodes may be introduced. To enable such advanced functions, the network will be based on bend-insensitive SMF. By means of optical switching or wavelength conversion, traffic flows (for wirebound and wireless service delivery) in the network may be steered in order to provide capacity on demand and thus improve network throughput and reduce energy needs, and in order to improve the support of user mobility. These advantages lead to enhanced operational efficiency of the network's resources, as they can be achieved without installing extra equipment. To create adequate granularity for the dynamic routing, a multitude of wavelength channels may be needed, in a dense wavelength division multiplexing (DWDM) scheme. The booming high-capacity needs for wireless services may be met by dynamically reconfigurable RoF-fed pico-cells. In particular, dynamically beam-steered optical wireless techniques will be introduced which can deliver on-demand high bandwidth services to densely spaced users in a flexible and energy-efficient way.

Overall, the evolution path of indoor networks is expected to follow a trend of increasing network capacity, reduced power consumption, and increased networking flexibility. This trend can uniquely be supported by fiber-based network solutions and their optically supported wireless last-meter links.

2.13 Concluding Remarks

Indoor networks face an exploding demand for more capacity, versatility, and flexibility. Optical fiber is optimally suited to meet this demand, thanks to its large bandwidth, low losses, and the extra dimension (namely wavelength) it can offer. It can handle a wide variety of services in a single network infrastructure. For the residential home area, in particular, point-to-point network topologies are attractive with large-core PMMA-POF as the fiber medium which provides ease of installation and not only performance-wise but also cost-wise already can outperform present mature (Cat-5e) electrical indoor network solutions. Bend-insensitive silica SMF and MMF are less easy to handle but require less space, offer huge bandwidth, very low losses, and readily support optical power splitting and (wavelength) routing functions, which in particular are beneficial in indoor networks in larger (semi-)professional buildings. The severely restricted bandwidth in large-core POF links requires powerful dispersion-robust modulation techniques such as DMT, and for less severely restricted bandwidths multilevel PAM combined with powerful equalization techniques. Gigabit Ethernet POF transceiver modules are entering the market.

In an alliance with wireless links, an indoor fiber network can provide a wide range of services with widely varying bandwidth, while supporting mobility of the users. Radio pico-cell solutions can effectively be supported by RoF techniques and can considerably increase the network's capacity for the delivery of wireless services while reducing the power consumption, in particular when applied in conjunction with optical techniques for radio beam steering and dynamic capacity allocation to the pico-cells.

OWC techniques are emerging to free up the radio spectrum; they can provide the ultimate capacity for wireless connectivity without EMI disturbances. In particular, beam-steered OWC techniques (as discussed in the remainder of this book) can offer huge wireless capacity for broadband services at high user densities and with enhanced energy efficiency. Thus, they can take away a heavy burden from RF wireless networks giving them more room to handle lower-bandwidth intermittent service delivery as required in the Internet of Things with its many diverse devices handling a plethora of services (housekeeping, house control, environment and safety control, etc.).

Recently, indoor fiber network proposals nick-named fiber-to-the-room (FttR) are gaining interest in standardization organizations such as the Telecommunications Sector of the International Telecommunication Union (ITU-T), the European Telecommunications Standards Institute (ETSI), and the China Communication Standards Association, with specific interest on introducing passive optical network (PON) technology well-known in FttH access networks into passively split indoor optical fiber networks to every room and on centralized networks combining Wi-Fi with fiber feeder links [2-68].

References

[2-1] A.M.J. Koonen, and E. Tangdiongga, "Photonic home area networks," *IEEE/OSA J. Lightw. Technol.*, vol. 32, no. 4, Feb. 2014, pp. 591–604, https://doi.org/10.1109/JLT.2013.2283145.

[2-2] A. Ng'oma, H. Yang, and R. Wagner, "The future of home networking,", *Proc. ECOC 2012*, Amsterdam, Sep. 16–20, 2012, Paper Mo.1.G.4.

[2-3] A. Ng'oma, "Home networks: Role of fiber and wireless technologies," *OFC 2023*, in Workshop S2C "Perennial Bandwidth at Home: LiFi or FiWi?," San Diego, Mar. 5, 2023.

[2-4] Cisco Annual Internet Report 2018-2023 [on-line]. Available: https://www.cisco.com/c/en/us/solutions/collateral/executive-perspectives/annual-internet-report/white-paper-c11-741490.html

[2-5] A.M.J. Koonen, H.P.A. van den Boom, E. Ortego Martinez, et al., "Cost optimization of optical in-building networks", *Opt. Express*, vol. 19, no. 26, pp. B399–B405, Nov. 2011, https://doi.org/10.1364/OE.19.00B399.

[2-6] O. Ziemann, J. Krauser, P. E. Zamzow, and W. Daum, *POF Handbook: Optical Short Range Transmission Systems*, 2nd ed. Berlin: Springer-Verlag, 2008.

[2-7] Bend-insensitive multimode OM4 fiber [on-line]. Available: https://www.corning.com/optical-communications/worldwide/en/home/products/fiber/optical-fiber-products/clearcurve-multimode-fiber.html

[2-8] Bend-insensitive single-mode fiber [on-line]. Available: https://www.thefoa.org/tech/ref/fiber/BIfiber.html

[2-9] ESKA plastic optical fiber [on-line]. Available: https://www.pofeska.com/pofeskae/

[2-10] Y. Koike, T. Ishigure, and E. Nihei, "High-bandwidth graded-index polymer optical fiber," *J. Lightwave Technol.*, vol. 13, no. 7, July 1995, pp. 1475–1489.

[2-11] Multi-core plastic optical fiber [on-line]. Available: https://i-fiberoptics.com/asahi-multi-core-cable.php

[2-12] T. Ishigure, Y. Aruga, Y. Koike, "High-bandwidth PVDF-clad GI POF with ultra-low bending loss," *J. Lightwave Technol.*, vol. 25, no. 1, pp. 335–345, Jan. 2007.

[2-13] M. Asai, Y. Inuzuka, K. Koike, et al., "High-bandwidth graded-index plastic optical fiber with low-attenuation, high-bending ability, and high-thermal stability for home-networks," *J. Lightwave Technol.*, vol. 29, no. 11, pp. 1620–1626, June 2011.

[2-14] H. Yang, S.C.J. Lee, E. Tangdiongga, et al., "47.4 Gb/s transmission over 100 m graded-index plastic optical fiber based on rate-adaptive discrete multitone modulation", *J. Lightwave Technol.*, vol. 28, no. 4, pp. 352–359, Feb. 2010.

[2-15] A. Argyros, R. Lwin, S.G. Leon-Saval, et al., "Low loss and temperature stable microstructured polymer optical fibers", *J. Lightwave Technol.*, vol. 30, no. 1, pp. 192–197, Jan. 2012.

[2-16] S. Randel, F. Breyer, S.C.J. Lee, J.W. Walewski, "Advanced modulation schemes for short-range optical communications", *IEEE J. Sel. Top. Quantum Electr.*, vol. 16, no. 5, pp. 1280–1289, Sep. 2010.

[2-17] C.M. Okonkwo, E. Tangdiongga, H. Yang, et al. "Recent results from the EU POF-PLUS project: Multi-gigabit transmission over 1 mm core diameter plastic optical fibers," *J. Lightwave Technol.*, vol. 29, no. 2, pp. 186–193, Feb. 2011.

[2-18] A. Nespola, S. Straullu, P. Savio, et al., "A new physical layer capable of record gigabit transmission over 1 mm step index polymer optical fiber," *J. Lightwave Technol.*, vol. 28, no. 20, pp. 2944–2950, Oct. 2010.

[2-19] S. Loquai, R. Kruglov, B. Schmauss, et al., "Comparison of modulation schemes for 10.7 Gb/s transmission over large-core 1 mm PMMA polymer optical fiber," *J. Lightwave Technol.*, vol. 31, no. 13, pp. 2170–2176, July 2013.

[2-20] D. Visani, C.M. Okonkwo, Y. Shi, et al., "3x2N-QAM constellation formats for DMT over 1-mm core diameter plastic optical fiber," *IEEE Photon. Technol. Lett.*, vol. 23, no. 12, pp. 768–770, June 2011.

[2-21] Y. Shi, E. Tangdiongga A.M.J. Koonen, et al., "Plastic optical fiber based in-home optical networks," *IEEE Comm. Mag.*, pp. 186–193, June 2014.

[2-22] IEEE standard for ethernet amendment 9: Physical layer specifications and management parameters for 1000 Mb/s operation over plastic optical fiber [on-line]. Available: https://standards.ieee.org/ieee/802.3bv/6023/

[2-23] Technical paper GSTP-OVHN overview G.hn technology [on-line]. Available: https://www.itu.int/dms_pub/itu-t/opb/tut/T-TUT-HOME-2021-3-PDF-E.pdf

[2-24] M.V. Ramana Murty, Optical link over GI POF [on-line]. Available: https://www.ieee802.org/3/dh/public/Sep_2022/murty_3dh_01_220912.pdf

[2-25] B.L. Dang, R.V. Prasad, and I. Niemegeers, "On the MAC protocols for radio over fiber indoor networks,", *Proc. ICCE*, Hanoi, Oct. 10–11, 2006, pp. 112–117.

[2-26] R. Griffin, P.M. Lane, and J.J. O'Reilly, "Radio-over-fiber distribution using an optical millimeter-wave/DWDM overlay," *Proc. OFC/IOOC 1999*, San Diego, Feb. 21–26, 1999, Paper WD6-1.

[2-27] A.M.J. Koonen, and M. Garcia Larrode, "Radio-over-MMF techniques – part II: microwave to millimeter-wave systems", *J. Lightwave Technol.*, vol. 26, no. 15, pp. 2396–2408, Aug. 2008, https://doi.org/10.1109/JLT.2008.927182.

[2-28] Y. Shi, D. Visani, C. M. Okonkwo, et al., "First demonstration of HD video distribution over large-core POF employing UWB for in-home networks," *Proc. OFC/NFOEC 2011*, Los Angeles, Mar. 6–10, 2011, Paper OWB5.

[2-29] S.T. Abraha, E. Tangdiongga, A. Crivellaro, et al., "Accurate ranging/localization technique using IR-UWB for smart fiber-wireless in-house networks," *Proc. ECOC 2012*, Amsterdam, Sep. 16–20, 2012, Paper P6.03.

[2-30] S.T. Abraha, H. Yang, C.M. Okonkwo, et al., "Novel generation and transmission of 2 Gbps impulse radio ultra-wideband over MMF for in-building networks application," *Proc. OFC/NFOEC 2010*, San Diego, Mar. 21–25, 2010, Paper OML4.

[2-31] S. Zou, H. Chen, F. Huijskens, et al., "Demonstration of fully functional MIMO wireless LAN transmission over GI-MMF for in-building networks," *Proc. OFC/NFOEC 2013*, Anaheim, Mar. 17–21, 2013, Paper JTh2A.08.

[2-32] P. Guignard, J. Guillory, P. Chanclou, et al., "Multi-format home networks using silica fibres," *Proc. ECOC 2012*, Amsterdam, Sep. 16–20, 2012, Paper Mo.1.G.5.

[2-33] E. Tangdiongga, Y. Shi, C. Okonkwo, et al., "Recent achievements in POF transmission," *Proc. ECOC 2012*, Amsterdam, Sep. 16–20, 2012, Paper Mo.2.G.1.

[2-34] Y. Shi, C.M. Okonkwo, E. Tangdiongga, and A.M.J. Koonen, "Bi-directional ultra-wideband services over large-core POF-PON based home networks," *Proc. OFC/NFOEC 2013*, Anaheim, Mar. 17–21, 2013, Paper NTu3J.4.

[2-35] Y. Shi, C. Okonkwo, D. Visani, et al., "Distribution of broadband services over 1-mm core diameter plastic optical fiber for point-to-multipoint in-home networks," *J. Lightwave Technol.*, vol. 31, no. 6, pp. 874–881, Mar. 2013.

[2-36] C.R.B. Corrêa, K.A. Mekonnen, F.M. Huijskens, et al., "Multi-gigabits per second spatial multiplexing transmission using passive OFE and WDM-over-POF," *Proc. ECOC 2022*, Basel, Sep. 18–22, 2022, Paper We3F.4.

[2-37] A. Meijerink, C.G.H. Roeloffzen, L. Zhuang, et al., "Phased array antenna steering using a ring resonator-based optical beam forming network", *Proc. 13th Annual Symposium of the IEEE/CVT Benelux Chapter*, Liège, Nov. 23, 2006, pp. 7–12.

[2-38] Z. Cao, R. Lu, Q. Wang, et al., "Cyclic additional optical true time delay for microwave beam steering with spectral filtering," *Opt. Lett.*, vol. 39, pp. 3402–3405, June 2014.

[2-39] N.M. Tessema, Z. Cao, J.H.C. van Zantvoort, et al., "A tunable Si3N4 integrated true time delay circuit for optically-controlled K-band radio beamformer in satellite communication," *J. Lightwave Technol.*, vol. 34, no. 20, pp. 4736–4743, Oct. 2016.

[2-40] B.-M. Jung, J.-D. Shin, and B.-G. Kim, "Optical true time-delay for two-dimensional X band phased array antennas", *IEEE Photon. Technol. Lett.*, vol. 19, no. 12, pp. 877–879, June 15, 2007.

[2-41 A.M.J. Koonen, Z. Cao, and E. Tangdiongga, "Optically controlled 2D radio beam steering system," *Proc. MWP 2014*, paper WD-3, Sapporo, Oct. 20–23, 2014, pp. 389–391.

[2-42] S. Yegnanarayanan, P. D. Trinh, and B. Jalali, "Recirculating photonic filter: a wavelength-selective time delay for phased-array antennas and wavelength code-division multiple access," *Opt. Lett.*, vol. 21, no. 10, pp. 740–742, Nov. 1996.

[2-43] B. Jalali, and S.S. Yegnanarayanan, "Method and apparatus for a wavelength selective true-time delay for an optically controlled device," US Patent 5793907, Aug. 11, 1998

[2-44] B. Vidal, J. L. Corral, M. A. Piqueras, and J. Marti, "Optical delay line based on arrayed waveguide gratings' spectral periodicity and dispersive media for antenna beamforming applications," *IEEE J. Sel. Top. Quantum Electron.*, vol. 8, no. 6, pp. 1202–1210, Nov.–Dec. 2002.

[2-45] Z. Cao, *Radio beam steering in indoor fibre-wireless networks*, PhD thesis, Eindhoven Univ. of Technology, Apr. 15, 2015, Ch. 3.4.

[2-46] IEEE 802.11be, a.k.a. WiFi 7 [on-line]. Available: https://en.wikipedia.org/wiki/IEEE_802.11be

[2-47] A.M.J. Koonen, N.C. Tran, and E. Tangdiongga, "The merits of reconfigurability in WDM-TDM optical in-building networks," *Proc. OFC/NFOEC 2011*, Los Angeles, Mar., 2011, Paper JWA63.

[2-48] S. Zou, C.M. Okonkwo, Z. Cao, et al., "Dynamic optical routing and simultaneous generation of millimeter-wave signals for in-building access network," *Proc. OFC/NFOEC 2012*, Los Angeles, Mar. 6–10, 2012, Paper OTh3G6.

[2-49] A.M.J. Koonen, H.P.A. van den Boom, and G.-D. Khoe, "Broadband access and in-house networks – extending the capabilities of multimode fibre networks," *Proc. ECOC 2003*, Rimini, Sep. 21–25, 2003, Paper Mo3.1.5.

[2-50] A.M.J. Koonen, H.P.A. van den Boom, I. Tafur Monroy, and G.-D. Khoe, "High capacity multi-service in-house networks using mode group diversity multiplexing," *Proc. OFC/NFOEC 2004*, Anaheim, Feb. 22–27, 2004, Paper FE-H.

[2-51] H.S. Chen, H.P.A. van den Boom, and A.M.J. Koonen, "30-Gb/s 3×3 optical mode-group-division multiplexing system with optimized joint detection," *IEEE Photon. Technol. Lett.*, vol. 23, no. 18, pp. 1283–1285, Sep. 2011.

[2-52] S. Schöllmann, W. Rosenkranz, "Experimental verification of mode group diversity multiplexing over GI-POF at 21.4 Gb/s without equalization," *Proc. Plastic Optical Fiber Conference 2006*, Seoul, Sep. 11–14, 2006.

[2-53] P.J. Winzer, "Spatial multiplexing in fiber optics: The 10x scaling of metro/core capacities," *Nokia Bell Labs Techn. J.*, vol. 19, pp. 22–30, Sep. 2014, https://doi.org/10.15325/BLTJ.2014.2347431

[2-54] F. Träger (ed.), *Handbook of lasers and Optics*. Berlin: Springer-Verlag, 2012, chapter 25 Laser safety.

[2-55] Laser safety [on-line]. Available: https://en.wikipedia.org/wiki/Laser_safety

[2-56] V. Jungnickel, A. Forck, T. Haustein, et al., "Electronic tracking for wireless infrared communications," *IEEE Trans. Wireless Comm.*, vol. 2, no. 5, pp. 989–999, Sep. 2003.

[2-57] S.C.J. Lee, J.W. Walewski, S. Randel, et al., "Discrete multitone for novel application areas of optical communications", *Proc. IEEE/LEOS Summer Top. Meetings 2008*, Acapulco, July 21–23, 2008, Paper TuE3.3.

[2-58] J. Vucic, C. Kottke, S. Nerreter, et al., "513 Mbit/s visible light communications link based on DMT-modulation of a white LED," *J. Lightwave Technol.*, vol. 28, no. 24, pp. 3512–3518, Dec. 2010.

[2-59] G. Cossu, A.M. Khalid, P. Choudhury, R. Corsini, and E. Ciaramella, "3.4 Gbit/s visible optical wireless transmission based on RGB LED," *Opt. Express*, vol. 20, no. 26, pp. B501–B506, Dec. 2012.

[2-60] J. Grubor, S. Randel, K.-D. Langer, and J.W. Walewski, "Broadband information broadcasting using LED-based interior lighting," *J. Lightw. Technol.*, vol. 26, no. 24, pp. 3883–3892, Dec. 2008.

[2-61] D. O'Brien, H. Le Minh, G. Faulkner, et al., "High data-rate infra-red optical wireless communications: implementation challenges", in *Digest IEEE Globecom 2010 Workshop on Optical Wireless Communications*, Miami, Dec. 10, 2010, Workshop OWC03.

[2-62] J. Zeng, V. Joyner, J. Liao, et al., "A 5 Gb/s 7-channel current-mode imaging receiver front-end for free-space optical MIMO," *Proc. 52nd IEEE Int. Midwest Symp. Circuits Syst.*, Cancun, Aug. 2–5, 2009, vol. 1/2, pp. 148–151.

[2-63] D. O'Brien, R. Turnbull, H. Le Minh, et al., "High-speed optical wireless demonstrators: conclusions and future directions," *J. Lightwave Technol.*, vol. 30, no. 13, pp. 2181–2187, Jul. 2012.

[2-64] H. Le Minh, D. O'Brien, G. Faulkner, et al., "A 1.25-Gb/s indoor cellular optical wireless communications demonstrator," *IEEE Phot. Technol. Lett.*, vol. 22, no. 21, pp. 1598–1600, Nov. 2010.

[2-65] C.J. Henderson, D.G. Leyva, and T.D. Wilkinson, "Free space adaptive optical interconnect at 1.25Gb/s, with beam steering using a ferroelectric liquid-crystal SLM," *J. Lightwave Technol.*, vol. 24, no. 5, pp. 1989–1997, May 2006.

[2-66] A.D. Yalcinkaya, H. Urey, D. Brown, et al., "Two-axis electromagnetic microscanner for high resolution displays," *IEEE J. Microelectromech. Syst.*, vol. 15, no. 4, pp. 786–794, Aug. 2006.

[2-67] H. Chen, H.P.A. van den Boom, E. Tangdiongga, and A.M.J. Koonen, "30-Gb/s bidirectional transparent optical transmission with an MMF access and an indoor optical wireless link," *IEEE Phot. Technol. Lett.*, vol. 24, no. 7, pp. 572–574, Apr. 2012.

[2-68] D. Zhang, J. Zhu, X. Liu, et al., "Fiber-to-the-room: a key technology for F5G and beyond," *IEEE/OPTICA J. Opt. Comm. and Netw.*, vol. 15, no. 9, 9pp, Sep. 2023, 10.1364/JOCN.485070.

3

Optical Wireless Communication Technologies[1]

3.1 Introduction

Already for decades, the demand for wireless communication has been growing relentlessly. The growth is being fueled by broadband mobile services to smartphones, tablet computers, laptops, etc., and by interaction over the internet among the myriads of devices constituting the Internet of Things. These wireless traffic loads are mainly generated indoors, and are driving today's wireless networks such as those in the IEEE 802.11 suite into serious congestion. New radio spectrum bands are being opened, for example, sub-THz ones stretching the spectrum up to some 300 GHz. Nevertheless, the ongoing exponential traffic growth is quickly exhausting the capabilities offered by radio communication. However, in the vast frequency regions of the optical domain, a wealth of additional spectrum is available, without requiring licensing fees (see Fig. 3-1).

This yet unoccupied vast spectrum can be disclosed by optical wireless communication (OWC) techniques. Moreover, OWC enables the creation of smaller communication cells in the network, which is a vital key to unlock the path to exponential capacity growth. According to Cooper's law [3-1], the density of radio communications has increased by a factor of 10^6 in the past 45 years. To this factor, the opening-up of new radio spectrum has contributed with a factor of 25, frequency division techniques with a factor of 5, and more comprehensive modulation formats with a factor of 5, but the major factor (1600×) was contributed by the introduction of spatial division, in particular by smaller cells and/or more antennas per cell enabling spectrum re-use.

---

1 This chapter is largely based on A.M.J. Koonen, "Indoor optical wireless systems: Technology, trends, and applications," *IEEE/OSA J. Lightwave Technol.*, vol. 36, no. 8, pp. 1459–1467, Apr. 2018, https://doi.org/10.1109/JLT.2017.2787614.

Short-Reach Optical Wireless Communication: By Directed Narrow Beams, First Edition. Ton Koonen.

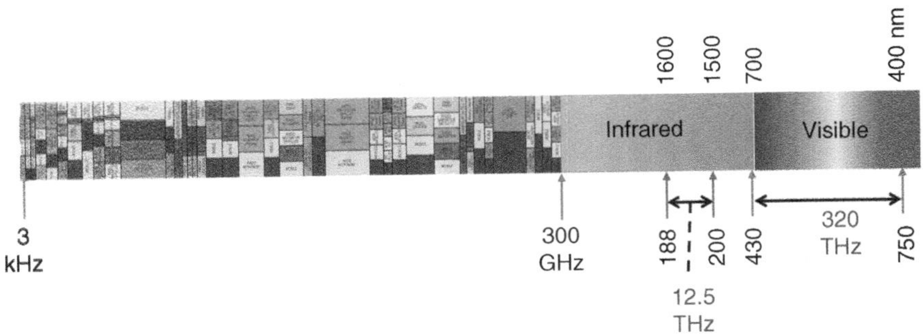

Fig. 3-1 The crowded radio spectrum: the optical domain offers huge extra spectrum

OWC can offload heavy traffic loads from congested radio-based wireless networks, thus making the room available which is desperately needed for the many intermittent low-capacity streams such as in the Internet of Things. However, OWC is not intended to fully replace radio wireless communication. For example, it cannot reach out-of-sight devices (such as smartphones, shielded within clothing), while radio techniques (below 60 GHz) can do that.

Actually, OWC by means of beams of light was already proposed by Alexander Graham Bell in 1880 [3-2]. He invented the "photophone" together with his assistant Charles Sumner Tainter on February 19, 1880, in his laboratory in Washington, DC (Fig. 3-2). A voice telephone message was sent by the photophone from the roof of the Franklin School to the window of his laboratory over a distance of about 213 meters. Bell stated that the photophone was "his greatest, most important invention." Of his 30 patents, 4 were about the photophone. The master patent was issued in December 1880 (US Patent 235,199 "Apparatus for Signalling and Communicating, called Photophone"). The photophone is regarded as the precursor of fiber–optic communication systems.

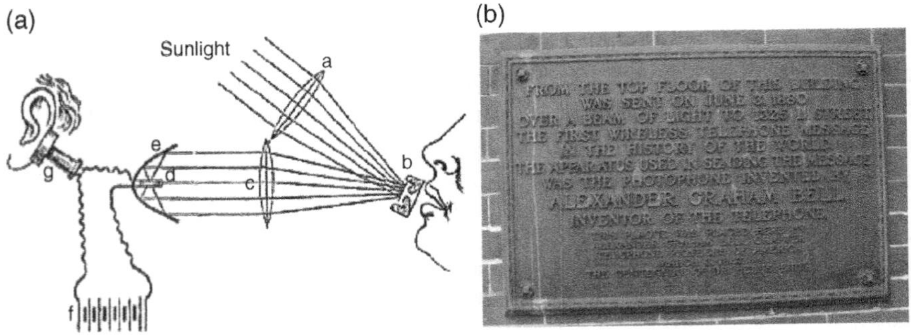

Fig. 3-2 Alexander Graham Bell's photophone. (a) Schematic of the photophone. (b) Plaque at Franklin School, Washington D.C. (Unknown author / Wikimedia Commons / Public domain)

This chapter intends to give a concise tutorial overview of the research domain of OWC. This field is very broad, with a large diversity of activities. The overview given here, therefore, cannot be exhaustive and mainly focuses on OWC techniques for short-range (indoor) communication. The key OWC functions, namely, transmission and reception of the optical beams and localization of the users, as well as the architecture of indoor OWC systems are treated in-depth in the respective Chapters 4–8.

After discussing the application domains of OWC in Section 3.2, techniques for visible light communication (VLC) and its network-based variant light fidelity (LiFi), and for beam-steered infrared light communication (BS-ILC), which are the two main methods for OWC, are discussed in Sections 3.4, 3.5, and 3.6, respectively. In Section 3.7, techniques for the localization of the users' devices (amongst others to support beam steering) are discussed. Subsequently, techniques for OWC receivers in order to increase the field-of-view (FoV) for easing receiver alignment and to enlarge the aperture for capturing more power are addressed in Section 3.8. In Section 3.9, the application of these techniques in hybrid bidirectional networks (i.e., combining optical wireless and radio wireless communication) and in all-optical bi-directional OWC networks is discussed. In Section 3.10, the characteristics of beam-steered OWC are compared with those of wide-beam OWC (i.e., VLC/LiFi), regarding EMI immunity, data capacity per device, latency, and energy efficiency. Beam-steered OWC is very strong in these aspects; it is less strong in the support of user/device mobility and may be hampered by non-line-of-sight (non-LoS) conditions. Photonic integration technologies will play an important role in enabling robust and compact indoor OWC systems. So far, photonic-integrated circuits (PICs) for communication have been mainly explored for optical fiber communication. Their potential for various OWC functions is discussed in Section 3.11. The chapter concludes with some considerations about the prospects of indoor OWC in Section 3.12.

3.2 OWC's Application Domains

OWC is not hampered by electromagnetic interference (EMI) nor does it generate interference for EM-sensitive devices, it is nearly impossible to jam and offers a high level of privacy (e.g., does not penetrate walls to get to the neighbors), its installation can be relatively easy (no wired infrastructure needed), and it gives access to a huge amount of spectrum which is license-free. Building on its ability to selectively cover small areas, by means of spatial re–use it can offer a large, aggregated capacity. These characteristics are advantageous in densely populated environments where people generate high traffic loads, such as in open office rooms, conference halls, exhibition halls, airplane cabins, train cabins, (virtual reality) gaming halls, etc. Its relative ease of installation (e.g., VLC may be integrated with illumination) and high privacy level make it attractive for residential services at home and in office buildings. Its EMI-insensitivity enables robust and highly reliable communication such as needed for industrial manufacturing robots, intra-data center top-of-the-rack interconnects, temporary (disaster-recovery) networks, intensive care hospital rooms, surgery rooms, etc. Its latency is substantially lower than that of fiber-based links, making it particularly attractive for machine-to-machine communication, for instance in intra-data center networks. Atmospheric conditions may hamper its outdoor use, in particular over longer reaches and in foggy conditions. Mobile backhauling for small radio cells (spaced several hundreds of meters), car-to-car and car-to-roadside communication are examples of attractive outdoor scenarios.

3.3 Indoor OWC Techniques

As illustrated in Fig. 3-3, two major categories of indoor OWC techniques may be distinguished: visible light communication (VLC, and its networked variant LiFi) and beam-steered infrared (IR) light communication (BS-ILC).

VLC (and more specifically LiFi [3-4]) gives access to a frequency spectrum of no less than 320 THz in the 400–700 nm range. VLC typically is deployed in

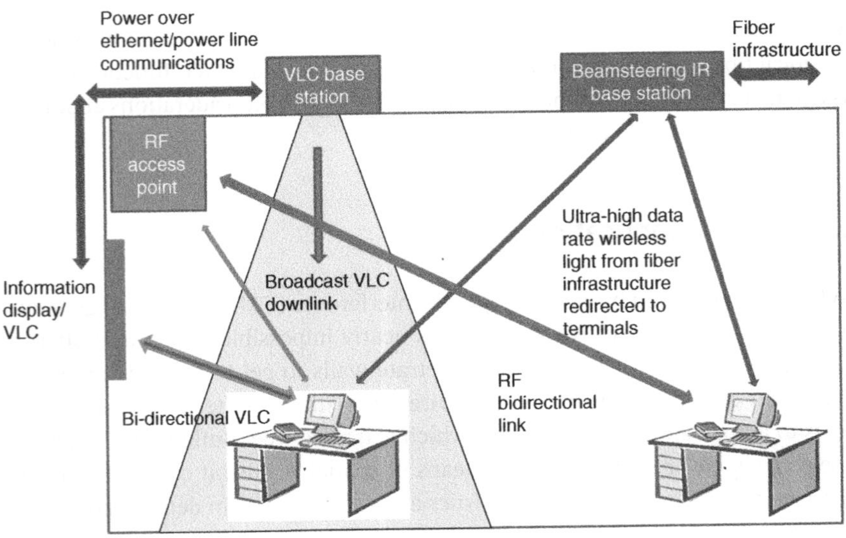

Fig. 3-3 Basic indoor OWC scenarios ([3-3]/D.C. O'Brien.)

conjunction with illumination, by modulating the data on the LED luminaries. The re-use of the existing infrastructure can lower the entry threshold for VLC. These LEDs have primarily been designed for generating light in the most efficient way and have a restricted modulation bandwidth. Moreover, the illumination systems typically serve a large area in which multiple devices have to share this limited VLC capacity by means of a suitable medium access control (MAC) protocol. The received light intensity goes down proportionally with the squared distance to the LED source, thus limiting the SNR and hence the reach. VLC requires that the illumination is switched on, which may not always be desired (e.g., when the room is already filled with bright daylight).

BS-ILC brings the light only there where and when it is needed. It operates independently from illumination systems and needs a separate infrastructure. Benefiting from well-established technologies in the S + C + L wavelength bands for fiber–optic communication, access can be obtained to a frequency spectrum of 20.9 THz in the 1460–1625 nm wavelength range. Multiple beams may each serve a user terminal individually within the room, without sharing. Thus, no MAC protocols to arrange shared access are needed, and each terminal can get a guaranteed capacity without conflict with other terminals. The directivity of a beam implies that the received power can be substantially higher than with VLC, enabling a higher SNR and thus a higher data rate and longer reach at better power efficiency. As a beam link offers a high bandwidth, for transporting high data rates no comprehensive spectrum-compressing signal processing is needed, thus minimizing the communication latency. Moreover, IR light does not reach the vulnerable retina in the human eye (see Fig. 2-26), which implies that relatively high transmitted power levels are allowed within the eye safety limits (up to 10 mW for $\lambda > 1.4\,\mu m$, according to IEC 60825, ANSI Z136 standards [3-5]). Furthermore, the photodiode's (PD) responsivity basically is proportional to wavelength[2] [3-6], which enables receivers to have a higher sensitivity in the IR domain than in the visible domain.

The IEEE 802.11 suite of radio standards is presently widely established and used for indoor wireless communications. Its basic characteristics are compared in a non-exhaustive way with exemplary OWC characteristics in Table 3-1. With BS-ILC, no capacity sharing is incurred, thus a guaranteed very high capacity per device can be provided. With both VLC and BS-ILC, however, line-of-sight (LoS) is needed. With its recently established IEEE 802.11bb standards, LiFi based on its multipoint networking can relax the LoS demands. With the IEEE 802.11 radio techniques (at least below 60 GHz) non-LoS devices can be reached also.

---

2 Responsivity $\mathcal{R} = \eta\, e\, \lambda / h\, c_0$ with η the quantum efficiency of the photodiode, e charge of an electron, λ wavelength of the light incident on the photodiode, h Planck's constant, c_0 speed of light in vacuum [3-6].

Table 3-1 OWC compared with IEEE 802.11 suite

	VLC	Li-Fi	BS-ILC	Wi-Fi IEEE802.11
Cap. per user	*Shared (P2MP)*; 9.5 Gbit/s over 1m	*Shared* (P2MP); 220 Mbit/s over few m (845 Mbit/s P2P)	*Unshared* (P2P); very high (up to 112 Gbit/s per beam)	*Shared*; 802.11n (2.4 GHz): <150 Mbit/s 802.11ac (5 GHz): <867 Mbit/s 802.11ad (60 GHz): <2.4 Gbit/s
Reach	Short-to-medium (few m) Needs LoS	Short-to-medium (few m)	Medium (<10 m)-long Needs LoS	Medium Does not need LoS (<60 GHz)
Energy consumption	High (Watts) if via illumination (needs to be <u>on</u>)	Low/moderate	Low (per beam, <10 mW), on-demand only	Base station >100 mW
Carrier freq., av. bandwidth	400–700 nm; spectrum width 320 THz; LED source few tens of MHz	Near-IR, 800–1000 nm	1460–1625 nm (S + C + L bands); spectrum width 20.9 THz, per channel ~35 GHz (AWGR steering)	2.4 GHz band; width 83.5 MHz channels 20 or 40 MHz 5 GHz band; width 575 MHz channels 20 ~160 MHz 60 GHz band; width 7 GHz; channels <320 MHz
Safety aspects, power limit	Penetrates eye; if collimated: <<1 mW	Reduced eye risk; if collimated, <1 mW	Does not penetrate eye; collimated <10 mW	EM hypersensitivity issues
Privacy	Good (contained by walls)	Good (contained by walls)	Good (contained by walls)	Open; needs encryption
Infrastructure	Share LED illumination, use electrical infra (w. PLC)	Use of electrical infra (w. PLC)	New indoor (fiber) infra	Electrical cable (Cat5) infra
Standardization	steps being made in IEEE 802.11 LC study group	IEEE 802.11bb standard ready	Not yet	Extensive, mature, keeps evolving Backward compatible
Utilization	Some products on market	Products on the market	Laboratory phase	Very widespread; more Wi-Fi devices than people on earth; >50% of internet traffic through Wi-Fi

P2MP: point to multipoint links; P2P: point-to-point link.

Products of the IEEE 802.11 Wi-Fi suite are abundantly available commercially. Empowered by the recent IEEE 802.11bb standard, LiFi products in various embodiments are entering the market. VLC is bringing some first products into the market, and BS-ILC is still largely in the laboratory research phase. Due to the difference in market maturity, cost comparisons are hard to make. Basically, from a capital expenditures (CapEx) perspective the IEEE 802.11 Wi-Fi suite, LiFi and VLC tend to incur lower costs, as they need little infrastructure or re-use existing infrastructure (by means of powerline communication, or power-over-ethernet techniques), while BS-ILC needs a new fiber backbone infrastructure (see Chapter 4). From the perspective of recurring operating expenditures (OpEx), such as powering, the BS-ILC may be more beneficial as it can be more energy-efficient due to its flexibility to offer capacity-on-demand where and when needed.

3.4 Visible Light Communication Techniques

LEDs which are primarily designed for illumination purposes typically deploy a blue LED chip plus a phosphor coating which generates white light (see Fig. 3-4a [3-7]). The slow decay time of the phosphor severely limits the bandwidth to a few MHz only. Some improvement is obtained when filtering the blue LED light at the receiver, but the bandwidth is still restricted (typically <20 MHz). Employing this blue-filtering and basic NRZ-OOK modulation, 40 Mbit/s transmission has been achieved at an illumination level of 700 lux (office areas require 200–800 lux) [3-8]. Spectrum-efficient advanced modulation schemes are to be used in order to realize higher data rates. Applying OFDM with 256 subcarriers and bit- and power-loading, in conjunction with a pre-equalizing filter before the LED driver and a blue filter before a PIN-TIA receiver, a net data rate of 2.0 Gbit/s over 1.5 meters free-space reach was achieved, and 0.79 Gbit/s over 3 meters [3-9].

Alternatively, white light may be produced by a multi-element LED, composed of tricolor elements (red + green + blue; see Fig. 3-4b [3-7]). Although each of the LED elements may have a limited bandwidth (typically some 15 MHz), by modulating them separately the capacity can be tripled. This wavelength division multiplexing (WDM) approach requires a more complex driver at the transmitter end, and three color-filtering stages in parallel at the receiver. Taking this approach and using DMT modulation with 512 subcarriers and color-filtering APD receivers, 3.4 Gbit/s was achieved over 10 cm (at 410 lux), and 2.2 Gbit/s over 30 cm (at 25 lux) [3-10]. With a quad-color LED (red + green + blue + yellow) and tailored carrierless amplitude phase (CAP) modulation (128-CAP for red, 32-CAP for green, 64-CAP for blue, and 128-CAP for yellow), an aggregate capacity of 8 Gbit/s over 1 meter was achieved (see Fig. 3-5) [3-11].

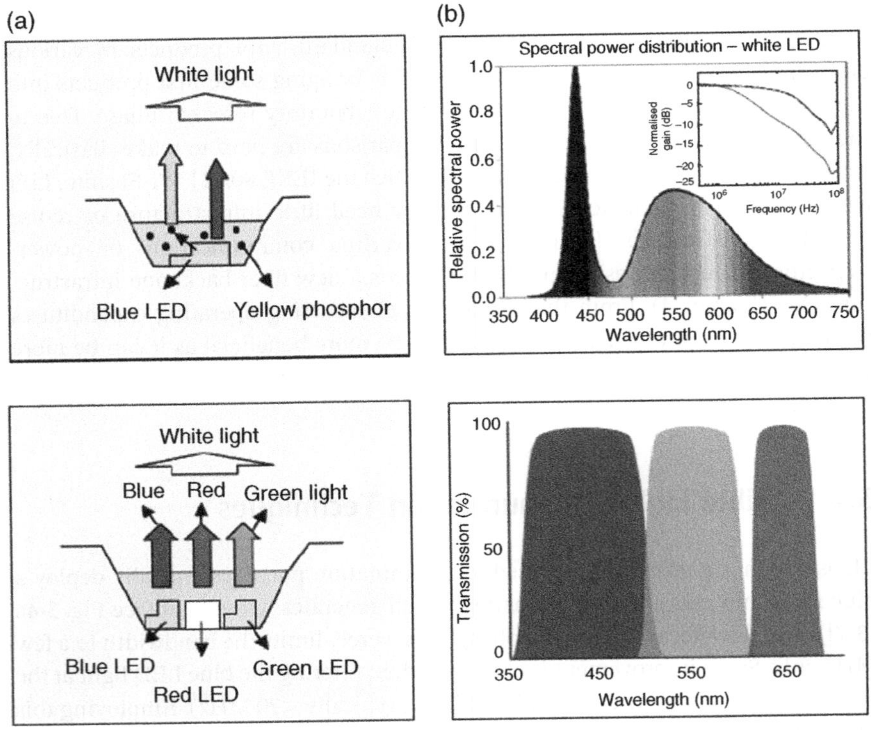

Fig. 3-4 LED types for VLC; (a) blue LED + phosphor, (b) RGB multielement LED (Langer et al. [3-7]/Fraunhofer HHI)

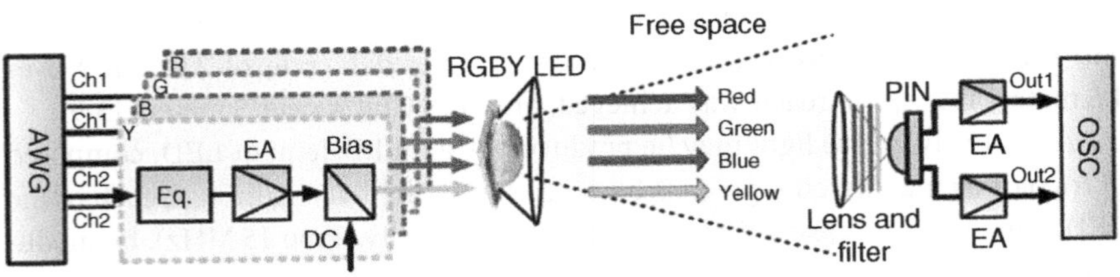

Fig. 3-5 8 Gbit/s quad-color VLC system using CAP modulation ([3-11]/Nan Chi)

The integration of lighting and communication functions brings particular challenges, as the consumer should not perceive any light flickering effects, and dimming the light over a wide range should be possible, while communication should continue to be supported. Within IEEE 802.15.7, standards are set for high-speed

VLC up to 96 Mbit/s. Among their physical layer techniques is the use of multicolor sources using color shift keying (CSK) [3-12]. Multicolor CSK schemes using three colors (RGB) and four colors (blue, cyan, yellow, and red) have been proposed; the four-dimensional (4D) coding scheme can have a higher efficiency and better reliability for a given multipath delay spread [3-13]. CSK uses intensity modulation of the colored LED elements individually, which implies that the resulting overall color varies quickly. These instantaneous color variations should not be perceivable by the human eye, so should be faster (i.e., <10 ms) than the critical flicker fusion threshold and the critical color fusion threshold. The transmission by the multicolor LED should be configured adequately in order that a specific constant color is perceived when time-averaged. This requires adequate LED-driving algorithms [3-14].

Yet another approach is to apply an array of micro-LEDs, which each due to their small active area can have a large bandwidth. Using an array of blue ($\lambda = 420$ nm) GaN μLEDs each having an active area diameter of only 42 μm, bandwidth of 400 MHz, and 5.7 mW output power, with PAM-4 modulation 3.8 Gbit/s (3.5 Gbit/s excluding 7% FEC) over 0.75 m was achieved, and 5.37 Gbit/s (5 Gbit/s excl. FEC) when using DC-biased optical OFDM (DCO-OFDM) with 512 bit-loaded subcarriers [3-15].

Spatial multiplexing can open another route toward a higher VLC capacity. In a MIMO system setup with a two-dimensional (2D) array of nine individually addressable Ø39 μm blue GaN μLEDs emitting at $\lambda = 450$ nm and with nine APDs (each $200 \times 200\ \mu m^2$, bandwidth >90 MHz) made in a 0.18-μm CMOS process, with PAM-8 modulation about 890 Mbit/s per channel was achieved over a free-space link of 50 cm, and an aggregate rate of 6.95 Gbit/s (excl. 7% FEC) [3-16].

VLC systems are becoming available commercially. Bidirectional point-to-point systems are offered providing 1 Gbit/s connectivity with low latency (<2 ms) [3-17].

3.5 Light Fidelity (LiFi) Techniques

Dubbed LiFi, VLC techniques which are basically point-to-point are extended into a fully networked system in an atto-cells network architecture, supporting seamless handover between these cells and bidirectional operation [3-4]. The LiFi-X system of PureLiFi supports 40Mbit/s full-duplex communication, using a USB dongle at the user's device [3-18]. Trulifi products of Signify offer data rates of 220/160 Mbit/s in down/up point-to-multipoint link, and up to 845 Mbit/s in a down/up point-to-point link [3-19]. Standardization of LiFi is well underway. The 802.11bb standard ([3-20], established by the IEEE 802.11 Light Communication

Task Group with efforts from PureLiFi and Fraunhofer Heinrich Hertz Institute) builds on the well-established IEEE 802.11 suite of standards and describes the communication protocols for LiFi devices. Different from VLC, the LiFi technology is not primarily aiming to re-use the existing illumination technology. The 802.11bb standard describes LoS networking using light in the near-infrared spectrum 800–1000 nm for data transport between 10 Mbit/s and 9.6 Gbit/s.

3.6 Beam-Steered IR Optical Communication Techniques

When working with narrow IR beams, accurate and independent 2D steering of multiple beams is needed in order to establish links to multiple (non-stationary) terminals. Several approaches can be followed to establish beam steering, which may be classified coarsely in those using active steering devices and in those using passive ones [3-21].

In active steering devices, the beam steering is accomplished by changing the state of the device upon the command from a control channel. For each beam to be steered individually, per beam a specific state needs to be set and a separate control channel is required. This leads to a complex comprehensive beam steering device if many beams need to be steered. The device needs to be powered locally, which increases installation, maintenance, and operation costs. Examples of active steering devices are those using spatial light modulators (SLMs) [3-22–3-24], and those using micro-electromechanical system (MEMS) mirrors [3-25]. Active beam steering concepts are discussed in more depth in Section 5.2.

In passive steering devices, the physical state of the device is not changed; no moving parts or other electrically controlled elements are needed. Also, no local powering is required, which eases the installation, operation, and maintenance. The beam steering may be accomplished by diffractive means such as gratings [3-27, 3-28]. By changing the wavelength of a data signal fed to the device also the direction is changed in which the signal's beam is steered. Moreover, multiple beams can readily be steered by a single diffractive device by just changing the wavelength of each data signal [3-21]. Hence, a relatively simple device suffices for steering many beams individually. Moreover, each wavelength acts both as a carrier of the data signal and as the control channel for the signal's beam steering. That is, the control channel is embedded in the data channel and no separate bookkeeping of which control channel belongs to which data channel is needed, thus significantly easing the network management and control functions. Passive beam steering concepts are discussed in more depth in Section 5.3 and following.

3.7 Localization

Localizing the user is needed in a beam-steered OWC system. It may also be useful in wide-beam OWC systems such as LiFi or VLC systems, when devices are to be tracked in for instance an industry 4.0 setting and warehouses.

Next to sending data for communication, a VLC system composed of (multiple) LED luminaries can send signals that contain the identity information of the LEDs and can assist in determining the position of a device. As compared to RF, visible light suffers less from multipath effects and can provide more accurate positioning. The location of the device may be determined by an algorithm using triangulation based on time-of-arrival (ToA, like in GPS) data of the light signals from various directions, on time difference of arrival (TDoA) data, on angle of arrival (AoA) data, or on received signal strength (RSS) data [3-29]. The ToA and TDoA methods require accurate synchronization. AoA can achieve the highest accuracy but is the most complex to realize. RSS is the most popular approach in VLC systems as it operates asynchronously and is simple to implement, although it needs well-assessed path loss models.

Optical camera communication (OCC) [3-30, 3-31] is being explored for low-to-moderate speed data transfer by LED lighting, and deploys an image sensor or camera at the user's device such as smartphone. The data speed is basically limited by the camera's frame rate. MIMO techniques can be exploited using an LED array and image sensor with pixel-to-bit mapping. By means of computer vision analysis techniques which enable extraction of the LED coordinates, OCC can also enable accurate device positioning.

Alternatively, localization may use an algorithm based on fingerprinting, or vision analysis (using a camera on the device). In [3-32], a 16×16 array of $100 \times 100\,\mu m^2$ GaN μLEDs emitting at $\lambda = 450\,nm$ is used to generate a time sequence of spatial illumination patterns. These patterns create a specific fingerprint optical signal sequence per site, which when detected by a simple photodetector can provide the device's location. Fast positioning within 10 ms was shown, with an accuracy better than 5 mm over a 40 cm distance; accuracy and area coverage can be improved when more emitters are used.

For beam-steered OWC, the position of the user's device needs to be known with an accuracy depending on the beam's width. The device needs to be localized in order to establish the communication link, and it also needs to be tracked in order to maintain the link when moving. To fully support the mobility of the device, the speed of the localization process should be at least as fast as the user's movements. When this speed is not sufficient, the process may still be able to support nomadic users, that is, users who have been settled down for some time and have established a connection, but subsequently decide to move, interrupt their connection, settle down somewhere else, get re-localized and re-establish their connection.

An example of such nomadic use may be the use of a laptop computer. The localization may be done by camera observation and thus be fast enough to support (not too fast) user mobility [3-34, 3-35]. The user positions registered by the camera need to be calibrated to the beam steering. If the localization is done by beam scanning (using the active or passive beam steering devices discussed in Section 3.6) combined with a feedback loop reporting successful localization, the process may not be fast enough to support full mobility, but can still satisfy the needs of nomadic mobility. By following the beam-scanning approach and using optical retro-reflector technology, the need for a pre-existing feedback loop is avoided, and auto-calibration of the localization to the beam steering is achieved as the scanning beam becomes the signal transport beam after completion of the localization [3-36]. The latter also implies that parallax errors are avoided (which would occur when the localization and the signal transmission are physically separated, such as when localization is done with a camera). In Chapter 6 localization techniques for beam-steered OWC systems are discussed in full depth.

3.8 OWC Receiver

An OWC receiver is a key part of the user's device. It should be compact and low-cost, it should not require tedious alignment with respect to the downstream transmitted optical beams, and it should capture enough optical power to enable a high downstream data capacity. Hence, it should have a wide field-of-view (FoV) and a large aperture.

However, increasing the active area of a photodetector is typically accompanied by a reduction of its bandwidth. The "conservation of etendue" law in physics states that etendue (also known as the $A \times \Omega$ product, the product of entrance area A and solid angle Ω subtended by a cone of light) is conserved when light travels through optical systems with perfect, loss-free reflections and refractions. Etendue can then remain constant or increase (e.g., by applying a diffuser), but not decrease [3-37]. By non-imaging optics, such as a compound parabolic concentrator (CPC, a reflective non-imaging optical element typically used for solar energy concentration [3-38]) put in front of a photodetector, the receiving aperture can be increased, see for instance the OWC system shown in Fig. 5-3. However, this comes at the expense of a reduced FoV, and the size of the CPC compromises the compactness of the receiver.

The law of conservation of etendue may be broken by converting the incoming light captured with a wide aperture (a slab waveguide doped with a fluorophore [3-39], or long fiber [3-40]) to a longer wavelength by means of phosphorescence. The converted light is then guided to a detector. Dispersion effects in those waveguides reduce the suitability for handling high data rates.

Another approach is to separate the light collection function from the light detection function, which allows the optimization of each function separately. For example, the light may be captured by a surface grating coupler (SGC) into a waveguide which brings it to a fast PD, co-integrated with the SGC and the waveguide in a PIC [3-41]. The receiving aperture may be scaled up by apodization of the SGC, and by creating an array of SGCs which jointly bring the collected light into the waveguide. However, the light that can be captured is limited by the acceptance angle of an SGC and its typical polarization dependence. This integrated cascaded aperture optical receiver concept is described in more detail in Section 3.11.2.

The effective detection area of the OWC receiver may be enlarged while maintaining adequate bandwidth by using a 2D array of fast PDs co-integrated with individual electrical preamplifiers followed by a summation amplifier (e.g., [3-42]). Such an angular diversity receiver (ADR) configuration is relatively complex, puts high requirements on the phase synchronization among the amplifiers, and has noise accumulation of the amplifiers, and an increased power consumption.

A novel concept using a 2D matrix of interconnected PDs has been proposed in order to increase the receiver's aperture and its FoV while maintaining the large bandwidth of a single PD [3-43–3-45]. This concept requires only a single electrical preamplifier, thus reducing the complexity of the receiver circuitry and avoiding the phase synchronization, noise accumulation and power consumption issues as encountered in an ADR. Its design and characteristics are discussed in depth in Chapter 7, and its potential for photonic integration is discussed in Section 3.11.2.

3.9 Bidirectional Wireless Networks

Many network concepts that have been reported in literature focus on high-capacity downstream transmission using OWC and provide the upstream path by means of (established) lower-capacity radio techniques. The capacity asymmetry of such a hybrid optical/RF wireless system is acceptable in many service scenarios, such as the downloading of large video files, but may be inadequate for symmetric capacity needs such as fast bidirectional file transfer between data servers or computers.

Within the European Research Council's project BROWSE – Beam-steered Reconfigurable Optical-Wireless System for Energy-efficient communication, we have investigated a hybrid OWC/radio network architecture as shown in Fig. 3-6 [3-26]. This architecture is discussed in detail in Chapter 4. An indoor fiber backbone network routes the optical signals on demand to the appropriate rooms. For the downstream OWC, each room has a diffractive pencil-radiating antenna (PRA), which can steer multiple narrow beams (so-called pencel beams) two-dimensionally in directions determined by their wavelengths. The upstream path from each mobile

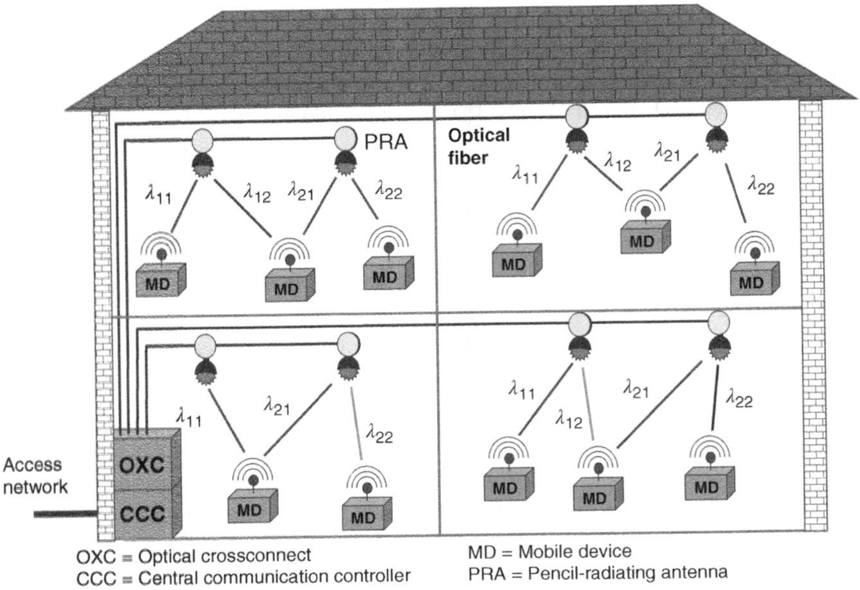

Fig. 3-6 Hybrid OWC/radio network using wavelength-controlled 2D infrared steered beams for downstream and 60 GHz radio beams for upstream (Koonen et al. [3-26]/© [2016] IEEE)

device (MD) is established with 60 GHz radio beams steered with phased array antennas. An integrated full-duplex hybrid system has been reported in [3-33] which provides 35 Gbit/s with OOK modulation downstream per infrared beam steered by a PRA composed of an 80-ports C-band AWGR with 35 GHz channel bandwidth driving a 2D fiber matrix and $f = 40$ cm condenser lens structure (for more details, see Section 6.1, Fig. 6-1). The beam diameter was 8.3 cm, and with a reach of 2.5 meters, an angular coverage of 17° × 17° was achieved. Upstream, 5 Gbit/s ASK was sent by 60 GHz radio beams, deploying two 30 dBi gain horn antennas at the device side and a 16 dBi antenna at the PRA side. The horn antennas were placed on a 2D electromechanical positioning stage, which enabled the device localization function. By applying PAM-4 modulation, the system's downstream throughput has been upgraded to 112 Gbit/s per beam [3-46].

Hybrid optical/RF wireless systems lack several of the key attractive benefits of beam-steered OWC techniques, notably their immunity for EMI, enhanced privacy, large bandwidth, and inherent potential for high user connection density. Fully optical bidirectional OWC systems are preferred in order to reap these benefits, that is, systems in which both the downstream and the upstream links are established by OWC techniques.

As reported in literature, bidirectional all-optical wireless indoor systems typically duplicate downstream links into upstream ones and thus establish point-to-point bidirectional links between each user and a separate ceiling unit (cf. those employing SLM [3-24] or MEMS [3-25] active beam steerers, see the examples given in Section 5.2). A more efficient system, however, consists of a common ceiling unit acting as a point-to-multipoint multicasting unit that connects to multiple users. Such a system, in particular for indoor communication according to the BROWSE architecture, has been reported in [3-47] and will be discussed in depth in Chapter 8.

3.10 Beam-Steered OWC versus Wide-Beam OWC[3]

Offering wireless connectivity in a wide variety of application domains (e.g., in indoor offices and residential settings, inside data centers, in industry 4.0 applications, in exhibition pavilions, and in mobile transport settings such as trains, busses, and airplanes) by means of OWC using narrow (infrared) beams can bring a number of advantages with respect to radio-based wireless communication and VLC/LiFi, as outlined below. In Chapters 4–8, we will focus on beam-steered communication using narrow infrared beams.

3.10.1 Electromagnetic Interference

Radio technologies using the ISM band or mm-wave frequencies may suffer from the electromagnetic interference (EMI) generated by the automated manufacturing equipment, such as arc-welding robots and heavily powered electromagnetic machines, and by jamming from neighboring sites. Optical wireless links are not affected by EMI nor jamming, hence can enhance link reliability in harsh Industry 4.0 environments.

3.10.2 Data Capacity per Device

Radio-based wireless links typically operate with relatively wide beam profiles. When closely spaced items have to be connected, for example, several arms of a robot, or a swarm of transporters, they may need to share the same beam, and thus have to share the data capacity of it. This complicates the communication with each one individually, by necessitating robust MAC protocols. By using higher radio frequencies and high-gain (e.g., horn-shaped) antennas, the radio beam

---

3 Extracted from A.M.J. Koonen, K.A. Mekonnen, Z. Cao, et al., "Beam-steered optical wireless communication for Industry 4.0," invited paper, *IEEE J. Sel. Top. Quantum Electron.*, vol. 27, no. 6, Nov./Dec. 2021, https://doi.org/10.1109/JSTQE.2021.3092837.

profile may be narrowed which alleviates the sharing issue, and more power can be delivered into the restricted aperture of the receiver. By means of simple optical elements such as lenses, however, narrow optical beams can be created with much less divergence (which is equivalent to much higher antenna gain). Such beams can create wireless connections at a much higher link power budget and spatial density, thus also avoiding the need for capacity sharing among closely spaced devices. A significantly higher connection bandwidth per device can be realized as compared with radio-based systems such as the ones of the IEEE 802.11 family.

Fig. 3-7 illustrates how an OWC system using narrow infrared beams may outperform the various IEEE 802.11 systems, notably at higher data rates per device and higher device densities. In IEEE 802.11 systems, the product of bitrate per device × number of devices per cell is limited by the sharing of the system's aggregate capacity. In the PHY layer, excluding multi-antenna MIMO techniques, the IEEE 802.11n (also known as Wi-Fi 4) offers 150 Mbit/s per channel, IEEE 802.ac (also known as Wi-Fi 5) up to 867 Mbit/s, and the latest IEEE 802.11ax (also known as Wi-Fi 6) up to 1201 Mbit/s [3-48]. Wi-Fi 6 can deploy up to 8 × 8 MIMO offering 9.6 Gbit/s in theory, and 5.9 Gbits in practice. LiFi-based OWC systems using a single luminary (as introduced in the market, with PHY rates of around 250 Mbit/s typically) also employ capacity sharing. These are based on the Gigabit

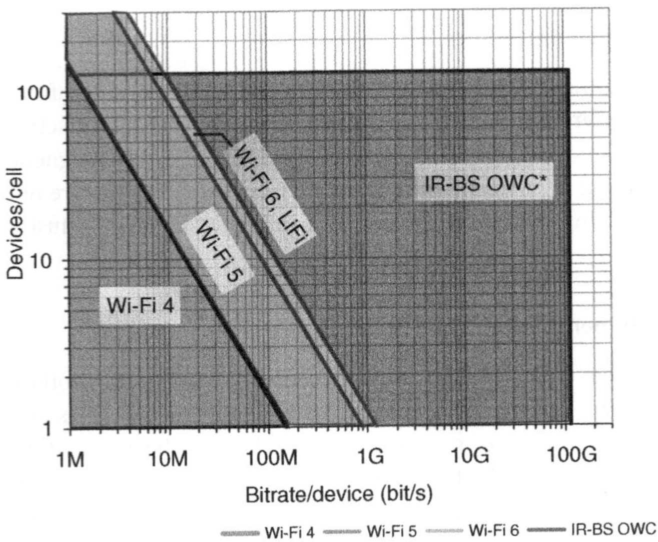

Fig. 3-7 Wireless connectivity using beam-steered OWC compared to using radio technologies (IEEE 802.11n also known as Wi-Fi 4, IEEE 802.11ac also known as Wi-Fi 5, IEEE 802.11ax also known as Wi-Fi 6; IR-BS OWC: infrared beam-steered optical wireless communication featuring 128 beams at 112 Gbit/s each)

Home Networking (G.hn) specification developed in ITU for wired home networks (telephone lines, coaxial cables, power lines, and plastic optical fiber) which also supports the Wi-Fi standards [3-49]; it limits the data rate to 1.6 Gbit/s. Bidirectional point-to-point VLC systems are offered with 1 Gbit/s capability [3-17], and according to the IEEE 802.11bb standard [3-20] LiFi in an 8×8 configuration may offer data rates up to 9.6 Gbit/s, comparable to Wi-Fi 6.

In contrast, an infrared beam-steered OWC (IR-BS OWC) system principally can offer a very high bandwidth per device, independent of the number of devices, where this number is only constrained by the maximum number of cells that can be addressed [3-50]. A capacity of up to 112 Gbit/s per beam has been demonstrated using PAM-4 modulation [3-46], and an ability to steer 128 beams individually [3-51].

3.10.3 Latency

In machine-to-machine communications, the latency in the link should be minimized to ensure the timely and reliable operation of robots individually and their interaction. In comparison with wired fiber–optic connections, wireless connections made through radio or optical waves have less propagation delay as the signals propagate at the speed of light in free space, whereas in fiber, the speed of light is reduced by the refractive index of its material ($n \approx 1.5$; note that hollow-core fibers are emerging from research which can largely avoid this retardation). Therefore, typically the latency in free space is 33% less than in fiber, which may make a noticable difference in the performance of high-speed machine-to-machine bidirectional communication, such as in data center rack-to-rack interconnects. Optical beams offer a much larger bandwidth than a radio link: each nm in the 1.5 μm wavelength range corresponds to 125 GHz of bandwidth wherein many multi-GHz channels can be accommodated. In the Wi-Fi family, within an overall bandwidth of 645 MHz in the 5 GHz band (5.180–5.825 GHz) multiple channels with bandwidths up to some 160 MHz can be used. In the mm-wave 60 GHz domain, an overall bandwidth of 7 GHz (57–64 GHz) is available; up to 4 channels of 2.16 GHz each can be bundled in one channel according to IEEE 802.11ay reaching 20–40 Gbit/s. Due to the restricted bandwidth available, radio systems have to apply comprehensive signal modulation techniques (e.g., OFDM with high-level QAM) in order to compress the signal's spectrum and enable high data rate transport. Such modulation techniques need sizable digital signal processing (DSP) both at the transmit and receive ends of the link, which adds to the latency. Moreover, the capacity sharing requires careful MAC processing, which also adds to the latency. For example, in the carrier-sense multiple access with collision avoidance (CSMA/CA, [3-52]) protocol, the latency occurs due to the waiting time required for a free channel in the listen-before-talk procedure.

In LiFi systems considerable signal processing time is needed for compressing the signal spectrum within the limited bandwidth available of the light sources and for overcoming the low SNR due to a low received signal intensity. The system's latency is increased further by the time needed for the MAC processing due to the beam sharing among the users' devices.

In an OWC system using narrow infrared beams, however, each beam offers a huge bandwidth and high link budget, and it does not need to be shared with other devices. This significantly relaxes the signal modulation efforts, avoids comprehensive MAC protocols, and thus minimizes the associated latency.

3.10.4 Energy Efficiency

Next to raising the capacity sharing constraint among a number of devices, the beams of RF (radio-based) wireless systems cover cells with a relatively large footprint, and therefore the signal energy which is captured per device by its limited aperture cannot be very large. The same holds for LiFi systems. This results in a lower SNR and therefore in addition to the sharing constraint puts a limit on the capacity per device that can be reached. Moreover, the beam's intensity decreases at least with the square of the distance to the receiver, thus restricting the reach and further tightening the achievable capacity. By means of a phased array antenna, RF beams can be shaped to much lower divergence and hence higher intensity, which alleviates the reach issue, but at the expense of a comprehensive multi-element antenna design with many power-consuming RF phase shifters and amplifiers [3-53]. The gain G_A of an antenna is the product of its directivity D_A and its radiation efficiency $\eta_{antenna}$ [3-54]. For example, for an antenna gain of 30 dBi (i.e., dB with respect to an isotropic antenna uniformly radiating into all directions), no less than 1000 hypothetical isotropic antenna elements are needed. A half-wave dipole antenna can be almost 10% efficient and have a gain of 2.15 dBi; for 30 dBi antenna gain, about 600 dipole antennas would be needed. Passive horn antennas may also deliver narrow RF beams, but are relatively large and may achieve gains up to 25 dBi [3-55]. The antenna's radiation efficiency $\eta_{antenna}$ is defined as the total power radiated by the antenna divided by the net power fed to the antenna by its transmitter [3-56]. It depends on the transfer efficiency of the interface and may depend on frequency. The directivity D_A is the ratio of the radiation intensity in a certain direction and the radiation averaged over all directions; for a hypothetical isotropic antenna, it is 1, so 0 dBi [3-57]. If the direction is not specified, the directivity refers to the axis of maximum radiation intensity. In a simplified far-field analysis, assuming a point source P emitting an omnidirectional uniform power density (such as an ideal isotropic antenna does) as shown in Fig. 3-8, the directivity D_A of a directional antenna may be interpreted as the ratio of the power emitted over the full sphere divided by the power emitted by the antenna in its angular aperture with half-angle θ_0; so

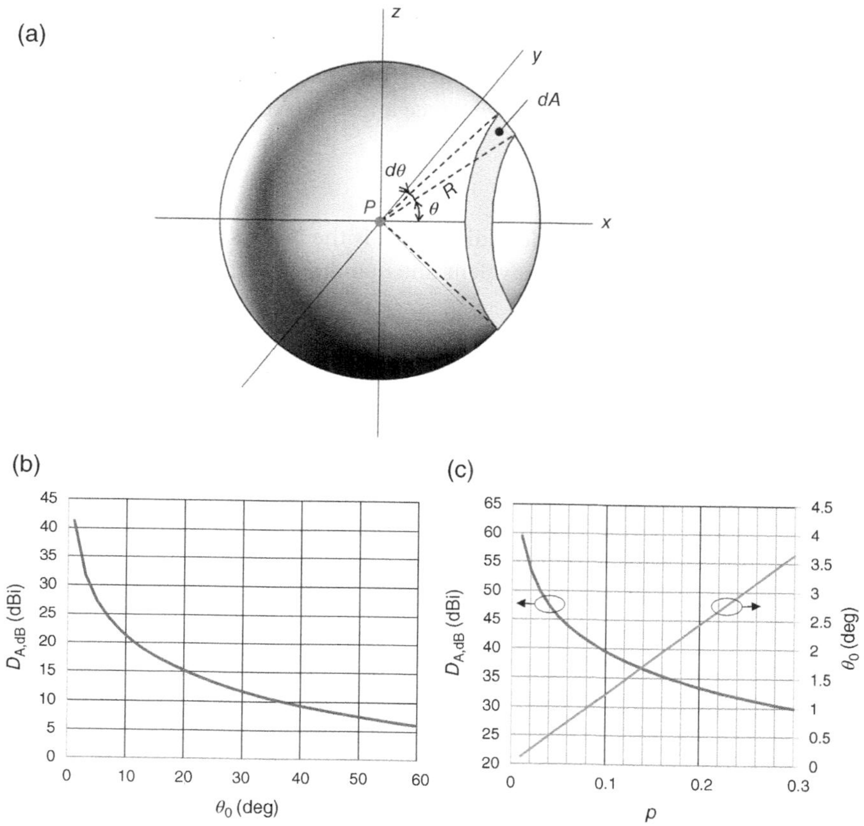

Fig. 3-8 Antenna directivity. (a) Modeling antenna directivity. (b) Antenna directivity $D_{A,dB}$ of RF antenna versus its aperture half-angle θ_0. (c) Antenna directivity $D_{A,dB}$ of lensed optical beam steerer versus defocusing parameter p

$$D_A = \frac{A_{tot}}{\Delta A} = \frac{4\pi R^2}{\int_{\theta=0}^{\theta_0} 2\pi R^2 \sin\theta \, d\theta} = \frac{4\pi R^2}{4\pi R^2 \sin^2\left(\frac{\theta_0}{2}\right)} = \frac{1}{\sin^2\left(\frac{\theta_0}{2}\right)} \tag{3-1}$$

In contrast to the RF wireless scenario where narrow beams require sophisticated antenna designs and/or comprehensive phased-array antenna structures, in the OWC scenario, narrow beams are readily created by employing a relatively simple single lens. Thus, by creating narrow dynamically-steered infrared beams each device can get an individual beam with a small footprint, which does not need to be shared and by which the reach is extended. By using just a single passive lens adequately corrected for spherical aberration, a highly collimated beam can be obtained, which has a very large equivalent antenna gain. No power-hungry electronics like in an RF-phased array antenna are needed. For example, in the 2D optical beam steerer based on an AWGR and a 2D fiber array reported in Section 5.5 and in [3-51], we used a lens in slightly defocused ($p = 21\%$) operation, which yields beams with a small divergence angle θ_0 (see Eq. (5-27))

$$\tan\theta_0 = 2p\,\frac{\lambda}{\pi w_0} \tag{3-2}$$

where w_0 is the SMF's mode field radius and λ is the beam's wavelength. Hence, the beam steerer provides an equivalent antenna directivity.

$$D_\mathrm{A} \approx \frac{1}{p^2}\left(\frac{\pi w_0}{\lambda}\right)^2 \tag{3-3}$$

A high antenna directivity can easily be realized by only deploying a passive lens. For example, for typical values $w_0 = 4.5\,\mu\mathrm{m}$ and $\lambda = 1.5\,\mu\mathrm{m}$, with $p = 0.21$ a $D_{\mathrm{A,dB}} = 33.0$ dBi is readily achieved (where $D_{\mathrm{A,dB}} = 10^{10}\log(D_\mathrm{A})$).

Virtually, the capabilities of a narrow optical beam are much like the ones of a fiber, but without needing a fiber. In indoor applications without atmospheric disturbances, the loss in the link that conveys the signal energy from the OWC transmitter to a device (and vice versa) is, therefore, less than in an RF wireless link, hence the SNR can be better and the attainable capacity much higher. Conversely, for a specific capacity needed much less energy is needed. Moreover, the beams can be activated and pinpointed to the respective devices only where and when needed, and no signal energy is wasted, thus further improving the energy efficiency of a beam-steered OWC system. In radio systems, measures have been introduced also to improve energy efficiency by directing radio beams. In Wi-Fi standard IEEE 802.11ba, the wake-up radio protocol has been introduced in which the user device is equipped with a primary connectivity radio (PCR) chain handling the high data rate and a low-power companion connectivity radio (CCR) chain waking up the PCR as soon as it detects data arriving [3-58]. In IEEE 802.11ax, the target wake time (TWT) is introduced where access devices can determine when and how frequently they will wake up to send or receive data, thus reducing their power consumption [3-59].

3.11 Photonic Integrated Circuits for Optical Wireless Communication[4]

The fast increase in demand for wireless connectivity, fueled by the growing bandwidth needed by mobile services as well as by the staggering numbers of wireless devices to be connected, is driving radio wireless networks into saturation. OWC

4 This section is largely based on T. Koonen, and Z. Cao, "Photonic integrated circuits for optical wireless communication," *Proc. ACP 2021*, Shanghai, Oct. 24–27, 2021, Tutorial paper M5D1.

has shown to have great potential to solve this imminent congestion. It can offer huge amounts of additional bandwidth; the visible domain (400–700 nm) brings an extra 320 THz, and the fiber–optic infrared domain (1500–1600 nm) an extra 12.5 THz. Moreover, when using narrow optical beams very small picocells can be created, which enables dense high-capacity coverage by spatial division multiplexing (SDM). SDM is uniquely capable of meeting exponentially growing capacity demand [3-60].

Two major directions in OWC may be discerned: one using very wide beams which do not need extra steering, and one using narrow beams which need to be pinpointed to users accurately [3-21]. LiFi systems which are typically built on top of illumination systems (using LED-based luminaires) belong to the first direction. Because such wide beams typically cover multiple users, these users will need to share the capacity of the luminary with its limited bandwidth and thus can be provided only with modest capacity. By means of multiple input multiple output (MIMO) techniques, similar to those known in advanced radio wireless systems) applied at the transmitter and receiver ends, the capacity per user may be increased. The second direction encompasses systems that typically use narrow infrared light beams, which are steered to users individually. Such systems may be considered as "LiFi 2.0" [3-62]. Each user gets his own beam, a "virtual fiber," and thus can get very high capacity without sharing and with high privacy, whereas no signal energy is wasted, and optimum system energy efficiency is achieved.

The first commercial introduction of LiFi systems is already taking place. True mass deployment of OWC systems in the direct environment of users will be largely dependent on scalable manufacturability, power efficiency (in particular for battery-fed mobile devices), and cost-effectiveness. PIC technologies play a crucial role in this. In this section, we will focus on PIC structures for OWC transmitters and OWC receivers.

3.11.1 Photonic Integrated OWC Transmitters

Wide-beam LED-based OWC systems are typically limited in reach and data capacity per user, as intensity scales with the squared inverse of the distance between user and luminary. Illumination LEDs have been optimized for light efficiency, but not for modulation bandwidth. Hence, several research groups have studied dedicated LED technologies for OWC, notably gallium-nitride (GaN) micro-LEDs which due to their small active area operate at high current density, and can offer high modulation bandwidth as well as high output power density [3-61]. Such μ-LEDs can be made in large arrays, and thus jointly provide increased total output power. Moreover, such array structures can enable to drive μLEDs individually, and thus multiply the data capacity. The array can be mounted on top of an array of CMOS driver circuits, and thus optical signals from the individual μLEDs can be directly generated from digital electrical input signals,

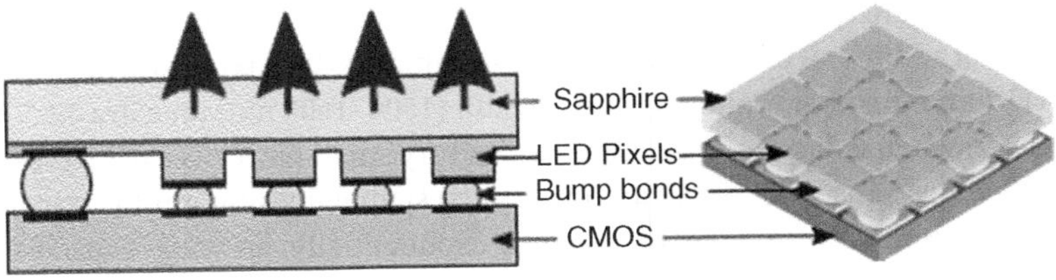

Fig. 3-9 μ-LED array bump-bonded to CMOS control chip ([3-61]/The Royal Society/CC BY 4.0)

see Fig. 3-9. Such an arrayed μLEDs approach opens the way toward effective OWC MIMO systems. As a mixture of the data signals from the individual μLEDs will arrive at each PD at the receiver side, additional signal processing is required both at the system's transmitter and receiver end to do the data channel separation. Up to 7 Gbit/s transmission was shown with a 3 × 3 array of 39 μm μLEDs [3-61]. With 10 μm blue (450 nm) μLEDs, a modulation bandwidth of 1.8 GHz was reported. Using DCO-OFDM modulation with bit- and power-loading up to 256-QAM, a data rate of 7.7 Gbit/s was achieved [3-63]. Operation in larger arrays of beyond 10,000 μLEDs is foreseen (which will bring associated management complexity issues).

Alternatively, in a beam-steered OWC system typically many narrow infrared beams are emitted, and each provides high data capacity to an individual user [3-50]. Each beam is pointed to a user who has been localized before (e.g., by means of a passive optical retroreflector array at the user device and an upstream power detector at the beam emission site (see [3-36], and Chapter 6), or visible LED tags and an upstream video camera [3-64]). By using passive diffractive modules and remote wavelength-tunable laser diodes, 2D beam steering has been demonstrated, for instance, with a high port count arrayed waveguide grating router (AWGR) of which the output ports were arranged in a 2D fiber matrix positioned in front of a large aperture lens [3-51] (see Section 5.5). A laboratory demonstrator has been built which can provide up to 128 beams. Real-time streaming of high-definition video signals embedded in 10 Gbit/s data has been shown. Moreover, optical phased arrays (OPAs) have been reported for beam-steering and -shaping, in conjunction with phase tuning in the OPA's arms and wavelength tuning of the laser diodes [3-65]. Next to OWC, they also find applications in LiDAR (Light Detection And Ranging). In [3-66], an OPA-based hybrid silicon PIC is reported which hosts 164 optical components (tunable laser diodes, optical amplifiers, PDs, phase tuners, and a photonic crystal lens; see Fig. 3-10). It can steer a beam over 23° × 3.6° with a beam divergence of 1° × 0.6°.

A compact PIC that can provide 2D beam steering using the cascaded-gratings concept has been reported in [3-67]; see Fig. 3-11. An arrayed waveguide grating

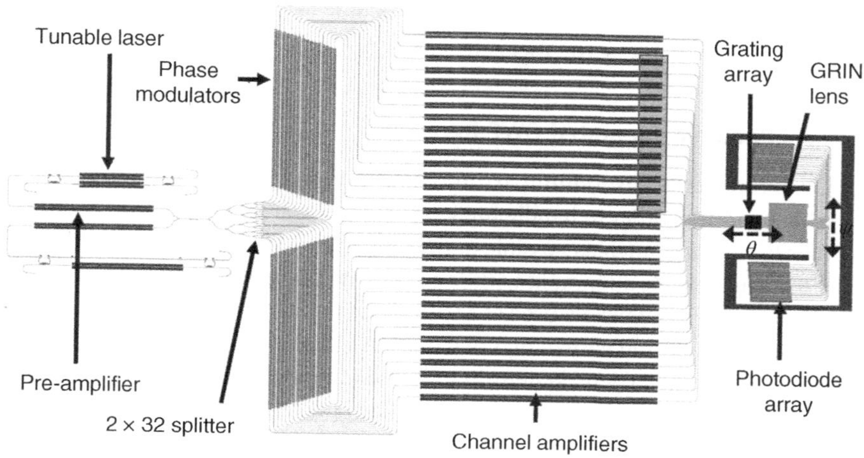

Fig. 3-10 OPA-based hybrid silicon PIC for 2D beam steering ([3-66]/J. C. Hulme)

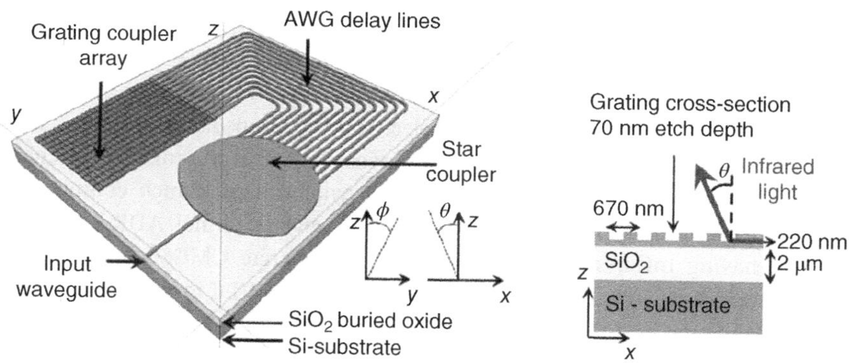

Fig. 3-11 2D beam steering photonic integrated circuit using λ-tuning ([3-75]/ © [2011] IEEE)

was co-integrated with an array of waveguide grating couplers in a passive silicon-on-insulator technology; input-to-free-space losses were 8.9–10 dB. 2D steering was achieved over 15° × 50° by wavelength tuning over 100 nm; with a beam's angular width of 4°, 50 resolvable spots could be generated in 2D. In [3-68], 2D beam steering by means of an active PIC implemented in indium-phosphide (InP) technology has been reported. While using an on-chip sampled grating distributed Bragg reflector (SG-DBR) laser diode with a wavelength tuning range beyond 30nm, by tuning the wavelength as well as tuning electro-optically, the local phase in the branches of an arrayed waveguide grating router 2D steering of a single beam over 3.5° × 10° was achieved.

3.11.2 Photonic Integrated OWC Receivers

An OWC receiver needs to fulfill an ambitious range of criteria: it should have a large bandwidth in order to handle high data rates, it should have a large aperture in order to catch as much as possible optical power, as well as have a wide FoV in order to avoid delicate alignment with respect to the beam. It should consume only a little power (as it is typically used at a battery-fed mobile device), and it should be cheap (so be simple and easy to manufacture). A large aperture may be created by using a large-diameter lens system that focuses the incoming beam on the PD. However, the basic physics' law of etendue states that a large aperture will imply a small FoV. A large diameter PD can help, but typically a large active area also brings a high junction capacitance, hence a reduced bandwidth. By using multiple PDs which jointly perform the photodetection the total active area can be effectively increased.

Two ways have been reported to circumvent the increased capacitance which would be associated with this: an angular diversity receiver (ADR), and a receiver deploying a 2D matrix constellation of PDs. In an ADR, multiple PDs are used, where each one is looking in a different direction and thus the PDs jointly cover a larger FoV. Each PD is followed by a separate electrical transimpedance amplifier, and the outputs of these amplifiers are summed. To ensure adequate summing, the characteristics of these amplifiers need to be well synchronized, which complicates the ADR design. As shown in Fig. 7-2 [3-42], a seven-channel ADR has been reported, having InGaAs MSM PDs followed by separate CMOS preamplifiers, which can handle 5 Gbit/s. In [3-69] an ADR is analyzed for VLC in conjunction with MIMO techniques. In [3-70], a densely integrated 19×32 2D array of 20–30 μm pixels was reported which offered a bandwidth of over 10 GHz. In a 6×6 array of 30 μm PDs, a bandwidth of 20 GHz was reached, and a free space 10 Gbit/s optical link over 1.5 meters with relaxed alignment was demonstrated [3-71].

As an alternative to the ADR approach, a 2D constellation of a matrix of PDs in combination with a single lens having a large numerical aperture has been proposed [3-43] (see Fig. 3-12, discussed in full detail in Chapter 7). Such a matrix has only a single output and thus requires only a single preamplifier and a single lens. This alternative solution considerably simplifies the receiver design. It can be shown that a 2D constellation of $M \times M$ PDs yields a capacitance that equals the capacitance of a single PD, whereas it has an M^2 times larger light-capturing area and generates M times larger photocurrent (assuming that junction capacitance is dominant, and the fill factor is large) [3-44]. A FoV of more than ±10° (half-angle) has been demonstrated with a 4×4 array of Ø150 μm active area

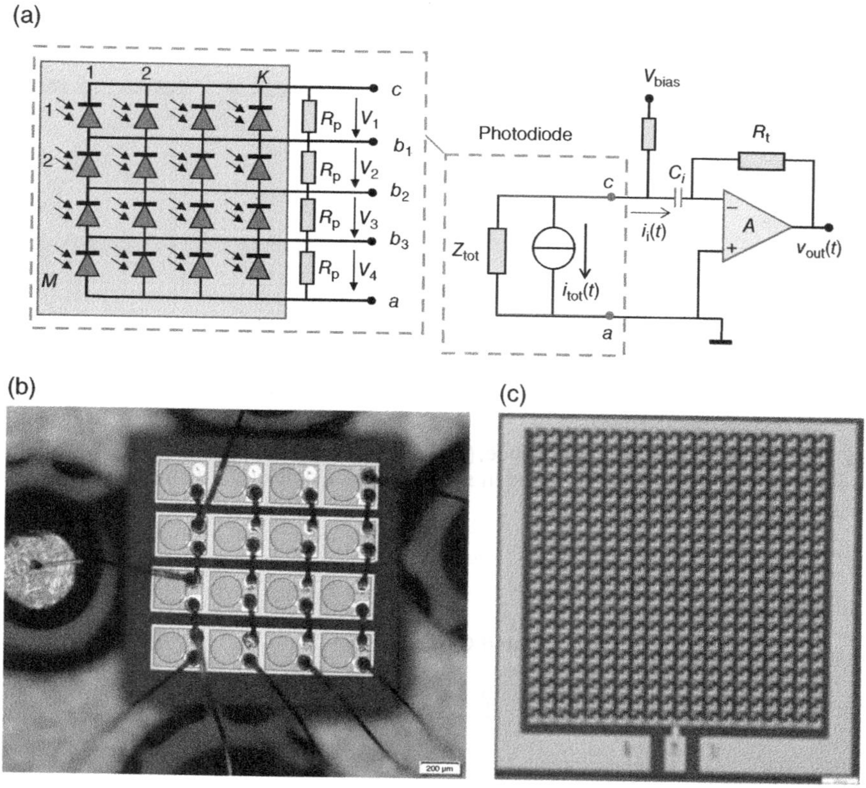

Fig. 3-12 OWC receiver with large FoV employing a 2D matrix of photodiodes [3-44, 3-45]. (a) Circuit diagram of TIA receiver with PD matrix ([3-44]/© [2021] IEEE). (b) 4 × 4 matrix of photodiodes, each Ø150 µm active area, by wire-bonding four 1 × 4 PD arrays ([3-44]/IEEE). (c) 20 × 20 matrix of Ø30 µm photodiodes, by integration on III–V substrate ([3-72]/IEEE)

PDs. In a laboratory system demonstrator, real-time optical wireless streaming of Gbit/s Ethernet video signals to an OWC receiver module connected to a laptop was shown [3-45].

The 2D PD matrix concept is well scalable while retaining the bandwidth of a single PD. With a 20 × 20 matrix with Ø30 µm PDs transmission of 25 Gbit/s NRZ modulated signals over an OWC link of 20 m has been demonstrated [3-72]. The PD matrix configuration may also be extended in a third dimension by a multi-stacked PIN PD structure, as shown in Fig. 3-13 [3-73]. Vertical pinholes with a diameter of 5 µm were applied, yielding a fill factor of 50%. Data transmission in an OWC link up to 20 meters at 20 Gbit/s non-return-to-zero (NRZ) OOK modulation has been shown.

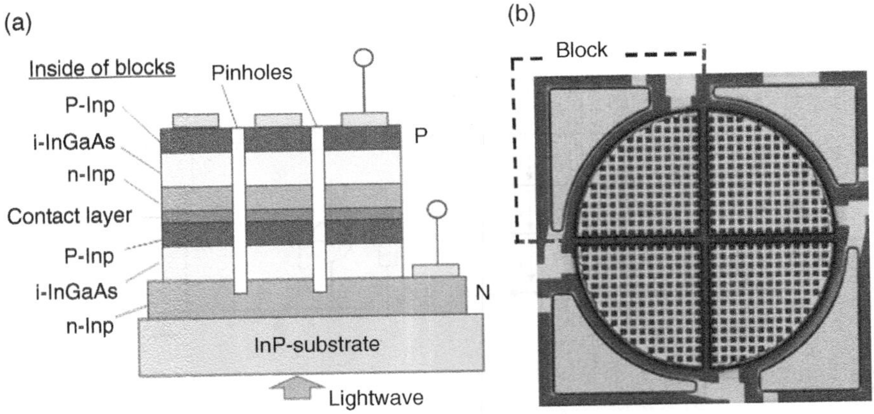

Fig. 3-13 Multi-stacked PIN photodiode; (a) two-stacked PIN photodiode block and (b) four two-stacked PD blocks interconnected in series (overall diameter 0.3 mm) ([3-73]/ Toshimasa Umezawa)

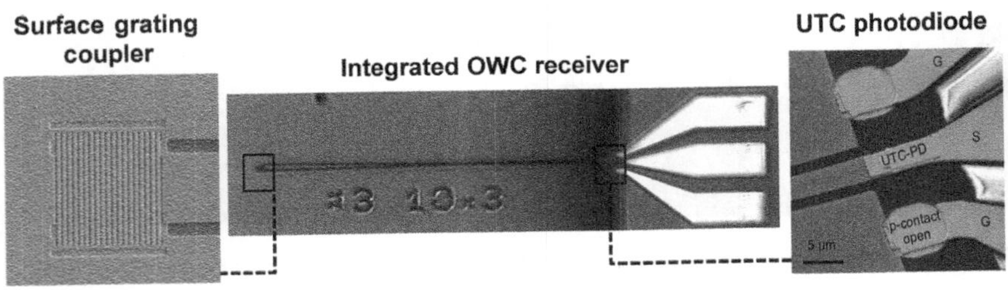

Fig. 3-14 Integrated cascaded aperture optical receiver ([3-41]/IEEE)

As briefly introduced in Section 3.8, the light collection function may be separated from the light detection function, which enables the optimization of both functions independently and can increase the receiver's aperture. Using InP photonic integration technology, a wide-area SGC collecting the incident light has been monolithically integrated with an InP waveguide feeding a co-integrated fast PD. A $10 \times 10\ \mu m^2$ SGC was coupled via a narrow waveguide to a $3 \times 10\ \mu m^2$ uni-travelling-carrier (UTC) PD which had an electrical bandwidth exceeding 67 GHz (see Fig. 3-14) [3-41]. With this integrated SGC-waveguide-PD device, free-space 40 Gbit/s NRZ-OOK data transmission has been demonstrated, as well as operation at 200 Gbit/s in WDM operation at five wavelengths. Without compromising the bandwidth, the light-collecting aperture can be increased further by enlarging the SGC by means of apodization, and by creating an array of SGCs plus an on-chip light combiner.

3.12 Concluding Remarks

OWC techniques hold great potential to resolve the imminent exhaustion of indoor radio-based wireless communication resources. The optical spectrum, both in the visible and in the infrared domain, offers ample extra bandwidth without licensing issues, and thus great opportunities for carrying large volumes of wireless traffic. OWC systems can take away much load from radio wireless systems, for example, by offloading much of the heavy burden put by broadband data transport in these networks. It will not fully replace radio wireless communication (e.g., it cannot support services without LoS), but it will complement and coexist with it. First VLC/LiFi products are entering the market in appealing applications [3-74, 3-75]. Beam-steered infrared communication techniques have shown their ability to offer ultra-high data rates (beyond 100 Gbit/s per beam), with high energy efficiency, and at high user densities with high privacy. Considerable work is still to be done to elevate these techniques from their research status (see Chapters 4 to 8) to market maturity, for example, regarding robust beam-steering OWC transmitters, compact low-cost OWC receivers, symmetric bidirectional OWC systems, scaling up the coverage area and reach, efficient localization, and co-existence scenarios with radio-based wireless technologies. Photonic integration is a key technology to achieve cost-effective high-performance OWC systems in large market volumes.

As OWC technologies have not matured yet, cost comparisons with other technologies are hard to make. From a general viewpoint, VLC/LiFi systems may be preferred regarding CapEx due to their lower infrastructure costs, whereas beam-steered OWC systems may be more beneficial regarding OpEx due to their lower energy consumption.

References

[3-1] Coopers law [on-line]. Available: http://www.arraycomm.com/technology/coopers-law/

[3-2] The photophone of Alexander Graham Bell [on-line]. Available: https://en.wikipedia.org/wiki/Photophone, https://www.loc.gov/item/today-in-history/june-03

[3-3] D.C. O'Brien, "Optical wireless communications: current status and future prospects," *Proc. IEEE Summer Topicals*, Newport Beach, Jul. 11–13, 2016.

[3-4] H. Haas, L. Yin, Y. Wang, and C. Chen, "What is LiFi?," *J. Lightwave Technol.*, vol. 34, no. 6, pp. 1533–1544, Mar. 2016.

[3-5] Laser safety [on-line]. Available: https://en.wikipedia.org/wiki/Laser_safety

[3-6] Photodiode responsivity [on-line]. Available: https://en.wikipedia.org/wiki/Responsivity

[3-7] V. Jungnickel, "Optical wireless in 5G," *IEEE Summer Topicals*, Newport Beach, Jul. 11–13, 2016, Tutorial.

[3-8] J. Grubor, S.C. Jeffrey Lee, K.-D. Langer, et al., "Wireless high-speed data transmission with phosphorescent white-light LEDs," *Proc. ECOC 2007*, Berlin, Sep.16–20, 2007, Paper PD3.6.

[3-9] X. Huang, S. Chen, Z. Wang, et al., "2.0-Gb/s Visible light link based on adaptive bit allocation OFDM of a single phosphorescent white LED," *IEEE Photon. J.*, vol. 7, no. 5, Oct. 2015, https://doi.org/10.1109/JPHOT.2015.2480541.

[3-10] G. Cossu, A.M. Khalid, P. Choudhury, et al., "3.4 Gbit/s visible optical wireless transmission based on RGB LED," *Opt. Express*, vol. 20, no. 26, p. B501–B506, Dec. 2012.

[3-11] Y. Wang, L. Tao, X. Huang, et al., "8-Gb/s RGBY LED-based WDM VLC system employing high-order CAP modulation and hybrid post equalizer," *IEEE Photon. J.*, vol. 7, no. 6, Dec. 2015, https://doi.org/10.1109/JPHOT.2015.2489927.

[3-12] S. Rajagopal, R.D. Roberts, and S.-K. Lim, "IEEE 802.15.7 visible light communication: modulation schemes and dimming support," *IEEE Commun. Mag.*, vol. 50, no. 3, pp. 72–82, Mar. 2012.

[3-13] R. Singh, T. O'Farrell, and J.P.R. David, "An enhanced color shift keying modulation scheme for high-speed wireless visible light communications," *J. Lightwave Technol.*, vol. 32, no.14, pp. 2582–2592, Jul. 2014.

[3-14] A.E. Aziz, K. Th. Wong, and J.-C. Chen, "Color-shift keying—how its largest obtainable "minimum distance" depends on its preset operating chromaticity and constellation size," *J. Lightwave Technol.*, vol. 35, no. 13, pp. 2724–2733, Jul. 2017.

[3-15] R.X.G. Ferreira, E. Xie, J.J.D. McKendry, et al., "High bandwidth GaN-based micro-LEDs for multi-Gb/s visible light communications," *Photon. Technol. Lett.*, vol. 28, no. 19, pp. 2023–2026, Oct. 2016.

[3-16] S. Rajbhandari, A.V.N. Jalajakumari, H. Chun, et al., "A multi-Gigabit/sec integrated multiple-input multiple-output VLC demonstrator, " *J. Lightwave Technol.*, vol. 35, no. 20, pp. 4358–4365, Oct. 2017.

[3-17] 1 Gbit/s VLC system of Fraunhofer HHI [on-line]. Available: https://www.hhi.fraunhofer.de/en/departments/pn/products-and-solutions/1-gbit-s-lifi-system.html

[3-18] LiFi-XC of Purelifi [on-line]. Available: http://purelifi.com/lifi-products/lifi-x/

[3-19] Trulifi of Signify [on-line]. Available: https://www.signify.com/global/innovation/trulifi

[3-20] IEEE 802.11bb LiFi standard [on-line]. Available: https://www.ieee802.org/11/Reports/tgbb_update.htm

[3-21] A.M.J. Koonen, "Indoor optical wireless systems: Technology, trends, and applications," *IEEE/OSA J. Lightwave Technol.*, vol. 36, no. 8, pp. 1459–1467, Apr. 2018, https://doi.org/10.1109/JLT.2017.2787614.

[3-22] Spatial light modulators [on-line]. Available: https://holoeye.com/products/spatial-light-modulators/

[3-23] A. Gomez, K. Shi, C. Quintana, et al., "Beyond 100-Gb/s indoor wide field-of-view optical wireless communications," *Photon. Technol. Lett.*, vol. 27, no. 4, pp. 367–370, Feb. 2015.

[3-24] A. Gomez, C. Quintana, G. Faulkner, and D.C. O'Brien, "Point-to-multipoint holographic beamsteering techniques for indoor optical wireless communications", *Proc. SPIE vol. 9772,* Broadband Access Communication Technologies X, 2016, 7p.

[3-25] K. Wang, A. Nirmalathas, C. Lim, et al., "Full-duplex gigabit indoor optical wireless communication system with CAP modulation," *IEEE Photon. Technol. Lett.*, vol. 28, no. 7, pp. 790–793, Apr. 2016.

[3-26] A.M.J. Koonen, C.W. Oh, K. Mekonnen, et al., "Ultra-high capacity indoor optical wireless communication using 2D-steered pencil beams," *J. Lightwave Technol.*, vol. 34, no. 20, p. 4802–4809, Oct. 2016, https://doi.org/10.1109/JLT.2016.2574855.

[3-27] M. Born, and E. Wolf, *Principles of Optics: Electromagnetic Theory of Propagation, Interference and Diffraction of Light*, 7th ed., 1999, Cambridge University Press, ISBN 0521642221.

[3-28] P.M. van den Berg, "Diffraction theory of a reflection grating," *Appl. Sci. Res.*, vol. 24, pp. 261–293, Dec. 1971.

[3-29] T.-H. Do, and M. Yoo, "An in-depth survey of visible light communication based positioning systems," *Sensors*, vol. 16, no. 5, art. no. 678, 2016 [on-line]. Available: http://www.mdpi.com/1424-8220/16/5/678.

[3-30] N. Saha, M.S. Ifthekhar, N. Tuan Le, and Y.M. Jang, "Survey on optical camera communications: challenges and opportunities," *IET Optoelectron.*, vol. 9, no. 5, pp. 172–183, 2015.

[3-31] N. Tuan Le, M.A. Hossain, and Y. Min Jang, "A survey of design and implementation for optical camera communication," *Elsevier Signal Process. Image Commun.*, vol. 53, pp. 95–109, 2017.

[3-32] J. Herrnsdorf, M.J. Strain, E. Gu, et al., "Positioning and space-division multiple access enabled by structured illumination with light-emitting diodes," *J. Lightwave Technol.*, vol. 35, no. 12, pp. 2339–2345, Jun. 2017.

[3-33] A.M. Khalid, P. Baltus, A. R. Dommele, et al., "Bi-directional 35-Gbit/s 2D beam steered optical wireless downlink and 5-Gbit/s localized 60-GHz communication uplink for hybrid indoor wireless systems," *Proc. OFC 2017*, Los Angeles, Mar. 19–23, 2017, Paper Th1E.6.

[3-34] A. Gomez, K. Shi, C. Quintana, et al., "A 50 Gb/s transparent indoor optical wireless communications link with an integrated localization and tracking system," *J. Lightwave Technol.*, vol. 34, no. 10, pp. 2510–2517, May 2016.

[3-35] N.Q. Pham, K.A. Mekonnen, E. Tangdiongga, et al., "Accurate indoor localization for beam-steered OWC system using optical camera," *Proc. ECOC 2019*, Dublin, Sep. 22–26, 2019, Paper P45.

[3-36] A.M.J. Koonen, K.A. Mekonnen, F.M. Huijskens, et al., "Fully passive user localization for beam-steered high-capacity optical wireless communication system," *J. Lightwave Technol.*, vol. 38, no. 10, pp. 2842–2848, May 2020, https://doi.org/10.1109/JLT.2020.2980428.

[3-37] Etendue [on-line]. Available: https://en.wikipedia.org/wiki/Etendue

[3-38] Non-imaging optics [on-line]. Available: https://en.wikipedia.org/wiki/Nonimaging_optics

[3-39] S. Collins, D.C. O'Brien, and A. Watt, "High gain, wide field of view concentrator for optical communications," *Opt. Lett.*, vol. 39, no. 7, p. 1756–1759, Apr. 2014.

[3-40] A. Riaz, and S. Collins, "A wide field of view VLC receiver for smartphones," *Proc. ECOC 2020*, Brussels, Dec. 6–10, 2020, Paper Tu2G.4.

[3-41] Z. Cao, L. Shen, Y. Jiao, et al., "200 Gbps OOK transmission over an indoor optical wireless link enabled by an integrated cascaded aperture optical receiver," *Proc. OFC 2017*, Los Angeles, Mar. 19–23, 2017, Post-deadline paper Th5A.6.

[3-42] J. Zeng, V. Joyner, J. Liao, et al., "A 5Gb/s 7-channel current-mode imaging receiver front-end for free-space optical MIMO," *Proc. IEEE MWSCAS 2009*, Cancun, Aug. 2–5, 2009, pp. 148–151.

[3-43] A.M.J. Koonen, K.A. Mekonnen, F.M. Huijskens, et al., "Novel broadband OWC receiver with large aperture and wide field of view", *Proc. ECOC 2020*, Brussels, Dec. 6–10, 2020, Paper Tu2G.6.

[3-44] T. Koonen, K. Mekonnen, Z. Cao, et al., "Beam-steered optical wireless communication for industry 4.0," *IEEE J. Sel. Top. Quantum Electron.*, vol. 27, no. 6, 10pp, Nov./Dec. 2021, https://doi.org/10.1109/JSTQE.2021.3092837.

[3-45] T. Koonen, K. Mekonnen, F. Huijskens, et al., "Optical wireless GbE receiver with large field-of-view," *Proc. ECOC 2021*, Bordeaux, Sep. 13–16, 2021, Paper Th1B.6.

[3-46] F. Gomez-Agis, S. van der Heiden, C.M. Okonkwo, et al., "112 Gbit/s transmission in a 2-D beam steering AWG-based optical wireless communication system," *Proc. ECOC 2017*, Göteborg, Sep. 17–21, 2017, Paper Th.2.B.1.

[3-47] T. Koonen, K.A. Mekonnen, F. Huijskens, and E. Tangdiongga, "Bi-directional all-optical wireless Gigabit Ethernet communication system using automatic self-aligned beam steering," *J. Lightwave Technol.*, vol. 41, no. 11, pp. 3446–3454, June 2023, https://doi.org/10.1109/JLT.2022.3231438.

[3-48] WiFi IEEE 802.11 standards [on-line]. Available: https://en.wikipedia.org/wiki/IEEE_802.11.

[3-49] Gigabit home networking specification G.hn [on-line]. Available: https://www.itu.int/dms_pub/itu-t/opb/tut/T-TUT-HOME-2021-3-PDF-E.pdf

[3-50] A.M.J. Koonen, K.A. Mekonnen, Z. Cao, et al., "Ultra-high capacity wireless communication by means of steered narrow optical beams," *Philos. Trans. R. Soc. A*, 19 pp, Mar. 2020, https://doi.org/10.1098/rsta.2019.0192.

[3-51] A.M.J. Koonen, F. Gomez-Agis, F.M. Huijskens, and K.A. Mekonnen, "High-capacity optical wireless communication using two-dimensional IR beam steering," *J. Lightwave Technol.*, vol. 36, no. 19, pp. 4486–4493, Oct. 2018, https://doi.org/10.1109/JLT.2018.2834374.

[3-52] CSMA/CA protocol [on-line]. Available: https://en.wikipedia.org/wiki/Carrier-sense_multiple_access_with_collision_avoidance

[3-53] P. Hannan, "The element-gain paradox for a phased-array antenna," *IEEE Trans. Antennas Propag.*, vol. 12, no. 4, pp. 423–433, July 1964.

[3-54] Antenna gain [on-line]. Available: https://en.wikipedia.org/wiki/Gain_(antenna)

[3-55] Horn antenna [on-line]. Available: https://en.wikipedia.org/wiki/Horn_antenna

[3-56] Radiation efficiency [on-line]. Available: https://en.wikipedia.org/wiki/Radiation_efficiency

[3-57] Antenna directivity [on-line]. Available: https://en.m.wikipedia.org/wiki/Directivity

[3-58] D.-J. Deng, S.-Y. Lien, C.-C. Lin, et al., "IEEE 802.11ba wake-up radio: Performance evaluation and practical designs," *IEEE Access*, vol. 8, pp. 141547–141557, Aug. 2020.

[3-59] Target wake time in IEEE 802.11ax protocol [on-line]. Available: https://www.commscope.com/blog/2018/802.11ax-fundamentals-target-wake-time-twt/

[3-60] T. Koonen, "Optical technologies to disclose the spatial diversity dimension in systems and networks," *Proc. ACP 2020*, Beijing, Oct. 24–27, 2020, Plenary talk, paper M1A1.

[3-61] A.D. Griffiths, J. Herrnsdorf, J.J.D. McKendry, et al., "Gallium nitride micro-light-emitting diode structured light sources for multi-modal optical wireless communications systems," *Philos. Trans. R. Soc. A*, 15 pp, Mar. 2020, https://doi.org/10.1098/rsta.2019.0185.

[3-62] M.D. Soltani, H. Kazemi, E. Sarbazi, et al., "Terabit Indoor Laser-Based Wireless Communications: LiFi 2.0 for 6G," *arXiv:2206.10532*, June 2022.

[3-63] L. Mareta, A. Lagrangea, and L. Dupré, "Ultra-high speed optical wireless communications with gallium-nitride microLED," *Proc. SPIE*, Mar. 5, 2021, https://doi.org/10.1117/12.2576092.

[3-64] N.Q. Pham, K. Mekonnen, E. Tangdiongga, et al., "User localization and upstream signaling for beam-steered infrared light communication system," *Photon. Technol. Lett.*, vol. 33, no. 11, pp. 545–548, June 2021.

[3-65] M.J.R. Heck, "Highly integrated optical phased arrays: Photonic integrated circuits for optical beam shaping and beam steering," *Nanophotonics*, vol. 6, no. 1, pp. 93–107, 2017.

[3-66] J.C. Hulme, J.K. Doylend, M.J.R. Heck, et al., "Fully integrated hybrid silicon two dimensional beam scanner," *Opt. Express*, vol. 23, no. 5, pp. 5861, Mar. 9, 2015.

[3-67] K. Van Acoleyen, W. Bogaerts, and R. Baets, "Two-dimensional dispersive off-chip beam scanner fabricated on silicon-on-insulator," *Photon. Technol. Lett.*, vol. 23, no. 17, p. 1270–1272, Sep. 2011.

[3-68] W. Guo, P.R.A. Binetti, C. Althouse, et al., "InP photonic integrated circuit with on-chip tunable laser source for 2D optical beam steering," *Proc. OFC 2013*, Los Angeles, Mar. 17–21, 2013, Paper OTh3I-7.

[3-69] C. He, S. Cincotta, M.M.A. Mohammed, and J. Armstrong, "Angular diversity aperture (ADA) receivers for indoor multiple-input multiple-output (MIMO) visible light communications (VLC)," *IEEE Access*, vol. 7, pp. 145282–145301, Oct. 2019.

[3-70] T. Umezawa, Y. Yoshida, A. Kanno, et al., "Integrated high-speed photodetector array for SDM communications," *Proc. ACP 2019*, Chengdu, Nov. 2–5, 2019, Paper M3H.4.

[3-71] T. Umezawa, Y. Yoshida, A. Kanno, et al., "FSO receiver with high optical alignment robustness using high-speed 2D-PDA and space diversity technique," *J. Lightwave Technol.*, vol. 39, no. 4, pp. 1040–1047, Feb. 2021.

[3-72] T. Umewaza, A.Q. Matsumoto, K. Akahane, et al., "400-pixel high-speed photodetector for high optical alignment robustness FSO receiver," *Proc. OFC 2022*, San Diego, Mar. 6–10, 2022, Paper M4I.3.

[3-73] T. Umezawa, S. Nakajima, A. Bekkali, et al., "Multi-stacked large-aperture high-speed PIN-photodetector for mobile-FSO communication," *Proc. ECOC 2023*, Glasgow, Oct. 1–5, 2023, Paper We.D.2.

[3-74] Signify's Trulifi, a new range of LiFi systems [on-line]. Available: https://www.signify.com/en-in/innovation/trulifi

[3-75] pureLiFi's LiFi@Home product family [on-line]. Available: https://www.purelifi.com/products/lifihome/

4

Beam-Steered Indoor OWC System[1]

4.1 Introduction

Much research effort has been reported in visible light communication (VLC) and LiFi, which can piggyback on existing LED illumination systems [4-1, 4-2]. VLC/LiFi systems are typically omnidirectional, implying capacity sharing of the light source by multiple user terminals. Also, the light intensity at the receiver inherently goes down with the square of the distance to the source, which after the optoelectrical conversion in the photodetector implies that the electrical signal-to-noise ratio (SNR) at the receiver decreases with the fourth power of the distance to the source, and thus the link reach is quite limited.

Links with a larger reach can be realized with directional collimated optical beams. Accurate agile two-dimensional (2D) beam steering is needed in order to address the (moving) user terminals; the concepts for this approach were introduced by Riza in [4-3]. Optical beam 2D steering systems have been reported using actively controlled devices such as micro-electromechanical system (MEMS)-actuated mirrors [4-4, 4-5], spatial light modulators (SLMs) [4-3, 4-6], and an optical switch for coarse steering followed by an SLM for fine steering [4-7]. Each beam may require a separate steering device, which makes scaling up to many beams challenging. It has been shown that by operating an SLM in a spatial division mode, a limited number of beams can be supported in a point-to-multipoint scheme [4-8]. Moreover, beam steering systems that use passive static diffractive devices have been reported, where the steering is controlled by

1 This chapter is largely based on A.M.J. Koonen, C. W. Oh, K. Mekonnen, et al., "Ultra-high capacity indoor optical wireless communication using 2D-steered pencil beams," *IEEE/OSA J. Lightwave Technol.*, vol. 34, no. 20, pp. 4802–4809, Oct. 2016, https://doi.org/10.1109/JLT.2016.2574855, and also on US patent 9246589 B2, WO patent 2013107853, priority date Jan. 20, 2012, "Two-dimensional optical beam steering module," by Antonius Marcellus Jozef Koonen, Peter G.M. Baltus, and Antonio Liotta.

changing the wavelength of the light beam fed to the device via fiber by remotely located tunable laser diodes. The steering is determined by the wavelength of the channel carrying the data; thus, no separate control channel for the steering is needed, which eliminates the need for tedious bookkeeping of which control channel belongs to which signal channel. The wavelength is actually serving as an embedded control channel. One-dimensional (1D) steering by means of a plane reflection grating was introduced in [4-9]. This concept was demonstrated with high-speed free-space communication for the first time in [4-10], showing transmission at 1 Gbit/s over 2 meters reach using on–off keying (OOK) modulation. Similarly, transmission at 2.5 Gbit/s OOK over 60 cm was shown in [4-11], and at 10 Gbit/s OOK over 2.5 meters in [4-12]. Steering of multiple beams is supported by using multiple wavelengths. A system using multiple narrowband Bragg gratings stored in photo-thermo-refractive (PTR) glass volumes was reported in [4-13]; the number of beams supported equals the number of Bragg gratings. Here, multi-beam generation via a single aperture was proposed using multiwavelength selection. The speed of the beam steering can be considerably faster than when using actively controlled devices, as the wavelength tuning time of a laser diode (with wavelength settling times less than 1 μs) can be much smaller than the electromechanical tuning time of a MEMS mirror or the tuning time of an SLM (both in the ms-range). 2D optical beam steering solutions by means of a photonic integrated circuit have been reported in [4-14, 4-15]. These circuits showed relatively high losses and no free-space data transmission was applied.

In this chapter, we introduce in Section 4.2 an indoor optical wireless communication system based on passive diffractive beam steerers according to the BROWSE concept proposed by us in [4-16–4-20] and discuss its design boundaries which define functional details of the key OWC modules to be elaborated in Chapters 5–8. Via an indoor fiber network, in every room, 2D steering of directional narrow optical infrared pencil beams is deployed for the wireless paths, which are each aimed at a specific user terminal. Thus, the full ultrahigh capacity of such a beam is available to that terminal, and neither congestion nor interference with other terminals occur. Each beam actually may be regarded as a "virtual fiber," offering the connectivity power of a single fiber (and even more than that, as its latency is 50% less, and waveguide dispersion is absent). In Section 4.3, the 2D beam steering function is discussed using a novel concept deploying wavelength tuning in combination with a pair of crossed diffraction gratings. Adding wavelengths results in additional beams for steering; thus, the system can readily be scaled to many beams. To the best of the authors' knowledge, this has been the first wavelength-controlled 2D beam-steered system demonstrated for ultrahigh-speed multi-beam communication. The pencil beam's directivity effectively alleviates the reach limitations of VLC and saves energy by spatial allocation of capacity on demand, offering capacity only when and where it is needed. The upstream wireless paths from the terminals in a bidirectional wireless system may implemented in a hybrid optical/RF fashion by high-capacity radio wireless channels in the 60 GHz band, which are transported over the indoor backbone fiber network by means of radio-over-fiber techniques employing recovery of the unmodulated optical carrier from a downstream channel, as described in Section 8.2. The upstream RF paths may be impacted by external EMI, and/or induce crosstalk to other users when not sufficiently confined. Alternatively, fully building on the key strengths of optical wireless communication, the bidirectional system may be implemented as an all-OWC system by also establishing the upstream links with optical beams, see Chapter 8. The chapter concludes by summarizing the main research directions for the key beam-steered OWC modules to be elaborated in Chapters 5–8.

4.2 The BROWSE System Concept

As discussed in Chapter 3, OWC by means of 2D-steered infrared beams is an attractive technique to provide broadband services wirelessly to users individually even at high user densities. We have proposed the concept of this system shown in Fig. 4-1 (first reported in [4-16]) which we have investigated in the project "BROWSE – Beam-steered Reconfigurable Optical-Wireless System for

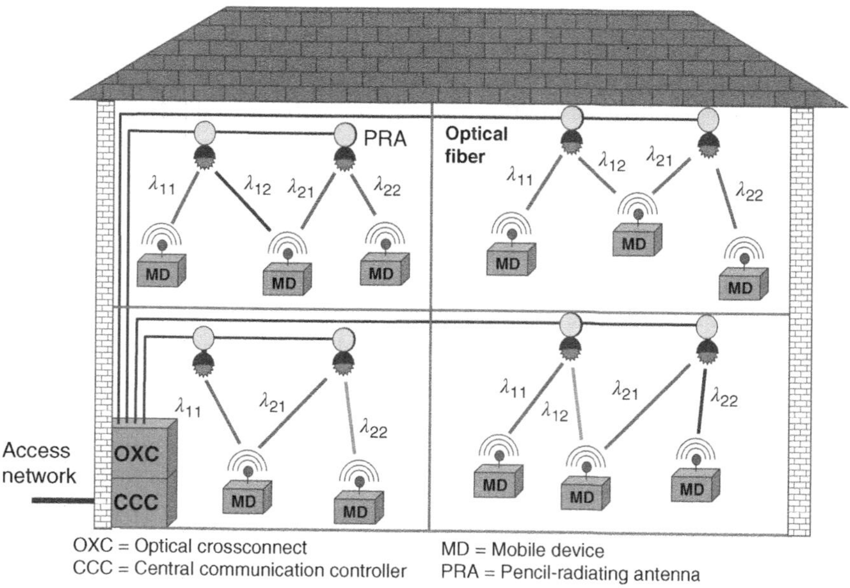

Fig. 4-1 Free-space indoor optical communication by pencil beams

Energy-efficient communication," funded by the European Research Council within the Advanced Investigator program. Inside a building, each room is equipped with one or more pencil-radiating antennas (PRAs). Each PRA can emit multiple optical pencil beams and contains passive diffractive optics. The capacity foreseen per beam is at least 1 Gbit/s, and capacities of 10 s of Gbit/s are within reach. Multiple PRAs per room are foreseen in order to provide full coverage of the room, and to enable circumvention of possible line-of-sight path blockings. The circumvention is realized by re-routing the optical signal to another PRA by means of the optical cross-connect (OXC). Inside a PRA, a passive diffractive module steers a beam in two angular dimensions (i.e., 2D steering), by just varying the wavelength of the beam. The wavelength of each beam thus acts as an embedded control channel, which simplifies the management of the indoor traffic flows as it avoids the need to implement a separate control network and the need for tedious bookkeeping of which control channel belongs to which signal channel. We have proposed two options for realizing such a diffractive module: one using a pair of crossed diffraction elements [4-18], or an alternative one using an arrayed waveguide grating router (AWGR) with its output ports grouped in a 2D matrix [4-19]. Full design details are given in Sections 5.4 and 5.5, respectively.

For the diffractive module using a pair of crossed diffraction elements, we propose to deploy two elements (e.g., reflection gratings, see Fig. 4-2a and Section 5.4 for more details) where one element has relatively low diffractive power, and the other one has a high diffractive power. These elements together form a fully passive PRA module, which steers a beam in two dimensions according to its wavelength. The beam is continuously swept in one dimension by changing the wavelength over less than the large free spectral range (FSR) of the low-dispersive element, whereas it is swept multiple times in the orthogonal dimension when it traverses multiple small FSRs of the other highly dispersive element. As a result, an area is scanned in two dimensions line by line, by just varying a single parameter, namely the wavelength λ of the beam. The wavelength of the beam thus also acts as a control channel of the beam steering, embedded in the data channel; hence, a separate control channel is avoided, which relaxes network management and control.

For the diffractive module using an AWGR with its output fibers arranged in a 2D matrix, wavelength tuning of the input signal yields steering of the beam in both of the two dimensions with discrete steps (Fig. 4-2b, see Section 5.5 for more details).

We propose to use infrared wavelengths beyond 1.4 μm, where eye safety regulations allow higher beam powers than in the visible range; up to 10 mW is allowed according to ANZI Z-136 and IEC 60825 standards [4-21]. Moreover, beyond 1.5 μm there are already many mature devices readily available, designed for optical fiber communication systems. Multiple beams can be handled individually by just adding wavelengths, making this concept readily scalable to large

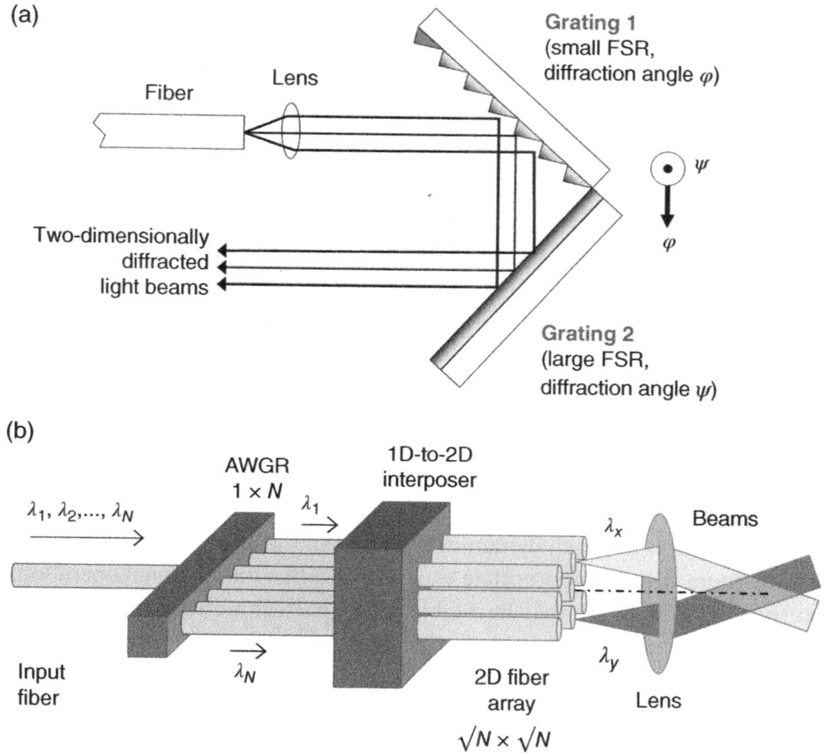

Fig. 4-2 Two-dimensional diffractive optical beam steering module using (a) a pair of crossed gratings (FSR: Free Spectral Range), or (b) an arrayed waveguide grating router with output ports in a 2D matrix

numbers of beams. The wavelength-tunable sources are hosted jointly in the central communication controller (CCC; see Fig. 4-1), and are dynamically connected upon demand to the PRAs in the respective rooms by means of an OXC switch and the point-to-point (P2P) fiber feeder backbone network. Optionally, one may share fiber lines in a point-to-multipoint (P2MP) topology by multiplexing wavelength channels groupwise into bands at the OXC and demultiplexing these bands near the PRAs (cf. a WDM-PON). However, in order to ensure that this wavelength-partitioning leaves enough wavelength tuning range per PRA, this would require an ultra-wide wavelength tuning range per tunable source. For example, 30 nm wide tuning by an integrated sampled-grating distributed Bragg reflector (SG-DBR) laser diode has been reported in [4-15]. Conceptually, wider tuning may be achieved by putting multiple tunable lasers each covering a distinct range in parallel. To limit the wavelength tuning requirements, we have chosen the P2P fiber network topology in our system concept.

The indoor OWC system concept using 2D-steerable IR beams we propose is shown in Fig. 4-1 [4-18]. In each room, multiple pencil-beam radiating antennas (PRA-s) are installed. A PRA contains passive diffractive optics and can emit multiple beams, each of which is 2D steered under the control of its wavelength. The wavelength of each beam thus acts as an embedded control channel, which simplifies the management of the indoor traffic flows as it avoids the need to implement a separate control network. By using multiple PRA-s per room, line-of-sight blocking can be circumvented, as one PRA can take over the duties of another one. Each PRA is fully passive; it contains no moving parts and does not need electrical power or electrical control signals. Its optical data signals are fed remotely via the indoor fiber backbone network from a central site in the building, where all network management and control functions are located. This co-location also eases the management, maintenance, and upgrading of the whole network. At this CCC, tunable transceivers together with a transparent OXC module take care that the user data streams are each dynamically routed to the appropriate PRA in the appropriate room with the appropriate wavelength. For a bidirectional operation, the return path for each user may be established by RF techniques such as 60 GHz radio-over-fiber techniques or alternatively may be established by optical wireless links too, as is described in detail in Chapter 8.

4.3 2D Beam-Steering Using Diffractive Means

For specifying the functionality of the PRA, we assume that it has to cover an area of $L \times L$ while it is positioned at a height (H) with respect to that area. It has to address the area by means of 2D steering of a beam having a diameter D_{beam} (see Fig. 4-3). The number of steering steps needed is $N = (L/D_{\text{beam}})^2$. Given that for diffractive steering of a beam a wavelength tuning range $\Delta\lambda$ is available, the wavelength tuning step size is $\delta\lambda = \Delta\lambda/N = \Delta\lambda\,(D_{\text{beam}}/L)^2$. Hence, a lower bound for the beam diameter is given by

$$D_{\text{beam}} > L\sqrt{\frac{\delta\lambda}{\Delta\lambda}} \tag{4-1}$$

When assuming that the PRA is launching (nearly) collimated beams out of a circular aperture, the beam spot projected in the user plane is also circular. The receiver at the user may be tilted over an angle ε_{rx} with respect to the user plane, and the angle of beam incidence α on the receiver is

$$\alpha = \varphi_c + \varepsilon_{\text{rx}} \tag{4-2}$$

where φ_c is the angle at which the beam is launched from the PRA, see Fig. 4-3b.

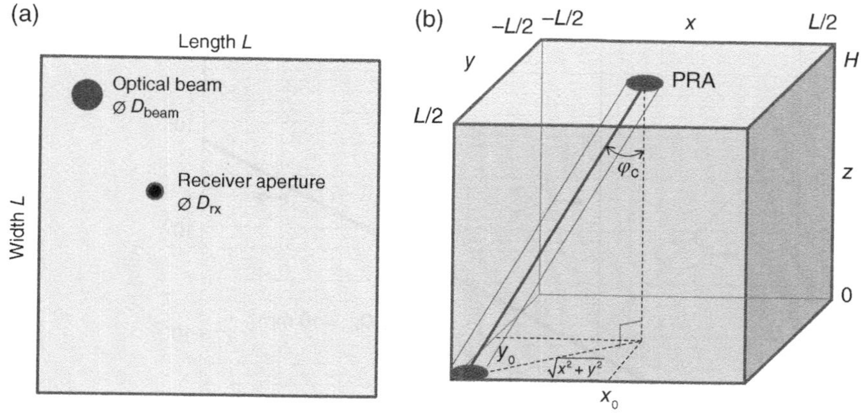

Fig. 4-3 Scanning the room with a collimated optical beam: (a) scanned user area and (b) scanning of a room

The maximum incidence angle α_{max} which the receiver can handle is specified as its FoV α_{FoV}. The required FoV of the receiver α_{FoV} for the coverage area is

$$\alpha_{FoV} \geq \text{atan}\left(\frac{L}{H\sqrt{2}}\right) + \varepsilon_{rx} \tag{4-3}$$

In a first-order analysis, we assume that the FoV of the receiver exceeds the maximum incidence angle α_{max}, and thus the received power is not restricted by the FoV. Such an approximation holds for a lensless receiver where the received light is incident directly on the active area of a photodiode. When the receiver in the user terminal has an aperture with diameter D_{rx}, the PRA is fixed in the middle of the ceiling at height H, and the power intensity in the beam is uniform across its diameter D_{beam}, the fraction T of the beam's power P_{beam} which is captured by the receiver with aperture diameter D_{rx} is

$$T = \frac{P_{rx}}{P_{beam}} = \left(\frac{D_{rx}}{D_{beam}}\right)^2 \cos\varepsilon \quad \text{if } D_{rx} \leq D_{beam} \tag{4-4}$$

$$T = 1 \qquad \text{if } D_{rx} > D_{beam}$$

where P_{rx} is the power captured by the receiver which implicitly gives an upper bound to the beam diameter D_{beam} in order to guarantee an adequate minimum received power level $P_{rx,min}$. Fig. 4-4 shows how D_{beam} can be determined as a

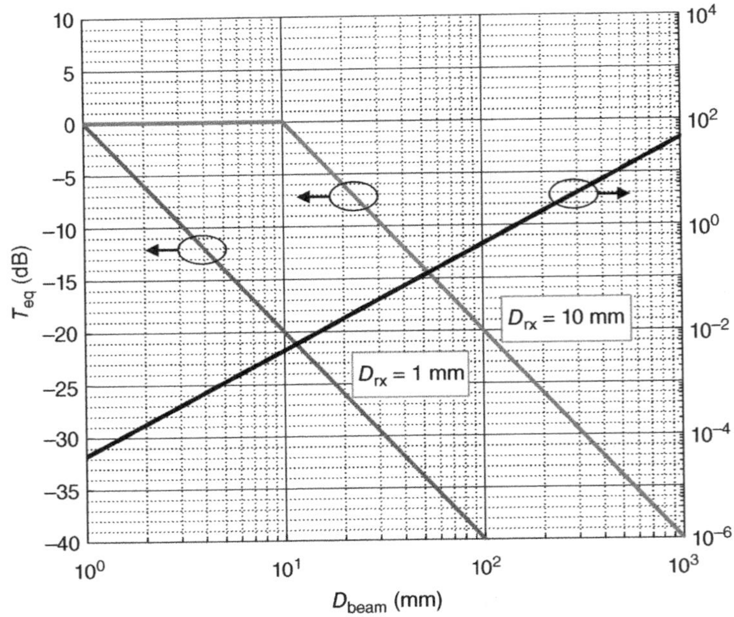

Fig. 4-4 Choosing the beam diameter D_{beam} for a lensless receiver (for covered user area $L \times L$ with $L = 1.5$ meters, and receiver aperture $D_{rx} = 1$ mm or $D_{rx} = 10$ mm)

compromise between a minimum wavelength tuning step and a minimum receiver sensitivity. If the beam-to-receiver power coupling fraction T_{eq} should be at least −20 dB (so for a beam power of +10 dBm at least −10 dBm is captured by the receiver's aperture), and if an area with $L = 1.5$ meters should be covered with a total available wavelength tuning range $\Delta\lambda = 100$ nm, the beam diameter should be $D_{beam} \leq 10$ mm and the tuning step size $\delta\lambda = 4.4$ pm when the receiver's aperture diameter $D_{rx} = 1$ mm. The tuning step size limits the bandwidth BW of the OWC link, as $BW << \delta\lambda c_0/\lambda^2$. Therefore, a narrow beam complicates the required accuracy for the beam steering, the required stability of the wavelength tuning, and the user localization process, as well as limits the link's bandwidth. For example, with $\delta\lambda = 4.4$ pm, the bandwidth $BW << \delta\lambda c_0/\lambda^2 \approx 587$ MHz in the 1.5 μm wavelength window. When a smaller beam-to-receiver power coupling fraction T_{eq} is allowed, the system requirements are relaxed. For example, a $T_{eq} > -30$ dB is achieved with $D_{beam} \leq 10$ mm and $\delta\lambda = 46$ pm, allowing a bandwidth $BW << 6.1$ GHz. The system requirements are significantly relaxed further when the receiver's aperture is increased: for example, at $T_{eq} > -30$ dB an aperture $D_{rx} = 10$ mm enables a beam diameter $D_{beam} \leq 314$ mm and tuning step size $\delta\lambda = 4.4$ nm (i.e., a bandwidth $BW << 584$ GHz).

However, increasing the receiver's aperture by increasing the photodiode leads to a reduction of the receiver's bandwidth as the junction capacitance of a photodiode scales with its active area. Alternatively, while maintaining the photodiode's size, the aperture may be increased by applying intermediate optics, such as a compound parabolic concentrator (CPC), a condenser lens, etc. Unfortunately, due to the basic physics law of the conservation of etendue, these aperture-enlarging optics inevitably also restrict the FoV α_{FoV} of the receiver. In Chapter 7, it is analyzed how the FoV can be increased by applying a certain degree of defocusing, that is, by putting the photodiode not in the focal plane of the lens but somewhat closer to it by a distance x.

When assuming a uniform beam and ideal thin lens without any lens aberrations, if the incident beam diameter D_{beam} exceeds the diameter D_{lens} of the lens with focal length f, the FoV of the receiver is (cf. Eq. (7-39))

$$\alpha_{\mathrm{FoV}} = \operatorname{atan}\left(\frac{|pD_{\mathrm{lens}} - D_{\mathrm{PD}}|}{2f(1-p)}\right) \tag{4-5}$$

where D_{PD} is the diameter of the active area of the photodiode, and the defocusing parameter $p = x/f$ $(0 \leq p < 1)$. Enlarging p not only enables to increase in the FoV but also reduces the fraction of the beam power coupled to the photodiode. According to Eq. (4-3), given the area to be covered and assuming a negligible tilt of the receiver (i.e., $\varepsilon_{\mathrm{rx}} = 0$), the required defocusing p follows:

$$p \geq \frac{D_{\mathrm{PD}} + \frac{L}{H} f\sqrt{2}}{D_{\mathrm{lens}} + \frac{L}{H} f\sqrt{2}} \tag{4-6}$$

As introduced in Chapter 7, a large photodetector area can be created by building a matrix of photodiodes while preserving the bandwidth of the individual small photodiodes. This allows an increase in the defocusing p and thus an increase in the FoV, but is accompanied by a decrease of the beam-to-photodetector power coupling. A suitable compromise between the FoV and the received optical power needs to be defined, depending on the system's requirements. According to eq. (4-6), the required defocusing p implies an equivalent beam-to-photodiode power coupling fraction T_{eq} (cf. eq. (7-40))

$$T_{\mathrm{eq}} = \frac{\eta}{M}\left(\frac{D_{\mathrm{PDmatrix}}}{pD_{\mathrm{beam}}}\right)^2 \leq \frac{\eta}{M}\left(\frac{D_{\mathrm{lens}}}{D_{\mathrm{beam}}} \frac{1 + \frac{L}{H}\frac{f}{D_{\mathrm{lens}}}\sqrt{2}}{1 + \frac{L}{H}\frac{f}{D_{\mathrm{PDmatrix}}}\sqrt{2}}\right)^2 \tag{4-7}$$

In Eq. (4-7), a matrix configuration of $M \times M$ photodiodes is assumed, which has a fill factor η (i.e., the aggregate active area of all the photodiodes in the matrix divided by the total area within the circle circumscribing the matrix, see Section 7.3.1).

Similarly, when the beam diameter D_{beam} is smaller than the lens diameter D_{lens}, the FoV of the receiver is

$$\alpha_{\text{FoV}} = \text{atan}\left(\frac{|p\,D_{\text{beam}} - D_{\text{PDmatrix}}|}{2f(1-p)}\right) \tag{4-8}$$

and the required defocusing p is

$$p \geq \frac{D_{\text{PDmatrix}} + \frac{L}{H} f\sqrt{2}}{D_{\text{beam}} + \frac{L}{H} f\sqrt{2}} \tag{4-9}$$

which yields an equivalent beam-to-photodetector power coupling

$$T_{\text{eq}} = \frac{\eta}{M}\left(\frac{D_{\text{PDmatrix}}}{pD_{\text{beam}}}\right)^2 \leq \frac{\eta}{M}\left(\frac{1 + \frac{L}{H}\frac{f}{D_{\text{beam}}}\sqrt{2}}{1 + \frac{L}{H}\frac{f}{D_{\text{matrix}}}\sqrt{2}}\right)^2 \tag{4-10}$$

Fig. 4-5 shows how the beam diameter D_{beam} may be chosen in order to cover the $L \times L$ user area when a lensed receiver is applied. It also shows the optical spectral bandwidth $\delta\lambda$ available per beam, which corresponds with a maximum link bandwidth available for communication $BW_{\max} \approx \delta\lambda c_0/\lambda^2$. Two lenses have been assumed: an ideal thin lens with a diameter (Ø) of 25 mm and focal length (f) of 5 mm, and an ideal thin lens with a diameter of 50 mm and focal length of 10 mm. Fresnel lenses (which can have a large D/f number, so a large aperture/focal length ratio and thus a large FoV, see Chapter 7) with these parameters have been used in our OWC laboratory testbed. With the Fresnel lenses, the beam-to-PD matrix coupling has been numerically analyzed by means of ray tracing in Section 7.3.2, wherein a 4×4 PD matrix is assumed with Ø150 μm photodiodes spaced at 17 μm, yielding a PD matrix diameter $D_{\text{PDmatrix}} = 0.859$ mm. By increasing the beam diameter, the required wavelength step size $\delta\lambda$ is increased, and thus the maximum available bandwidth $BW_{\max} \approx \delta\lambda c_0/\lambda^2$ grows. When a beam-to-PD matrix power coupling factor $T_{\text{eq}} > -30$ dB is needed, without a lens the beam diameter should be $D_{\text{beam}} < 9.5$ mm, and the wavelength tuning step $\delta\lambda = 4.0$ pm ($BW_{\max} \approx 0.53$ GHz). With the lenses, the beam diameter can be significantly larger, and thus the wavelength tuning step $\delta\lambda$ larger, which eases the beam steering process. With the Ø25 mm lens, $D_{\text{beam}} < 54$ mm and $\delta\lambda =$ 0.13 nm ($BW_{\max} \approx 17$ GHz), and with the Ø50 mm lens $D_{\text{beam}} < 60$ mm and

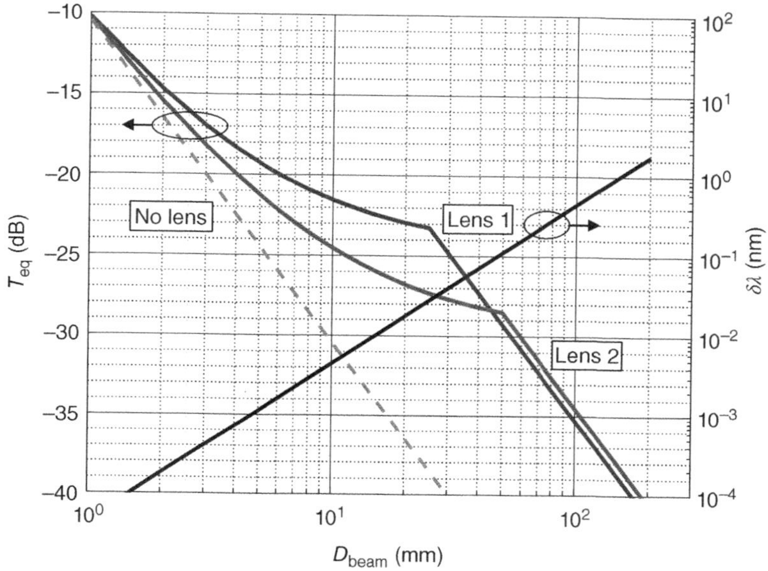

Fig. 4-5 Choosing the beam diameter D_{beam} for a lensed receiver with lens 1 (a thin lens with diameter Ø25 mm and $f = 5$ mm), and lens 2 (thin lens Ø50 mm and $f = 10$ mm), and for a lensless receiver (dashed curve). Assumptions: area size $L = 1.5$ meters, height $H = 2.5$ meters, and a 4 × 4 PD matrix with overall diameter $D_{PDmatrix} = 0.86$ mm and fill factor $\eta = 49\%$

$\delta\lambda = 0.16$ nm ($BW_{max} \approx 21$ GHz). Fig. 4-5 also shows that the power received by a lensless receiver equipped with the same PD matrix is significantly less for the same beam diameter.

It may therefore be concluded that for a practical robust OWC indoor system based on diffractive 2D steering of relatively narrow beams by means of wavelength tuning, it is preferred to maximize the diameter of these steered beams, and to facilitate this by using lensed receivers at the user which can provide an adequate FoV while also capturing a sufficiently large fraction of the incident optical beam to enable high-capacity communication. Such a receiver design approach is elaborated in more detail in Chapter 7.

4.4 The Indoor Fiber Backbone Network

A fiber backbone network is foreseen throughout the building in order to connect the PRA-s located in the various rooms, as shown in Fig. 4-1. A P2P fiber link connects each PRA to the CCC, where the tunable laser diodes are located. By means of an OXC, according to actual traffic demands these laser diodes are mapped to the appropriate PRA-s in the appropriate rooms. At the CCC, autonomic network management is located to intelligently locate and track the (roaming) users' mobile devices (MDs) and to control the OXC settings and the wavelength tuning accordingly. Machine learning techniques can be applied which gather user behavior characteristics, in order to speed up localization, tracking, and beam steering processes.

4.5 Concluding Remarks

Free-space optical communication deploying 2D steerable narrow optical pencil beams can offer the ultimate wireless capacity per user device. Diffractive beam steering techniques enable to emit multiple beams from a single ceiling unit (typically located at the ceiling in an indoor scenario), while each beam is individually steerable by changing its wavelength. Moreover, the optical signals that create the beams can be generated remotely, not at the ceiling unit, which facilitates maintenance, upgrading, and upscaling of the system. The proposed concept which was introduced in the ERC Advanced Grant BROWSE project employs a pair of passive crossed diffractive elements that enables scaling to large numbers of pencil beams, each individually steerable by tuning its wavelength in the 1.5 μm infrared regime. As described in detail in Section 5.4, angular beam steering over a 6° × 12° area (i.e., an area of $32 \times 64\,cm^2$ at 3 meters distance) was achieved by wavelength tuning from 1505 to 1630 nm. Further careful design of the diffractive stages augmented by angular magnification using a lens system can increase the area coverage. Downstream delivery of 32 Gbit/s per optical pencil beam in the 1.5–1.6 μm range over 3 meters reach using PAM-4 modulation with a simple free-space receiver has been demonstrated. With adaptive DMT modulation using 512 tones, a 42.8 Gbit/s gross bitrate has been achieved. In the return path, by using radio-over-fiber techniques, DMT modulation, and recovery of the downstream optical carrier, upstream delivery of 10 Gbit/s per upstream 60 GHz radio signal has been demonstrated.

Within the BROWSE project, we also explored an alternative diffractive 2D beam-steering technique based on an AWGR and 2D arrangement of its fiber output ports. This technique employs readily available AWGR modules which are widely used in optical fiber communication networks and do not involve the delicate alignment needed for the crossed-gratings technique. As discussed in Section 5.5, an indoor area coverage of 17° × 17° is readily feasible, and a data capacity of up to 112 Gbit/s per beam with PAM-4 modulation has been achieved.

For aligning the receiver aperture, automated electromechanical alignment measures can be applied which maximize the received power but need extra alignment time. Alternatively, the alignment effort can be relaxed by increasing the FoV of the receiver by means of a novel photodiode matrix constellation as will be discussed in Chapter 7.

References

[4-1] D. O'Brien, R. Turnbull, H. Le Minh, et al., "High-speed optical wireless demonstrators: Conclusions and future directions," *IEEE/OSA J. Lightwave Technol.*, vol. 30, no. 13, pp. 2181–2187, Jul. 2012.

[4-2] H. Haas, "Visible light communication," *Proc. of OFC 2015*, Los Angeles, Mar. 22–26, 2015, Paper Tu2G.5.

[4-3] N.A. Riza, "Reconfigurable optical wireless," *Proc. LEOS'99 12th Annual Meeting*, Vol. 1, San Francisco, Nov. 10–11, 1999, pp. 70–71.

[4-4] A.D. Yalcinkaya, H. Urey, D. Brown, et al., "Two-axis electromagnetic microscanner for high resolution displays," *IEEE J. Microelectromech. Syst.*, vol. 15, no. 4, pp. 786–794, Aug. 2006.

[4-5] K. Wang, A.T. Nirmalathas, C. Lim, and E. Skafidas, "Experimental demonstration of full-duplex optical wireless personal area communication system with 16-CAP modulation," *Proc. OFC 2015*, Los Angeles, Mar. 22–26, 2015, Paper M2F-7.

[4-6] A. Gomez, K. Shi, C. Quintana, et al., "Beyond 100-Gb/s indoor wide field-of-view optical wireless communications," *IEEE Photon. Technol. Lett.*, vol. 27, no. 4, pp. 367–370, 2015.

[4-7] N.A. Riza, and Z. Yaqoob, "Space-multiplexed optical scanner," *Appl. Opt.*, vol. 43, no. 13, pp. 2703–2708, May 2004.

[4-8] A. Gomez, C. Quintana, G. Faulkner, D.C. O'Brien, "Point-to-multipoint holographic beamsteering techniques for indoor optical wireless communications," *Proc. SPIE 9772, Broadband Access Communication Technologies X*, Feb. 2016, 7p.

[4-9] Z. Yaqoob, A.A. Rizvi, and N.A. Riza, "Free-space wavelength-multiplexed optical scanner," *Appl. Opt.*, vol. 40, no. 35, pp. 6425–6438, Dec. 2001.

[4-10] K. Liang, H. Shi, and S.J. Sheard, D.C. O'Brien, "Transparent optical wireless hubs using wavelength space division multiplexing," *Proc. SPIE 5550 Free-Space Laser Commun. IV*, Denver, CO, USA, Oct. 2004, 8p.

[4-11] C.W. Oh, F.M. Huijskens, Z. Cao, et al., "Toward multi-Gbps indoor optical wireless multicasting system employing passive diffractive optics," *Opt. Lett.*, vol. 39, no. 9, pp. 2622–2625, May 2014.

[4-12] C.W. Oh, E. Tangdiongga, and A.M.J. Koonen, "Steerable pencil beams for multi-Gbps indoor optical wireless communication," *Opt. Lett.*, vol. 39, no. 18, pp. 5427–5430, Sep. 2014.

[4-13] Z. Yaqoob, M.A. Arain, and N.A. Riza, "High-speed two-dimensional laser scanner based on Bragg gratings stored in photothermorefractive glass," *Appl. Opt.*, vol. 42, no. 26, pp. 5251–5262, Sep. 2003.

[4-14] K. Van Acoleyen, W. Bogaerts, and R. Baets, "Two-dimensional dispersive off-chip beam scanner fabricated on silicon-on-insulator," *IEEE Photon. Technol. Lett.*, vol. 23, no. 17, pp. 1270–1272, Sep. 2011.

[4-15] W. Guo, P.R.A. Binetti, C. Althouse, et al., "InP photonic integrated circuit with on-chip tunable laser source for 2D optical beam steering," *Proc. of OFC 2013*, Los Angeles, Mar. 19–21, 2013, Paper OTh3I-7.

[4-16] A.M.J. Koonen, C.W. Oh, and E. Tangdiongga, "Reconfigurable free-space optical indoor network using multiple pencil beam steering," *Proc. OECC/ACOFT 2014*, Melbourne, July 2014, Paper Tu3F-1.

[4-17] A.M.J. Koonen, C.W. Oh, K. Mekonnen, and E. Tangdiongga, "Ultra-high capacity indoor optical wireless communication using steered pencil beams," *Proc. MWP 2015*, Paphos, Oct. 25–29, 2015, Paper WeC-5.

[4-18] A.M.J. Koonen, C.W. Oh, K.A. Mekonnen, et al., "Ultra-high capacity indoor optical wireless communication using 2D-steered pencil beams," *IEEE/OSA J. Lightwave Technol.*, vol. 34, no. 20, pp. 4802–4809, Oct. 2016, https://doi.org/10.1109/JLT.2016.2574855.

[4-19] A.M.J. Koonen, F. Gomez-Agis, F.M. Huijskens, et al., "High-capacity optical wireless communication using two-dimensional IR beam steering," *IEEE/OSA J. Lightwave Technol.*, vol. 36, no. 19, pp. 4486–4493, Oct. 2018, https://doi.org/10.1109/JLT.2018.2834374.

[4-20] A.M.J. Koonen, K.A. Mekonnen, F.M. Huijskens, and E. Tangdiongga, "Bi-directional all-optical wireless Gigabit Ethernet communication system using automatic self-aligned beam steering," *J. Lightwave Technol.*, vol. 41, no. 11, pp. 3446–3454, June 2023. https://doi.org/10.1109/JLT.2022.3231438.

[4-21] Laser safety. [Online] Available: https://en.wikipedia.org/wiki/Laser_safety

5

Beam-Steering Techniques for Indoor OWC

5.1 Introduction

An essential task of the indoor beam-steered optical wireless network as introduced in Section 3.6 is to provide independent accurate steering in two dimensions of each optical beam. At the ceiling unit, multiple beams need to be steered downstream (DS) toward multiple user devices individually. Conversely, to establish bidirectional optical wireless communication, each user device needs to steer a beam upstream (US) to the ceiling unit. Several approaches can be followed to establish beam steering, which may be classified coarsely in those using active steering devices and in those using passive ones.

5.2 Active Beam-Steering Devices

In active beam-steering devices (e.g., micro-electromechanical system (MEMS) mirrors, or spatial light modulators (SLMs)), the state of the device is changed using a separate control channel, and a change in the state implies a change in the beam's direction. Per beam, a separate controllable state is needed, which complicates the scaling of the steering up to many beams. Local electrical powering and maintenance of the device are needed. For downstream communication, the device is typically mounted on the ceiling of the room and may not be readily accessible, which complicates maintenance and upgrading.

An SLM processes a beam by spatially altering the intensity, phase, or polarization of the incident light. An SLM contains many pixels of which the phase delay can be programmed, and thus it can perform as a dynamically adaptable reflection or transmission grating [5-1, 5-2]. It is typically based on a liquid crystal nematic micro-display, with transmissive (with LCD, liquid crystal display) or reflective (with LCOS, liquid crystal on silicon; gives better performance) pixels

Short-Reach Optical Wireless Communication: By Directed Narrow Beams, First Edition. Ton Koonen.

[5-3], [3-22]. First concepts using an SLM were introduced in ref. [5-4], describing beam steering by a material in which multiple holograms were recorded (see Fig. 5-1). By using a specific spatial code on the beam that is impressed by means of an SLM, the beam is deflected by the corresponding hologram into a specific direction.

The number of programmable pixels of an SLM and their size determine the beam-steering characteristics. As the pixels are typically large with respect to the wavelength, the steering angle is small [5-5]. Moreover, a comprehensive driver module is needed to program the SLM.

An SLM having 512×512 pixels with 256 phase levels was used for steering a beam at $\lambda = 1.55\ \mu m$ (see Fig. 5-2) [5-1]. The maximum steering angle θ_{max} attainable

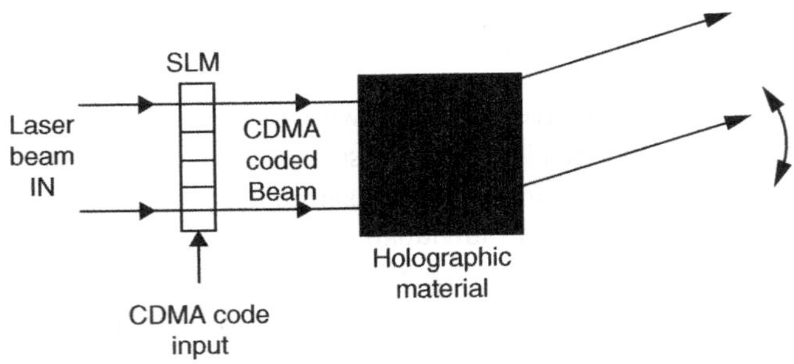

Fig. 5-1 2D beam steering by spatial coding. *Source:* Riza [5-4]. © [1999] IEEE

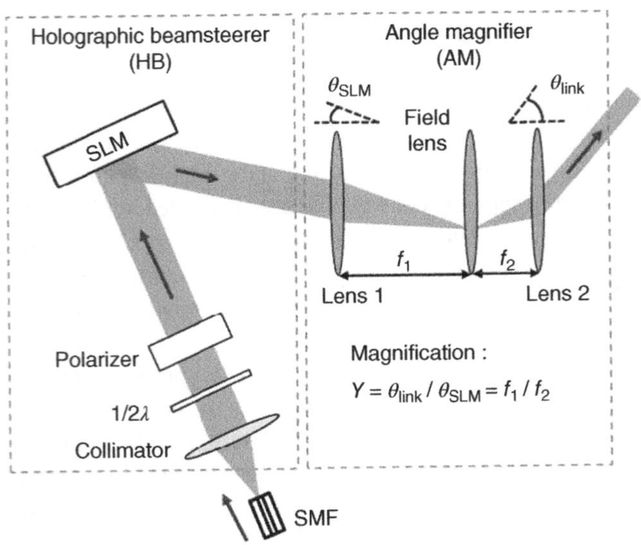

Fig. 5-2 2D beam steering with SLM and angle magnifier [3-23], [5-1]. © [1999] IEEE

by the SLM is determined by the SLM's pitch ξ; with a $\xi = 15\,\mu m$, it was only $\theta_{SLM,max} = \lambda/2\xi \approx 3°$. An angular magnifier composed of an afocal system with two lenses having focal lengths f_1 and f_2 was applied and can yield an angular magnification factor $\gamma_{mag} = \theta_{link}/\theta_{SLM} = f_1/f_2$, which inevitably goes together with a beam narrowing factor f_2/f_1. A field lens may be put in between to reduce lens aberrations. Using such an angular magnifier with effective $\gamma_{mag} = 10$, steering was achieved over $\theta_{link} = 30°$. Operating in WDM mode with three channels spaced by 10.1 GHz and carrying 16-QAM signals, transmission of 3×37.4 Gbit/s = 112 Gbit/s (incl. 7% FEC) over a 3-meters free-space link was achieved. If its aperture allows, an SLM may be operated in different zones each steering a beam [3-24], [5-6]. Thus, a (limited) number of beams can be steered by a single SLM, and optionally point-to-multipoint communication can be supported by sharing different zones.

By means of MEMS mirrors, 2D steering over relatively large angles can be achieved. In [3-25] [5-7], steering over >20° is reported, providing 10 Gbit/s with CAP-16 modulation over a 2-meters downstream link with 7 mW launched power at $\lambda = 1551$ nm (see Fig. 5-3). A compound parabolic concentrator (CPC) was used to increase the receiving aperture, followed by a fiber collimator and PIN photodiode receiver. In the US direction, a similar MEMS-steering mirror is deployed and by using a vertical cavity surface emitting laser (VCSEL) emitting 5 mW at $\lambda = 850$ nm 2 Gbit/s was transmitted.

Beam steering may also be implemented by means of an array of VCSEL-s. Low-divergence IR beams generated by a VCSEL array in conjunction with MIMO techniques have been proposed for meeting the high-density capacity demands in future 6G wireless networks [5-8] and have been coined as LiFi 2.0 [5-9].

5.3 Passive Beam-Steering Devices

Alternatively, the beam steering can be done by passive devices, which do not change state and have no moving parts. Hence, these devices do not need local powering nor a separate control channel. They have no failing risk and virtually need no maintenance.

The passive steering function may be implemented by a diffractive device, through which the beam steering is controlled by changing the wavelength of the optical data signal fed to the steering device [5-10]. The control is thus embedded in the data signal and no separate control channel is needed, which simplifies the network's architecture and the system's management, among others, by avoiding the tedious bookkeeping for monitoring which control channel belongs to which data channel. The optical data signals are fed to the steering device by means of an indoor fiber network and are generated by wavelength-tunable laser transmitters. For steering the downstream beams in each room, these transmitters may be

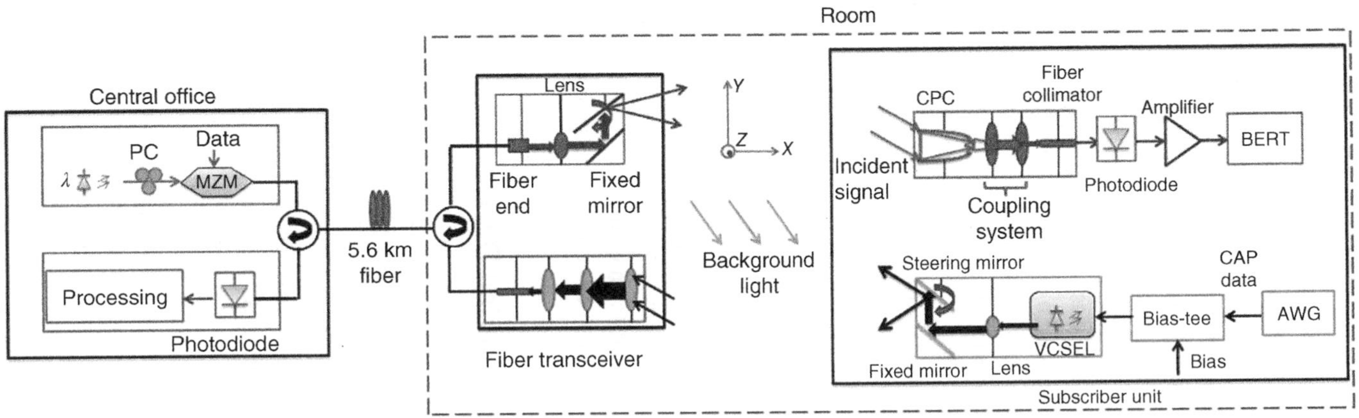

Fig. 5-3 2D beam steering with MEMS mirrors [3-25], [5-7]. © [1999] IEEE

jointly located and controlled in a centralized site, as proposed in the BROWSE system architecture (see Section 4.2), which eases maintenance, upgrading, and upscaling of the network. Laser tuning times are typically much less than the tuning times of a MEMS mirror or SLM, and therefore the beam-steering process can be much faster. If tuning times can be much shorter than the duration of the data packets, a single laser may be able to serve multiple users in a time division multiplexed mode, that is, in interleaved time slots, thus saving on the required number of tunable laser diodes. For instance, wavelength switching of a monolithic widely tunable laser from any one channel to another in the 100 GHz grid of the full C-band, within ±2.5 GHz of the ITU grid frequencies, has been reported in commercially available modules already in 2002. The switching time was less than 1 μs [5-11].

We propose two novel concepts for steering beams in two angular dimensions (2D) by wavelength-diffractive techniques: one based on crossed diffraction gratings presented in Section 5.4 and another one presented in Section 5.5 based on an arrayed waveguide grating router (AWGR). AWGR-s are already widely deployed in multiwavelength optical fiber networks, such as for wavelength-controlled traffic routing. In combination with wavelength tuning, these concepts are readily scalable to handle many beams individually by only a single steering module. The upscaling just requires that multiple wavelengths are fed to the device without the need to adapt it, where each wavelength independently determines the direction of its beam. These two concepts are therefore well suitable, for example, for steering multiple beams downstream from the central ceiling unit in an indoor scenario.

To establish the upstream (US) link in a bidirectional indoor optical wireless communication (OWC) system serving multiple users, from each user module a single beam has to be steered accurately toward the US receiver in the ceiling unit. Diffraction-based US beam steering is not preferred for cost reasons, as tunable laser diode transmitters are more complex. Hence, 2D beam steering by mechanical means using a laser diode operating at an arbitrary wavelength has been explored; its details are elaborated in Section 5.6.

The chapter is concluded in Section 5.7 by comparing the characteristics of the various beam-steering methods.

5.3.1 1D Steering by Diffraction Gratings

Diffraction gratings are well known for deflecting a light beam in one dimension (1D) at an angle ψ as a function of the beam's wavelength λ. For a reflective diffraction grating as illustrated in Fig. 5-4 where constructive interference occurs if neighboring ray path lengths differ by an integer multiple (the order m) of the wavelength λ, the beam steering is governed by the classical grating equation [3-27], [5-12]

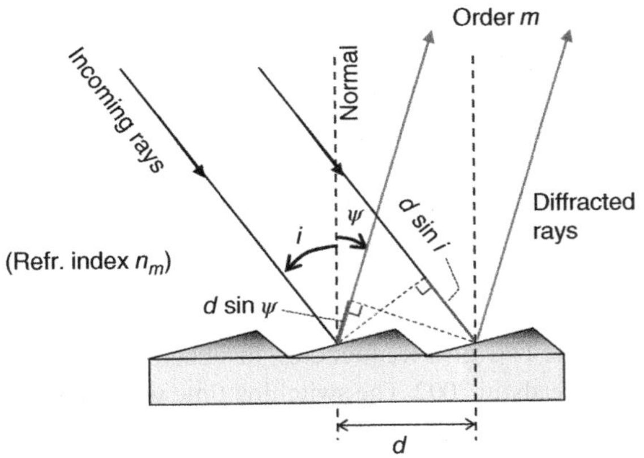

Fig. 5-4 Reflective diffraction grating

$$\sin\psi + \sin i = \frac{m\lambda}{n_m d} \tag{5-1}$$

where i is the beam's incident angle on the grating, ψ is the direction in which the diffracted rays constructively interfere in the order of interference m (both i and ψ are measured from the grating's normal, positive if counterclockwise), d is the grating's period, and n_m is the refractive index of the outside medium.

The beam is typically deflected in multiple orders m, as illustrated in Fig. 5-5. The achievable order m (an integer) is bounded by

$$\left\lceil \frac{n_m d}{\lambda}(\sin\theta_i - 1) \right\rceil \leq m \leq \left\lfloor \frac{n_m d}{\lambda}(\sin i + 1) \right\rfloor \tag{5-2}$$

where the function $\lceil x \rceil$ rounds x up to the smallest integer larger than x, and the function $\lfloor x \rfloor$ rounds x down to the largest integer smaller than x.

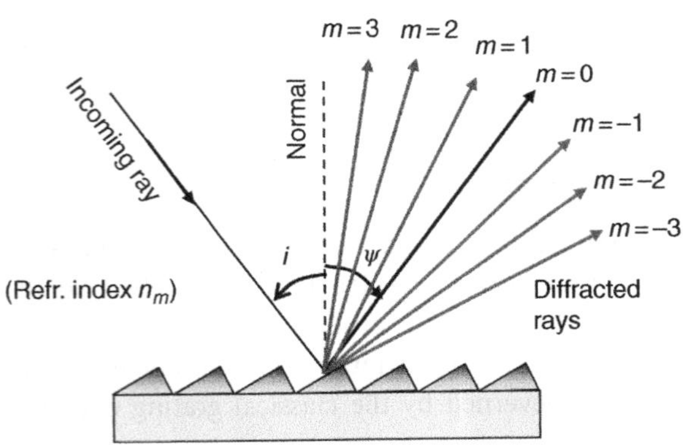

Fig. 5-5 Reflection grating operating in various orders m

The free spectral range (FSR) in order m is the wavelength range of $\Delta\lambda_{FSR,m}$ by which at an incident angle i the wavelength λ can be increased to arrive in the neighbor order $m-1$ at the same diffraction angle ψ. It also equals the wavelength range for which a beam in that order m does not enter a neighboring order. Hence, as illustrated in Fig. 5-6a, it follows with Eq. (5-1)

$$m\lambda = (m-1)\,(\lambda + \Delta\lambda_{FSR})$$

$$\Delta\lambda_{FSR,m} = \frac{\lambda}{m-1} \quad \text{for } m \geq 2 \tag{5-3}$$

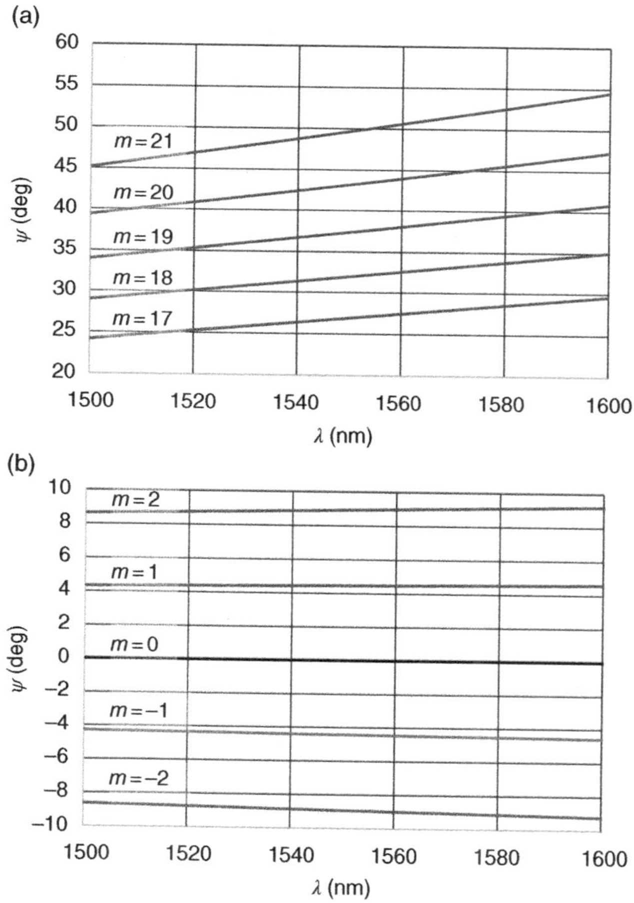

Fig. 5-6 Diffraction angle ψ versus wavelength λ for reflection grating operating in various orders m (grating with 50 grooves/mm). (a) Operating at higher orders m (incidence angle $i = 60°$). (b) Operating at low orders m (incidence angle $i = 0°$)

For $m \leq -1$ holds (cf. Fig. 5-6b)

$$\Delta\lambda_{\mathrm{FSR},m} = \frac{\lambda}{|m-1|} \tag{5-4}$$

The angular dispersion $\mathrm{d}\psi/\mathrm{d}\lambda$ is

$$\frac{\mathrm{d}\psi}{\mathrm{d}\lambda} = \frac{m}{n_m d \cos\psi} \tag{5-5}$$

The angular tuning range $\Delta\psi_m$ that can be achieved when tuning over one FSR in order m (with $m \geq 2$) from λ to $\lambda + \Delta\lambda_{\mathrm{FSR},m}$ is

$$\begin{aligned}\Delta\psi_m &= \psi(\lambda + \Delta\lambda_{\mathrm{FSR},m}) - \psi(\lambda)\\ &= \arcsin\left(\frac{m\,(\lambda + \Delta\lambda_{\mathrm{FSR},m})}{n_m d} - \sin i\right) - \arcsin\left(\frac{m\lambda}{n_m d} - \sin i\right)\end{aligned} \tag{5-6}$$

$\Delta\psi_m$ is maximum when $\psi_m\,(\lambda + \Delta\lambda_{\mathrm{FSR},m}) = \pi/2$, which implies that the maximum angular tuning range $\Delta\psi_{m,\max}$ which can be achieved in order m when tuning the wavelength over the full FSR $\Delta\lambda_{\mathrm{FSR},m}$ is given by

$$\cos\Delta\psi_{m,\max} = 1 - \frac{m}{m-1}\frac{\lambda}{d} \tag{5-7}$$

Moreover, $\psi_m\,(\lambda + \Delta\lambda_{\mathrm{FSR},m}) = \pi/2$ also implies $\sin\theta_i = \lambda/d\,(m^2/(m-1)) - 1$. As $\sin\theta_i \leq 1$, this yields an upper bound for the attainable $\Delta\psi_{m,\max}$ given by

$$\cos\Delta\psi_{m,\max} \geq 1 - \frac{2}{m} \tag{5-8}$$

In order to satisfy the two bounds for $\cos\Delta\psi_{m,\max}$, it follows for the maximum order $m_{\max}$

$$m_{\max} = \left\lfloor \frac{d}{\lambda}\left(1 - \sqrt{1 - \frac{2\lambda}{d}}\right)\right\rfloor \tag{5-9}$$

Hence, in order to achieve a relatively large angular tuning $\Delta\psi_{m,\max}$ within a small wavelength tuning range $\Delta\lambda_{\mathrm{FSR},m}$, a large order m and small grating groove spacing d are needed. For instance, when deploying a maximum wavelength tuning range $\Delta\lambda_{\mathrm{tun}} = 100$ nm (from $\lambda = 1.5$ to $1.6\ \mu$m), the maximum attainable angular tuning range $\Delta\psi_{m,\max} = 28.36°$ at order $m = 16$, which can be achieved with a reflection grating of $N = 75$ grooves/mm at a large incidence angle of $i = 66.93°$ (see Table 5-1).

It may be noted that the efficiency of diffraction gratings depends on the polarization state of the incident beam [5-13]; in particular, reflection gratings with

Table 5-1 Maximum angular tuning range $\Delta\psi_{m,max}$ using a reflection grating with N grooves/mm, achieved at $\lambda = 1500$ nm over a tuning range $\Delta\lambda_{tun}$ for the maximum order m_{max} and incidence angle i

N (gr/mm)	$\Delta\psi_{m,max}$ (deg)	Order m_{max}	Tuning range $\Delta\lambda_{tun}$ (nm)	i (deg)	$\Delta\psi_{m,max}$ bound (deg)
50	22.80	25	63	72.39	23.07
55	23.97	23	68	79.66	24.07
60	25.11	21	75	79.90	25.21
65	26.22	19	83	72.83	26.53
70	27.32	17	94	63.71	28.07
75	28.36	16	100	66.93	28.96
80	29.37	15	107	68.21	29.93

smaller groove spacings may have noticeably polarization-dependent performance at larger beam incidence angles.

1D beam steering with a plane reflection grating was introduced in [5-14] for fast optical scanning applications. It was applied for steering of 3.3 mm diameter beams transmitting 10 Gbit/s with OOK (on-off keying) modulation over 2.5 meters in [5-15], employing an echelle grating of 79 grooves/mm, with a blaze angle of 75°. An angular steering range of $\Delta\psi = 17.2°$ was achieved when tuning over 130 nm from $\lambda = 1500$ to 1630 nm in orders $m = 15$ and $m = 16$.

For the 2D beam steering using two crossed gratings, as discussed in depth in Section 5.4, tuning the wavelength over multiple orders of interference m is needed. Extrapolating Eq. (5-3), the tuning range $\Delta\lambda_{tun}$ which is needed to cover M orders and starts at λ_{min} in the highest order m_{max} is

$$\Delta\lambda_{tun} = \sum_{i = m_{max} - M}^{m_{max}} \Delta\lambda_{FSR,i} = \lambda_{min}\, M/(m_{max} - M) \tag{5-10}$$

Hence, the highest order in which the highly diffractive element needs to operate is

$$m_{max} = (L/D_{beam})\,(1 + \lambda_{min}/\Delta\lambda_{tun}) \tag{5-11}$$

For example, when we apply the above-mentioned $D_{beam} = 3$ cm, $\Delta\lambda_{tun} = 100$ nm and $\lambda_{min} = 1500$ nm, for an area size of $L = 1.5$ meters we find a highest order $m_{max} = 800$. Such a high order is far beyond the capabilities of regular diffraction gratings and needs specific multi-beam diffraction elements.

An element with high diffractive power operating in high order (and thus having a small FSR) is the arrayed waveguide grating (AWG), as shown in Fig. 5-7. Neighboring waveguides have an optical path length difference

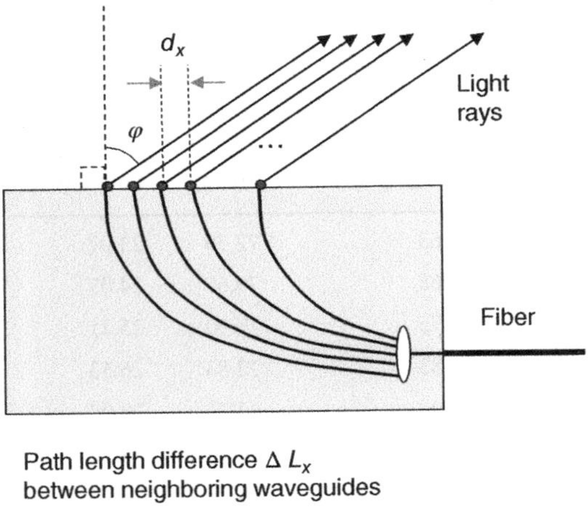

Fig. 5-7 Highly dispersive arrayed waveguide grating

ΔL_x and are spaced at the end facet by d_x. Multi-beam interference maxima occur in direction φ if

$$\Delta L_x + d_x \sin\varphi = m\lambda \tag{5-12}$$

which implies that the order m is bounded by

$$\frac{\Delta L_x - d}{\lambda} \leq m \leq \frac{\Delta L_x + d}{\lambda} \tag{5-13}$$

Typically, $\Delta L_x >> d > \lambda$, hence $m >> 1$ and a high angular dispersion $\mathrm{d}\varphi/\mathrm{d}\lambda$ can be achieved as

$$\frac{\mathrm{d}\varphi}{\mathrm{d}\lambda} = \frac{m}{d_x \cos\varphi} \tag{5-14}$$

From Fig. 5-7 follows that (with $m \geq 2$)

$$\Delta\lambda_{\mathrm{FSR},m} = \frac{\lambda}{m-1} = \frac{\lambda^2}{\Delta L_x + d_x\ \sin\varphi - \lambda} \tag{5-15}$$

Hence, by designing the AWG with a large internal optical path length difference ΔL_x a small FSR $\Delta\lambda_{\mathrm{FSR},m}$ at a high order m can be achieved. The angular dispersion characteristics for various orders m are illustrated in Fig. 5-8.

The angular tuning range $\Delta\varphi_m$ which is achieved in order m when tuning over one FSR from λ to $\lambda + \Delta\lambda_{\mathrm{FSR},m}$ is

$$\begin{aligned}\Delta\varphi_m &= \varphi(\lambda + \Delta\lambda_{\mathrm{FSR},m}) - \varphi(\lambda) \\ &= \arcsin\left(\frac{1}{d_x}\left(m\lambda\frac{m}{m-1} - \Delta L_x\right)\right) - \arcsin\left(\frac{1}{d_x}(m\lambda - \Delta L_x)\right)\end{aligned} \tag{5-16}$$

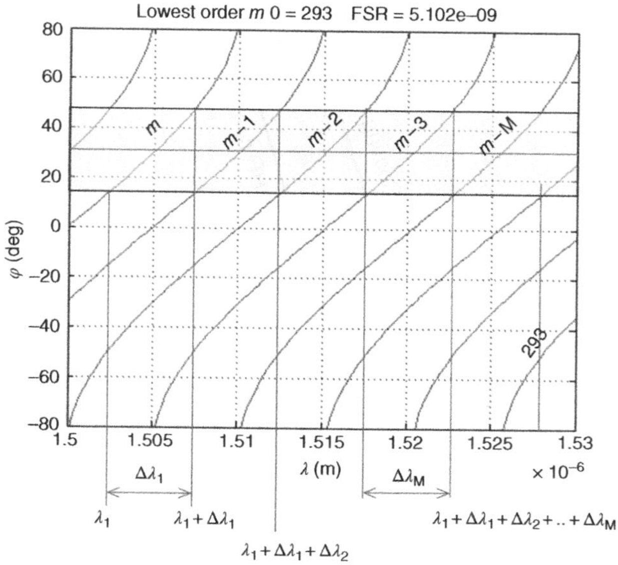

Fig. 5-8 Diffraction angle φ vs. wavelength λ for arrayed waveguide grating operating in high orders (with $\Delta L_x = 450\,\mu\text{m}$, $d_x = 3.3\,\mu\text{m}$)

$\Delta\varphi_m$ is largest when $\varphi(\lambda + \Delta\lambda_{\text{FSR},m}) = \pi/2$, which yields

$$\cos \Delta\varphi_{m,\,\max} = 1 - \frac{1}{m}\left(1 + \frac{\Delta L_x}{d_x}\right) \tag{5-17}$$

For example, when we deploy an AWG beam steerer with $\Delta L_x = 450\,\mu\text{m}$, $d_x = 3.3\,\mu\text{m}$, and $\lambda = 1.5\,\mu\text{m}$, an order $m \approx 300$ is achieved and an FSR $\Delta\lambda_{\text{FSR},m} \approx 5$ nm with a maximum angular tuning range $\Delta\varphi_{m,\max} \approx 56.9°$.

An alternative solution for obtaining a small FSR and high order m is the VIPA – Virtually Imaged Phased Array [5-16, 5-17], which relies on multi-beam interference generated by two reflective sides of a parallel plate, as shown in Fig. 5-9.

Interference maxima occur in the angular direction φ_o which satisfies the VIPA equation [5-16]

$$m\lambda = 2n_v t \cos(\varphi_i - \varphi_o) \tag{5-18}$$

where φ_i is the angle of incidence on the VIPA and n_v is the refractive index of the medium in the VIPA cavity. The interference order m is approximately

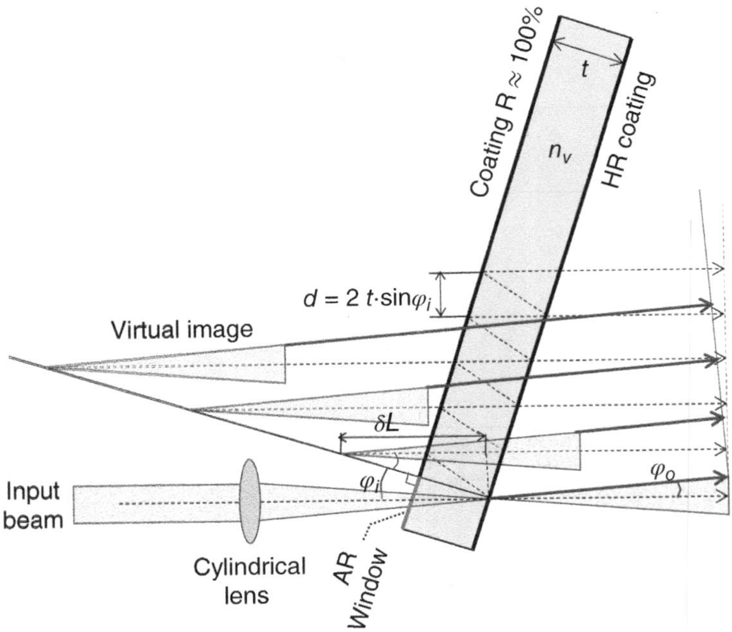

Fig. 5-9 Virtually Imaged Phased Array (HR: coating with high reflection (≥95%); AR: anti-reflection coating)

equal to twice the optical thickness of the plate (i.e., $2n_vt$) divided by the wavelength λ, hence a high order m can be achieved at reasonable plate thickness.

The angular tuning range $\Delta\varphi_{o,m}$ when tuning in order m over one FSR from λ to $\lambda + \Delta\lambda_{FSR,m}$ is (with $m \geq 2$)

$$\Delta\varphi_{o,m} = \varphi_o(\lambda + \Delta\lambda_{FSR,m}) - \varphi_o(\lambda) = \arccos\left(\frac{\lambda}{2n_vt}\cdot m\right) - \arccos\left(\frac{\lambda}{2n_vt}\frac{m^2}{m-1}\right) \quad (5\text{-}19)$$

The maximum $\Delta\varphi_{o,m}$ in order m is achieved when $m^2\lambda = 2n_vt\,(m-1)$, that is, when $\varphi_o(\lambda + \Delta\lambda_{FSR,m}) = \varphi_i$, and is

$$\Delta\varphi_{o,m_{max}} = \arccos\left(1 - \frac{1}{m}\right) \quad (5\text{-}20)$$

Hence, the maximum angular tuning range decays for large order m; for $m = 300$, we find $\Delta\varphi_{o,m\_max} = 4.68°$, and for $m = 800$, we find $\Delta\varphi_{o,m\_max} = 2.87°$.

5.4 2D Diffractive Beam Steerer Using Crossed Gratings[1]

As outlined in Section 4.3, our proposed line-by-line 2D scanning technique implies that $M = L/D_{\text{beam}}$ lines need to be scanned in order to fully cover the $L \times L$ area.

For 2D optical scanning, it has been proposed to create a number of tilted narrowband Bragg gratings in a photo-thermo-refractive (PTR) glass volume [5-18]. Each grating directs the beam having the wavelength which matches that grating into a specific direction. A limited number of beams can be steered; 2D scanning with $M \times N$ beams may be achieved by N PTR plates each containing M tilted gratings. Experimentally, a scan over 5° was shown with four tilted Bragg gratings at 816 lines/mm when tuning the wavelength over 80 nm.

For 2D steering which can be scaled to an arbitrarily large number of beams, we have introduced the use of two crossed diffraction gratings (see Fig. 4-2a and Fig. 5-10) [5-19]. The first grating has a small FSR and many orders of this grating are traversed when tuning the wavelength over the full system range $\Delta\lambda_{\text{tun}}$. The second grating has an FSR which is at least $\Delta\lambda_{\text{tun}}$. As a result, a full line-by-line

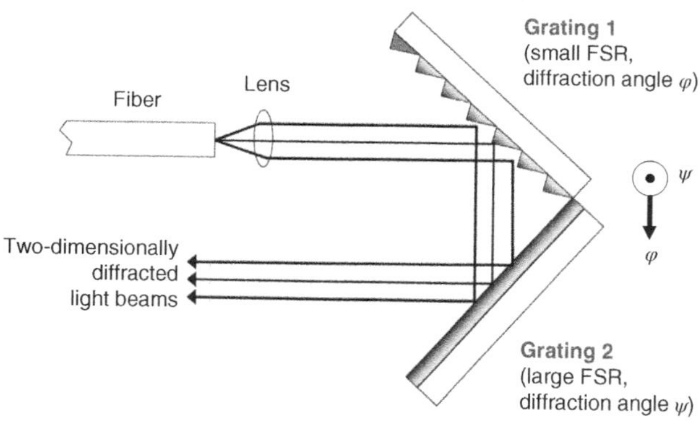

Fig. 5-10 2D beam steering with two crossed gratings. *Source:* [5-19]/A.M.J. Koonen

1 This section is largely based on A.M.J. Koonen, C.W. Oh, K.A. Mekonnen, Z. Cao, and E. Tangdiongga, "Ultra-high capacity indoor optical wireless communication using 2D-steered pencil beams," *IEEE/OSA J. Lightw. Technol.*, vol. 34, no. 20, pp. 4802–4809, Oct. 2016, https://doi.org/10.1109/JLT.2016.2574855. Related patent: US patent 9246589 B2, WO patent 2013107853, priority date Jan. 20, 2012, "Two-dimensional optical beam steering module," by Antonius Marcellus Jozef Koonen, Peter G.M. Baltus, and Antonio Liotta.

scanning of a 2D area takes place when tuning the wavelength of the beam over $\Delta\lambda_{tun}$: the beam is quickly steered along the angular diffraction direction φ of the first grating, and slowly along the orthogonal angular diffraction direction ψ of the second grating. An arbitrary number of beams can independently be steered by feeding an equal number of wavelengths.

5.4.1 Beam Steerer Design

When applying the crossed-gratings module, scanning one line is equivalent to passing through one FSR of the highly diffractive element, so the total wavelength tuning over $\Delta\lambda_{tun}$ should go across M orders (FSR-s) of this element.

For the overall crossed-gratings structure with the proposed arrayed waveguide (AWG) structure for grating 1 and a reflection grating for grating 2, it can be derived that the 2D angular diffracted intensity pattern I as a function of wavelength λ and the orthogonal angles φ and ψ when assuming a 2D AWG matrix of iso-tropically radiating apertures is [5-19]

$$I(\varphi,\psi,\lambda) = I_0\left[\frac{\sin\left(M\frac{\pi d_y}{\lambda}(\sin\psi - \sin i)\right)}{\sin\left(\frac{\pi d_y}{\lambda}(\sin\psi - \sin i)\right)}\right]^2\left[\frac{\sin\left(N\frac{\pi}{\lambda}(d_x\sin\varphi+\Delta L_x)\right)}{\sin\left(\frac{\pi}{\lambda}(d_x\sin\varphi+\Delta L_x)\right)}\right]^2 \quad (5\text{-}21)$$

where i is the angle of incidence on the reflection grating 2, d_y is the groove spacing, and M is the effective number of grooves of this grating; N is the effective number of waveguides in the AWG 1, and d_x is the spacing between the emitting end faces of the waveguides in this grating. The simulation results shown in Fig. 5-11 illustrate how the pencil beam can be steered across an (x, y)-plane on the floor of the room, where x and y are expressed in units as a fraction of the height H of the room (assuming that the PRA is mounted on the ceiling; $x/H = \tan\psi$ and $y/H = \tan\varphi$, refer to Fig. 4-3). The parameters assumed for the waveguide grating 1 are $N = 50$, $d_x = 6\,\mu m$, and $\Delta L_x = 450\,\mu m$ (yielding a $\Delta\lambda_{FSR1} = 5\,nm$). For grating 2, we choose a ruled Echelle grating (in Littrow mount) with $i = 75°$, $M = 50$, and $d_y = 12.05\,\mu m$ (83 grooves/mm). The wavelength λ of the pencil beam was stepwise tuned from $\lambda = 1.504$ to $1.574\,\mu m$, with a step size of 0.2 nm, so over a wavelength range covering 14 times $\Delta\lambda_{FSR1}$. Fig. 5-11 illustrates how area coverage by 2D beam steering deploying only simple λ-tuning can be achieved: angular coverage over $21° \times 20°$, which with $H = 2.5$ meters corresponds to a room area of 0.96×0.92 meters^2 in the (x, y) plane. The beam spots are elliptical with a longer axis in the ψ-direction than in the φ-direction, which is due to the higher angular dispersion $d\varphi/d\lambda$ of the AWG.

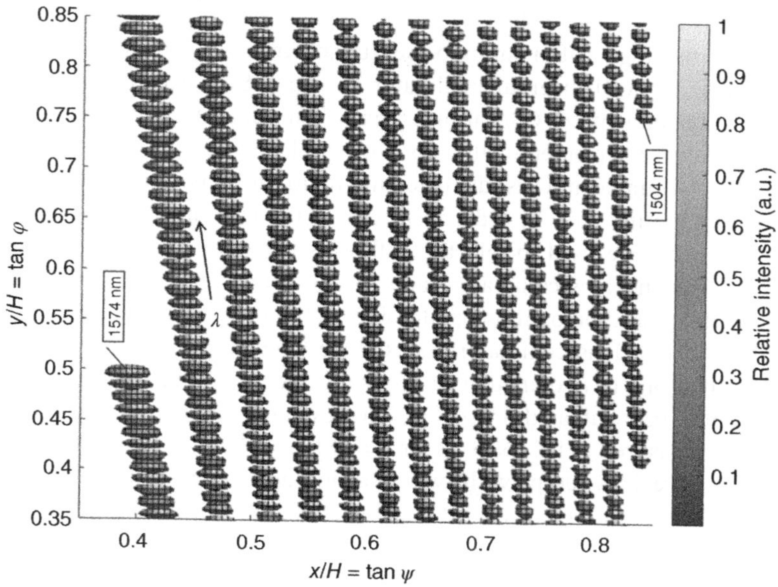

Fig. 5-11 2D scanning by λ-tuning of the room area (λ-step size 0.2 nm). *Source:* [5-10] © [2016] IEEE

5.4.2 System Experiments

As shown in Fig. 5-12, a laboratory system setup has been built for 2D beam-steering deploying an echelle reflection grating with 13.3 grooves/mm operating in order $m = 95$ at an incidence angle of 80.7° and wavelength of $\lambda = 1550$ nm, followed by a transmission grating with 1000 grooves/mm operating

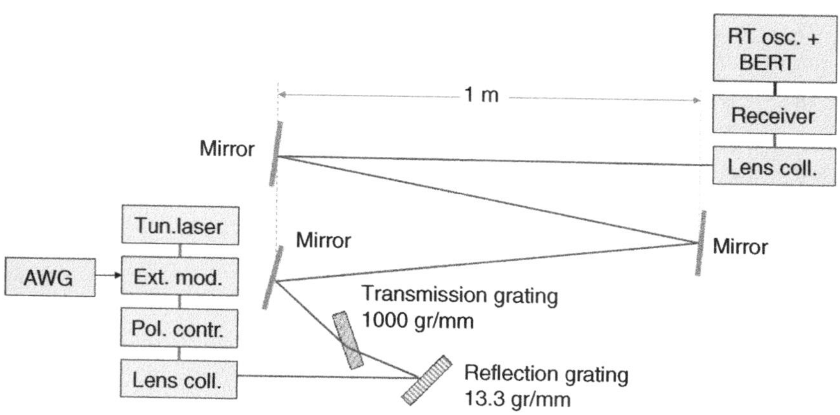

Fig. 5-12 2D beam-steered system experiment. *Source:* [5-10] © [2016] IEEE

in $m = 1$ and incidence angle of 49.9° [5-10]. The optical carrier is generated by a Keysight laser providing 13 dBm output power tunable from $\lambda = 1505$ to 1630 nm. It is followed by an MZI external modulator having 6–8 dB loss driven by a Keysight 20 GHz bandwidth 65 GS/s arbitrary waveform generator. After a polarization controller, via a triplet lens collimator with focal length $f = 18.4$ mm a pencil beam with a 3.3 mm beam waist diameter and up to 6 dBm power is launched onto the crossed grating pair. In order to fit within the laboratory space, three silver-coated mirrors are used to fold the path of 3 meters total length. At the receiver side, a second lens collimator captures the beam, and signal analysis takes place with a real-time oscilloscope to observe the eye pattern and analyze the BER. This second lens collimator needs to be aligned very carefully with respect to the first one in two angular dimensions, due to its very small field of view (FoV) (only 0.034° full angle). Total path losses in the link between the lens collimators are measured to be ≤6.15 dB for $\lambda = 1518$–1600 nm, which is 9.85 dB less than in our previous setups with two reflection gratings reported in [5-19, 5-20]. These path losses are mainly due to the losses of the two gratings; the three silver-coated mirrors contribute a reflection loss of only 0.11 dB each.

Fig. 5-13 shows the calculated 2D-steered beam positions when λ is varied from 1505 to 1630 nm, during which the order of the echelle grating varies from $m = 98$ to 90, corresponding to the nine scanning lines shown. The received optical powers measured in the encircled spots ranged between −0.15 and 1.2 dBm. Angular tuning over 6° in the φ-range and 12° in the ψ-range is achieved. The dispersion of

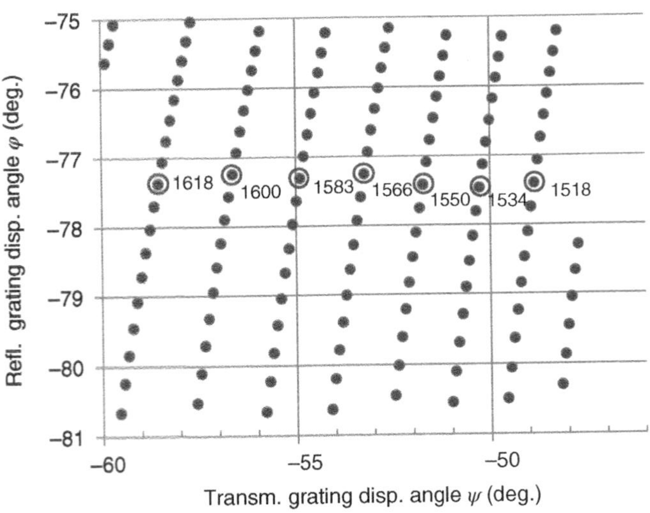

Fig. 5-13 2D beam scanning (simulated; encircled spots were measured at the wavelengths indicated). *Source:* [5-10]© [2016] IEEE

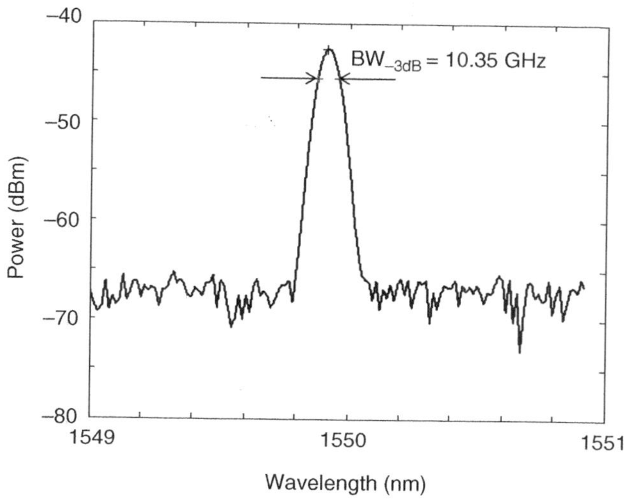

Fig. 5-14 Spectral characteristics of the 2D-steered pencil beam channel. Source: [5-10] © [2016] IEEE

the two gratings, together with the limited aperture of the receiver collimator, determines the spectral passband of the pencil beam channel. Using the broadband amplified spontaneous emission (ASE) of an erbium-doped fiber amplifier (EDFA), the −3 dB passband width was found to be 10.35 GHz, as shown in Fig. 5-14.

System BER measurements were done at $\lambda = 1520$ nm and $\lambda = 1550$ nm. With PAM-4 modulation, we have achieved 32 Gbit/s transmission at BER $< 4 \times 10^{-3}$ (i.e., below the HD-FEC – hard-decision forward error correction) limit [5-21]). A simple 10 GHz bandwidth receiver without extra equalization is used. Predistortion is applied at the arbitray waveform generator to spectrally shape the overall transmission channel. The BER performances of the link at $\lambda = 1520$ nm and $\lambda = 1550$ nm are shown in Fig. 5-15 for PAM-4 symbol rates ranging from 12 to 16 GBaud, both for the back-to-back scenario (i.e., connecting the fiber of the transmitting lens collimator directly to the fiber after the receiving lens collimator) and for the scenario including the free-space beam link. With respect to back-to-back measurements, at 16 GBaud, power penalties were 1.21 dB at $\lambda = 1520$ nm and 2.07 dB at $\lambda = 1550$ nm. The respective eye patterns are shown in Fig. 5-16.

In our previous system setup with two reflection gratings [5-20], we achieved transmission of 42.8 Gbit/s using discrete multitone (DMT) modulation with 512 tones and adaptive bit-loading per tone. A maximum loading of 7 bits/symbol (QAM-128) was observed, decreasing to 1 bit/symbol at the edge of the passband of the pencil beam channel. Excluding DMT overhead, the net transmitted data rate was 37.3 Gbit/s. With respect to the DMT solution, the PAM-4 solution

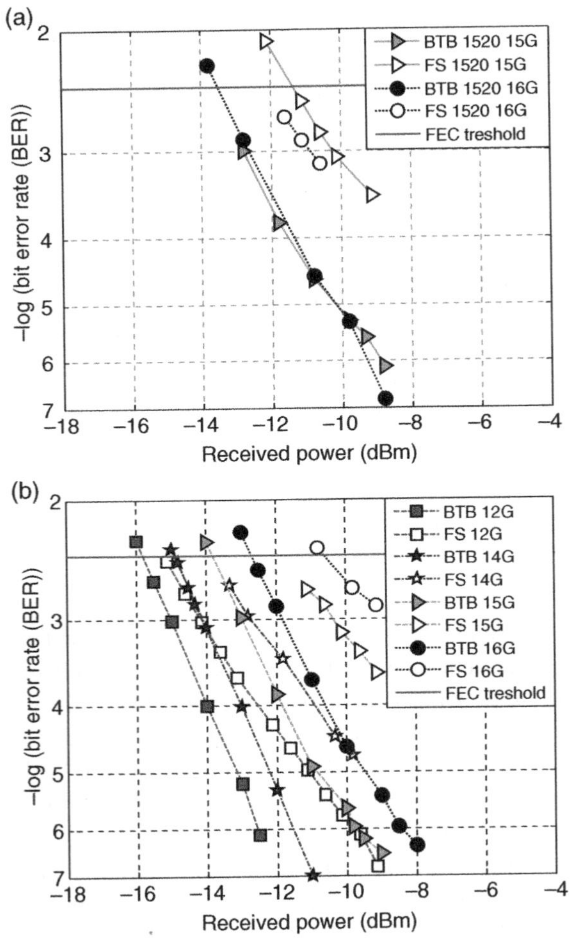

Fig. 5-15 BER performance of back-to-back (BTB) and free-space (FS) link at (a) $\lambda = 1520$ nm for 15 and 16 GBaud PAM-4, and at (b) $\lambda = 1550$ nm for 12–16 GBaud PAM-4. *Source:* [5-10] © [2016] IEEE

requires much less signal processing and thus offers the benefit of a much simpler receiver structure at the user's device. It also offers less power consumption and lower latency [5-22].

The crossed-gratings concept can also be realized with a VIPA as the element with a small FSR operating in high order m. As discussed in Section 5.3.1, a VIPA can be implemented by a glass plate with two parallel reflective sides, and its interference order m is about twice the optical thickness of the plate $n_v t$ divided by the wavelength λ [5-17]. With a VIPA of $t = 100$ μm thickness and refractive index

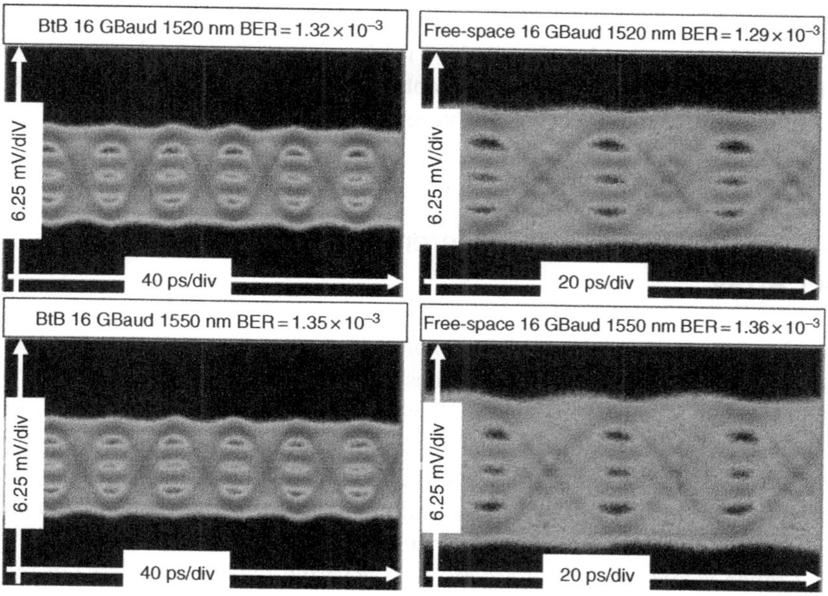

Fig. 5-16 Received PAM-4 eye patterns at 16 GBaud, measured in back-to-back and in the free-space pencil beam link at $\lambda = 1520$ nm (top row) and $\lambda = 1550$ nm (bottom row). *Source:* [5-10]/IEEE

of its cavity $n_v = 1.5$ (so $m = 200$ at $\lambda = 1500$ nm), in combination with a transmission grating of 500 lines/mm, 2D line-by-line scanning over $5° \times 9°$ when tuning from 1400 to 1600 nm can be achieved [5-16].

5.4.3 Concluding Remarks

Free-space optical communication deploying 2D steerable narrow optical pencil beams can offer the ultimate wireless capacity per user device. The proposed concept employs a pair of passive crossed diffractive elements which enables scaling to large numbers of pencil beams, each individually steerable by tuning its wavelength in the 1.5 μm infrared regime. Angular beam steering over a $6° \times 12°$ area (i.e., an area of 32×64 cm^2 at 3 meters distance) was achieved by wavelength tuning from 1505 to 1630 nm. Further careful design of the diffractive stages augmented by angular magnification using a lens system is needed to increase the area coverage. The efficiency of a reflection grating typically depends on the polarization state of the incident light; hence, some polarization control of the light carried by the input fiber (or the use of polarization-maintaining fiber in the feeder network) may be needed.

Downstream delivery of 32 Gbit/s per optical pencil beam in the 1.5–1.6 μm range over 3 meters reach using PAM-4 modulation with a simple free-space receiver has been demonstrated. With adaptive DMT modulation using 512 tones, a 42.8 Gbit/s gross bitrate has been achieved.

5.5 2D Diffractive Beam Steerer Using an AWGR and 2D Fiber Array[2]

In Section 5.4, we have proposed and explored an arrangement with two cross-aligned diffraction gratings, which provides 2D beam steering by wavelength tuning [5-10, 5-19]. The beam steering follows a line-by-line scanning process, in which by increasing the wavelength, the beam's footprint runs smoothly along a line; however, at the end of that line the footprint makes a stepwise jump to the next line. To achieve this, one of the gratings needs to operate in multiple high interference orders, and the other one in a single low order. The full wavelength tuning range comprises several orders (each covering the so-called FSR) of the high-order grating, but at most only one order of the low-order grating. This setup requires highly efficient gratings with low polarization dependency and careful, highly stable mechanical adjustment of these gratings.

As an alternative approach that requires less alignment effort and does not have a noticeable dependency on the polarization state of the incoming signals, we propose the beam-steering module concept, as shown in Fig. 5-17 [5-23–5-26] (also introduced in Fig. 4-2b). It uses an AWGR module with a large number of output ports. Such AWGR modules are readily available commercially, primarily for deployment in fiber-based multi-wavelength communication networks. The $N{\cdot}M$ output fibers of the AWGR are arranged in an $N \times M$ 2D fiber array, which is put in the object plane of a lens. The lateral position of a fiber in the object plane with respect to the lens' axis determines in which 2D direction its corresponding beam is emitted after the lens. As the fiber is selected by the AWGR by means of the wavelength of the data signal, discrete stepwise wavelength-controlled 2D beam steering is achieved. For operation with a wide FoV, the lens should have a small f/D number and should be well-corrected for having minimal off-axis lens aberrations and thus well-defined beams (e.g., an $f/0.95$ multielement camera lens with $f = 50$ mm is used in the laboratory setup; see Section 5.5.2). In combination

2 This section is largely based on A.M.J. Koonen, F. Gomez-Agis, F.M. Huijskens, and K.A. Mekonnen, "High-capacity optical wireless communication using two-dimensional IR beam steering," *J. Lightwave Technol.*, vol. 36, no. 19, pp. 4486–4493, Oct. 2018, https://doi.org/10.1109/JLT.2018.2834374. It is also based on patent PCT/EP2017/080716, priority date Nov. 29, 2016, "Two-dimensional optical beam steering module," by A.M.J. Koonen.

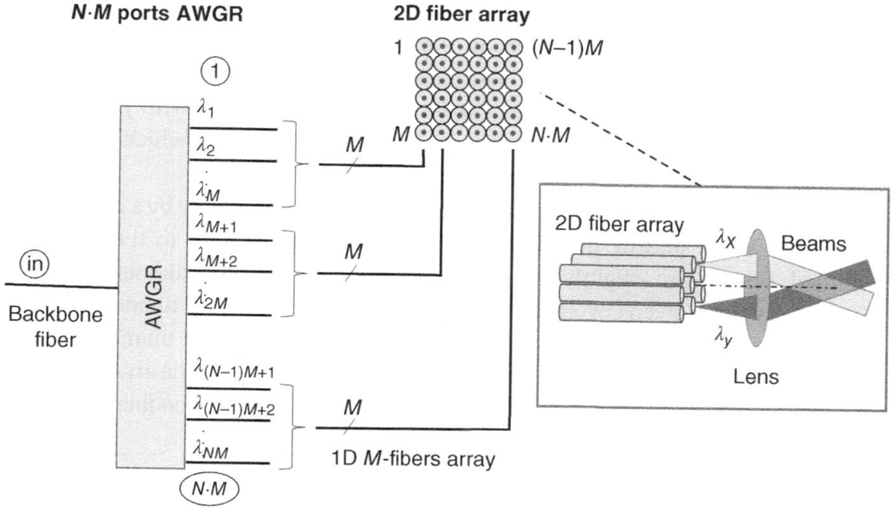

Fig. 5-17 2D steering of IR beams using a high port-count AWGR [5-23, 5-25]

with the lens focal length, the pitch between the fibers in the 2D array is chosen such that the beams create adjacent spots in the 2D user plane to be covered. By putting the 2D fiber array slightly closer to the lens than its focal plane, the beams are slightly diverging, and the module can be made more compact. The design of this beam-steering module is discussed in detail in the next section.

5.5.1 Beam Steerer Design

By putting the fiber array in the focal plane of the lens, collimated beams will be emitted with diameter $D_0 = 2f \tan\alpha$, where $\tan\alpha = \lambda/(\pi w_0)$ with mode field radius w_0 of the single-mode fiber. In the analysis (cf. [5-26]), paraxial geometric optics are used assuming an ideal aberration-free lens, that is, with no Petzval field curvature [5-27]. Each collimated beam creates a spot in the image plane with diameter $D_{\text{beam}} = D_0$. To cover a square image area of size $L \times L$ with no spacing between adjacent spots, $N = M$ and the number of AWGR ports needed is $N{\cdot}M = M^2 = (L/D_{\text{beam}})^2$. The spacing Δy of the fibers in the 2D array needs to be $\Delta y = L/[M(b/f - 1)]$, where $f = \pi w_0 L/(2\lambda M)$ is the lens focal length, and b is the distance between the image plane and the lens. Thus, for $D_{\text{beam}} = 3$ cm, $\lambda = 1.5$ µm, $w_0 = 4.5$ µm (so $\tan\alpha = 0.106$), $L = 1.5$ m and room height $b = 2.5$ m, we need an array of $M \times M = 2500$ fibers with $\Delta y = 1.80$ mm and $f = 14.1$ cm. For a given image area, a larger spot size D_{beam} will reduce the required AWGR port count; however, it also requires a larger aperture at the receiver in order to ensure that enough power is detected to handle the large data rate. Commercially available AWGR-s operating in C-band are typically offered with port counts up to 96. In previous experiments [5-24], we used an AWGR with 80 ports, so $M = 9$. To cover an area of only $0.75 \times 0.75\,\text{m}^2$ using a spot size $D_{\text{beam}} = 8.3\,\text{cm}$, a large condenser lens was needed with $f = 39.3\,\text{cm}$ and diameter $D_{\text{lens}} = 21\,\text{cm}$, and fiber spacing $\Delta y = 15.5\,\text{mm}$ which yielded a 2D fiber array with a total size of $12.4 \times 12.4\,\text{cm}^2$.

The size of such rather bulky setups can be reduced significantly by a defocusing approach: when putting the fiber array out of focus and closer to the lens, the emitted beams are slightly diverging, increasing the spot diameter D_{beam}. Conversely, for a given coverage area and a given acceptable spot diameter D_{beam}, defocusing results in a significantly smaller focal length (f) and diameter of the lens (D_{lens}). It also reduces the spacing (Δy) between the fibers in the array, thereby enabling a significant reduction of the size of the beam-steering module. The relative defocusing parameter p (with $0 \leq p < 1$) can be defined by

$$p = 1 - \frac{v}{f} \tag{5-22}$$

where v denotes the object distance between the array and the lens; defocusing by p implies that the fiber array is moved closer to the lens by a displacement $p\,f$. In order to cover an $L \times L$ area at distance b_0 from the lens by beam spots with equal diameter D_{beam} while the spots are contiguous (so with zero spacing between them), from Fig. 5-18 using paraxial geometric optics, it can be shown that the

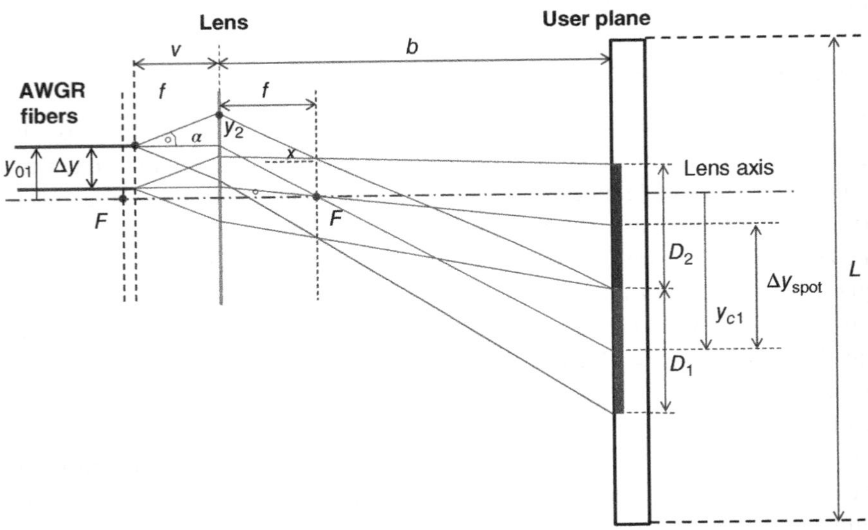

Fig. 5-18 Reducing the size of the beam-steering module by defocusing the fiber array ($D_1 = D_2 = D_{\text{beam}}$)

required lens focal length f, lens diameter D_{lens}, fiber spacing Δy, beam spot diameter D_{beam} as a function of distance b, beam divergence $\mathrm{d}D_{\text{beam}}/db$ and spacing between the centers of the spot Δy_{spot} are

$$f = \frac{1}{1-p}\left(\frac{L}{M2\tan\alpha} - pb_0\right) \tag{5-23}$$

$$D_{\text{lens}} = (M-1)|\Delta y| + 2f(1-p)\tan\alpha \tag{5-24}$$

$$\Delta y = 2f\tan\alpha\left(\frac{f}{b_0 - f} + p\right) = \frac{L}{M}\frac{L - 2p\,b_0 M\tan\alpha}{-L + 2b_0 M\tan\alpha} \tag{5-25}$$

$$D_{\text{beam}}(b) = 2\tan\alpha\,\{f + p\,(b-f)\} \tag{5-26}$$

$$\frac{\mathrm{d}D_{\text{beam}}(b)}{db} = 2p\tan\alpha \tag{5-27}$$

$$\Delta y_{\text{spot}} = \Delta y\left(\frac{b}{f} - 1\right) \tag{5-28}$$

In order to have well-defined spots in the user plane, the lens should have minimal field curvature errors (also known as Petzval curvature [5-27]) and hence should be well-corrected over a wide FoV. For example, in our OWC laboratory setup, we deployed a fast $f/0.95$ camera lens with $f = 50$ mm having 10 elements in 7 groups, which has been optimized for portrait photography [5-28].

Fig. 5-19 illustrates the impact of the defocusing on the module's size. As shown in Fig. 5-19a, for an 81-ports AWGR and coverage area of 0.75×0.75 meters^2 with room height $b_0 = 2.5$ meters, no defocusing (so $p = 0$) yields a lens with a focal length of $f = 39.3$ cm, fiber spacing of $\Delta y = 15.5$ mm, and lens diameter of $D_{\text{lens}} = 20.8$ cm for a beam spot size $D_{\text{beam}} = 8.3$ cm (which was used in [5-24, 5-29] to realize a beam which supports up to 35 Gbit/s OOK data transmission). However, using a defocusing of 12% (so $p = 0.12$), we find $f = 10.5$ cm, $\Delta y = 3.67$ mm, and $D_{\text{lens}} = 4.90$ cm only, while the spot size stays the same and hence the transmission capacity as well. Clearly, defocusing can reduce the size of the beam-steering module considerably without affecting the system's performance. A larger defocusing also decreases the lens speed needed, thereby increasing the f-number f/D: it increases from $f/D = 1.83$ for no defocusing ($p = 0$) to $f/D = 2.15$ for $p = 0.12$.

Similarly, for an AWGR with a larger port count, the coverage area can be increased while the defocusing keeps the module compact. As shown in Fig. 5-19b, when adopting the same beam spot size $D_{\text{beam}} = 8.3$ cm, for a 196-ports AWGR ($M = 14$) at room height $b_0 = 2.5$ meters, a 1.16×1.16 meters^2 area can be

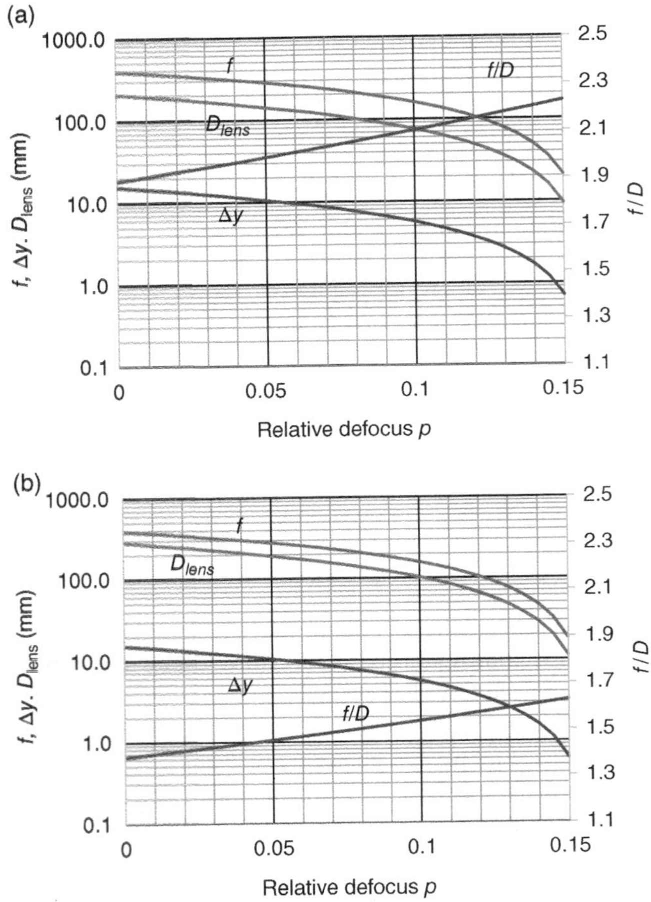

Fig. 5-19 Reducing the module size by defocusing, when opting for a beam diameter D_{beam} = 8.3 cm and applying a) AWGR with 81 ports (M = 9), or b) AWGR with 196 ports (M = 14)

covered with a defocusing factor $p = 0.135$ yielding $f = 61.2$ mm, $\Delta y = 2.08$ mm and $D_{lens} = 3.83$ cm.

Experiments in our laboratory setup with 9.5 dBm transmitted beam power and at the receiver a Keplerian telescopic 10× beam compressor with an aperture diameter of 5 cm have shown that a spot size $D_{beam} = 12.0$ cm allows a data transmission rate of more than 10 Gbit/s in OOK modulation format. For this spot size, with $M = 14$ and $b_0 = 2.5$ meters, an area of 1.68×1.68 m^2 (equivalent with an angular coverage of $18.6° \times 18.6°$) can be covered with $p = 0.21$ yielding

$f = 51.2$ mm, $\Delta y = 2.51$ mm, and $D_{lens} = 4.12$ cm (implying an $f/D = 1.24$ number of the lens, which is readily available, e.g., for camera lenses).

Another advantage of the defocusing approach is the better fill factor of the covered area. Fig. 5-20 shows how the beams from two neighboring fibers each diverge when defocusing is introduced and the distance b between the lens and the image plane increases. Without defocusing ($p = 0$), the beams are collimated and do not diverge. If the lens focal length f and fiber spacing Δy values have been set according to the equations (5-23) to (5-28) for a given lens-to-image plane

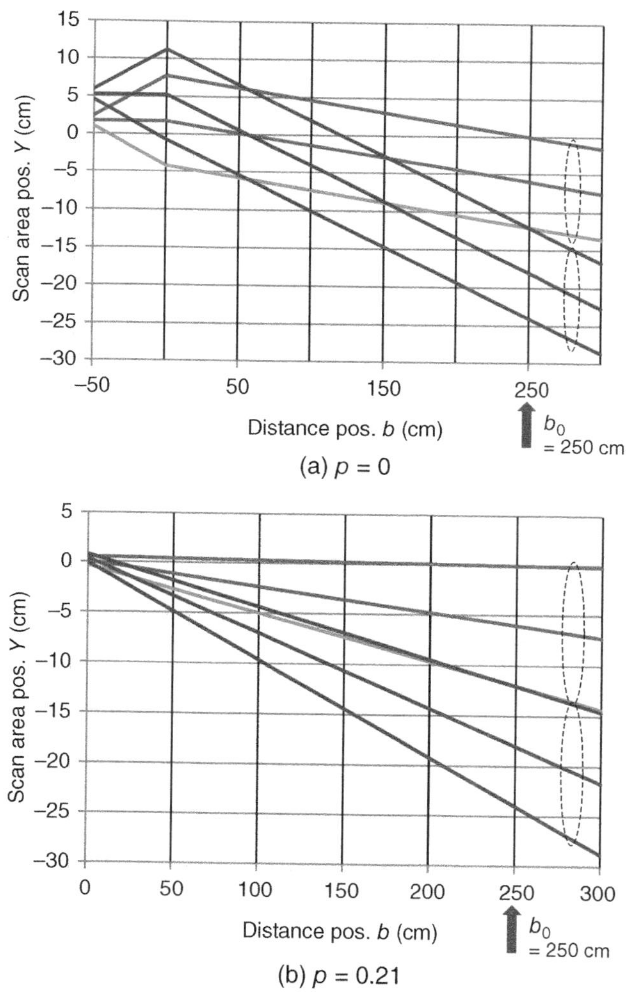

Fig. 5-20 Impact of defocusing on the beam profiles (a) without defocusing ($p = 0$) and (b) with defocusing $p = 0.21$. Assumption: room height $H = b_0 = 2.5$ meters, ideal lens at $b = 0$, $Y = 0$

distance b_0, the spots will exactly touch each other in the image plane where the distance b to the lens equals b_0. This is illustrated in the examples given in Fig. 5-20 when assuming $b_0 = 250$ cm and spot size $D_{beam} = 12$ cm: Fig. 5-20.a shows the case for a collimated beam, i.e., $p = 0$ where the beam's divergence $dD_{beam}/db = 0$, and Fig. 5-20b shows the case for a slightly diverging beam, where $p = 0.21$ and $dD_{beam}/db = 0.0445$ (implying a divergence half-angle of 1.28°).

Fig. 5-21 illustrates how the spots cover the image plane; with the e^{-2} intensity contours of the Gaussian spot footprints clearly shown. The spots exactly touch each other when $b = b_0$, as shown in Fig. 5-21a. In more detail, Fig. 5-22a and b show the coverage by beams with a uniform intensity profile and a Gaussian intensity profile, respectively, for $b = b_0$. When the image plane is moved to a

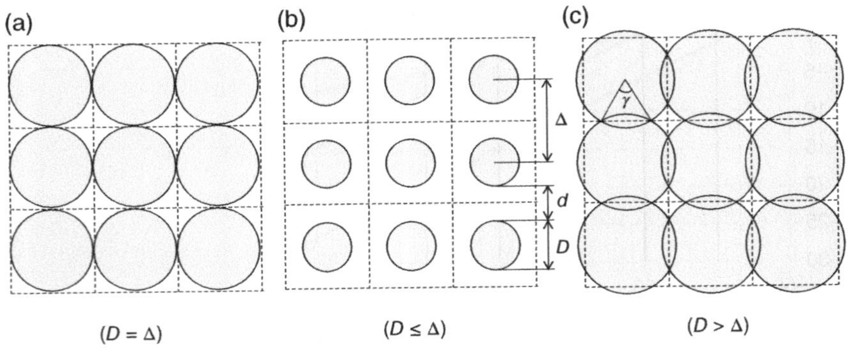

Fig. 5-21 Filling the image area by the beam spots

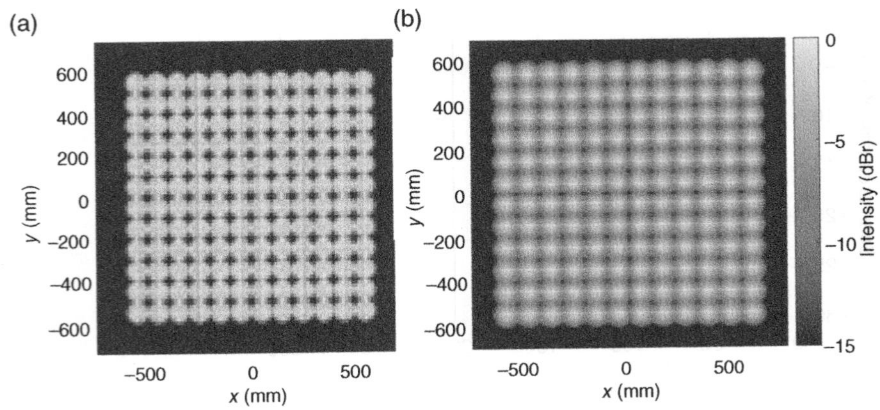

Fig. 5-22 Covering the user area by 144 beams (Ø100 mm beam footprint, ideal aberration-free thin lens $f = 50$ mm, room height $H = 2500$ mm (i.e., $b = b_0$; beam patterns by ray tracing, 1519 rays per beam). Note that the beams have different wavelengths; (a) area coverage by 12 × 12 spots of uniform beams, (b) area coverage by 12 × 12 spots of Gaussian beams (intensity in dB with respect to the maximum in the beams)

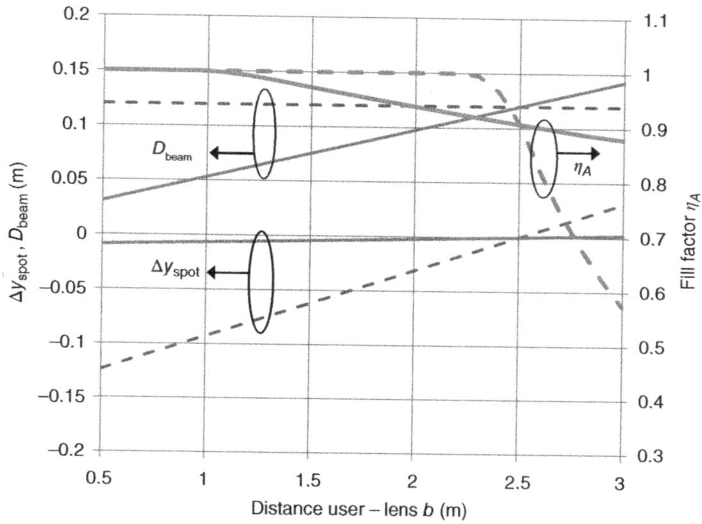

Fig. 5-23 Improving the area fill factor η_A by defocusing (with 12 × 12 uniform spots, and hexagonal fiber stacking; for beam diameter D_{beam} = 12 cm at $b = b_0$ = 2.5 meters; beam spot spacing Δy_{spot}; dashed curves: $p = 0$; solid curves: $p = 0.21$)

larger distance from the lens (i.e., when $b > b_0$), the spacing between the spot centers Δy_c increases ($\Delta y_c = \Delta y(b/f - 1)$). Without defocusing (i.e., $p = 0$), the beam spot diameter D_{beam} stays constant and becomes smaller than the spot spacing Δy_{spot}, hence, the area fill factor η_A (i.e., the summation of the footprint areas of all spots divided by the total coverage area) decreases when b increases (see Fig. 5-21b). For $b < b_0$, overlap between the spots occurs (see Fig. 5-21c). Fig. 5-22 illustrates the coverage of the user plane with a 12 × 12 matrix of spots from uniform beams and from Gaussian beams for $b = b_0$.

Fig. 5-23 shows how the beam spot diameter D_{beam}, the spacing between the beam spots Δy_{spot} and the area fill factor η_A, evolve for a hexagonal stacking of the fibers in the 2D array when the distance b of the image plane to the lens is increased, both for the collimated-beams case ($p = 0$) and the defocused case ($p = 0.21$). The design assumptions for this were a spot diameter $D_{beam} = 12\,\text{cm}$ at room height $b_0 = 2.5$ meters, and a 12×12 2D fiber array. Clearly, by means of defocusing, the dependency of the area fill factor η_A on the distance b to the lens is reduced.

5.5.2 System Experiments

Experimental demonstrations of the AWGR-based beam-steering concept have shown its versatility to provide ultra-high wireless data transmission to devices individually, with capacities per beam using ranging from 10 Gbit/s [5-24] up to 35 Gbit/s [5-29], and using PAM-4 up to 112 Gbit/s [5-30].

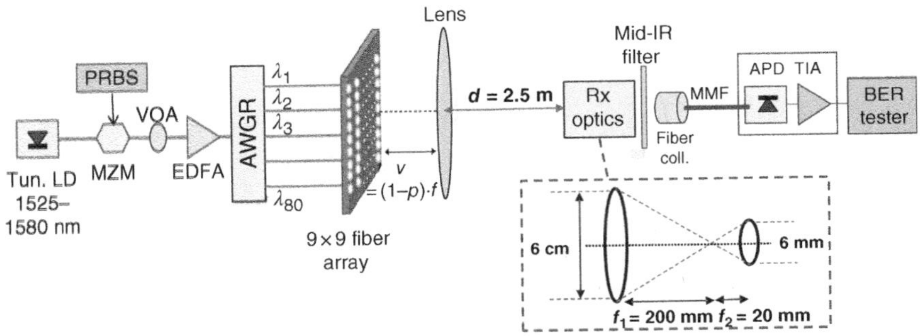

Fig. 5-24 Experimental setup applying the defocusing technique. *Source:* Koonen et al. [5-25] © [2017] IEEE

The first AWGR-based beam-steering laboratory experiment applying defocusing is shown in Fig. 5-24 [5-25]. It uses a laser diode tunable from 1525 to 1580 nm and emitting 0 dBm optical power, a PRBS $2^{23}-1$ generator of NRZ-OOK data driving a Mach Zehnder modulator (MZM), a variable optical attenuator followed by an erbium-doped fiber amplifier (EDFA), and a Gemfire 1×80 ports AWGR with channel center wavelengths from 1529.10 to 1569.80 nm, spaced by 50 GHz and −3 dB channel bandwidth of 0.28 nm (35 GHz). For the focused case (i.e., with $p=0$ and collimated beams), a condenser lens was needed with focal length $f=40$ cm, diameter $D_{lens}=20$ cm, and a fiber array pitch $\Delta y=15.0$ mm. The emitted power was 9.5 dBm, that is, below the 10 dBm eye safety limit for $\lambda > 1.4$ μm. At the receiver side, a telescopic lens system with 6 cm aperture was used for beam narrowing, followed by a mid-IR filter which suppresses ambient light and passes the IR beams, and subsequently a 6-mm aperture fiber collimator feeding the APD-TIA (avalanche photodiode plus transimpedance amplifier) receiver via a 50 μm core MMF (multimode fiber). For the defocused case, we took $p=0.12$, allowing a much smaller achromatic doublet lens with $f=10.0$ cm and $D_{lens}=50.2$ mm, and smaller fiber array with pitch $\Delta y=3.7$ mm. NRZ-OOK data transmission at 10 Gbit/s and at 20 Gbit/s over 2.5 meters in free-space were realized, within the AWGR's −3 dB bandwidth of 35 GHz. Steering was achieved over 17° × 17° when tuning over C-band. The performance was measured for a central spot (HS-1) and a next-neighbor spot (HS-2). As shown in Fig. 5-25, the BER curves measured at 20 Gbit/s for both HS-1 and HS-2 show nearly identical system performance for the focused and defocused cases, with a penalty of about 2 dB with respect to back-to-back operation (due to lens aberrations and slight misalignments). Hence, it can be concluded that the system performance is not reduced by the defocusing, whereas the system has become much more compact.

In a second laboratory experiment, shown in Fig. 5-26, 112 Gbit/s transmission per beam was achieved by deploying the same AWGR and using PAM-4 signal

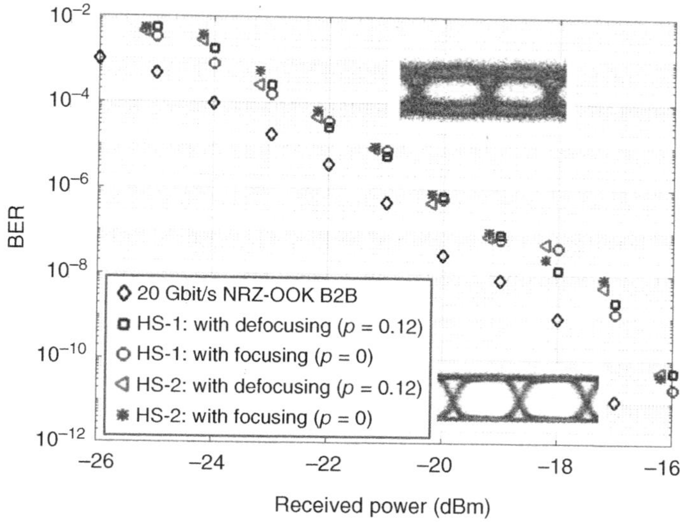

Fig. 5-25 20 Gbit/s performance without and with defocusing. *Source:* Koonen et al. [5-25] © [2017] IEEE

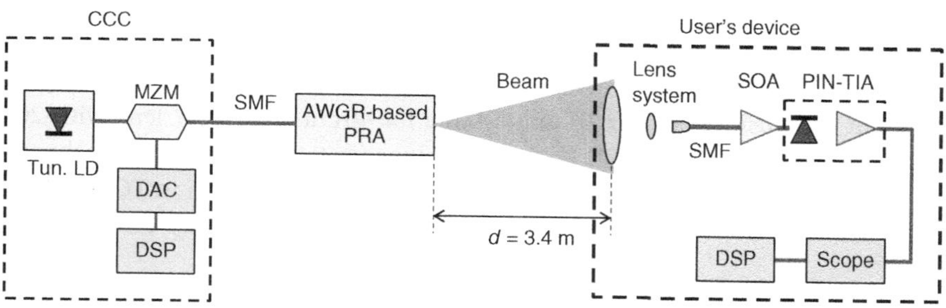

Fig. 5-26 AWGR-based beam-steered OWC system transmitting up to 112 Gbit/s PAM-4 per beam. *Source:* Adapted from [5-30]

modulation [5-30]. In order to overcome the bandwidth limitations of the AWGR (with a −3 dB bandwidth of 35 GHz) and the driver electronics, digital signal processing (DSP) was applied. This enabled spectrum compression by coding the data into a PAM-4 format and applying further equalization. PAM-4 signals with raised-cosine 0.2 roll-off waveforms at 42 GBaud and 56 GBaud were digitally generated using 19 taps and a 9-taps pre-compensation filter, and uploaded into a Keysight M8196A arbitrary waveform generator operating at 84 GS/s. The output of a tunable laser was modulated by the PAM-4 signals using a single-drive 40 GHz Mach-Zehnder modulator. After amplification in an EDFA, the signal is fed into the 80-ports AWGR and, via the 9 × 9 2D fiber array and an $f = 40$ cm focal length lens with a diameter of 20 cm, is launched into free-space as a collimated beam (without defocusing, i.e., with $p = 0$). The beam's power was +10 dBm, which is below the eye safety threshold. The beam's diameter was 8.5 cm, and it was coupled into a receiver at a distance of 3.4 meters. The receiver's lens system has an aperture of 50 mm and consists of: a first lens with $f = 50$ mm; a second lens with $f = 5$ mm; a third lens with $f = 4.51$ mm which couples the light into a 10 μm core single-mode fiber (SMF). The received power was −9.5 dBm. Further optical amplification was done by a linear semiconductor optical amplifier (SOA), providing a gain of 10 dB. Subsequently, the signal was detected by a 70 GHz PIN photodiode followed by a 35 GHz RF amplifier.

The BER performance was measured for 42 GBaud PAM-4 signals (84 Gbit/s) and 56 Gbit/s PAM-4 signals (112 Gbit/s; 100 Gbit/s net after removing 12% HD-FEC overhead). The measured BER curves are shown in Fig. 5-27. At 42 GBaud, for a received power of −7.5 dBm, the BER was 1.7×10^{-3} (below the HD-FEC limit), and at 56 GBaud, the BER was below the HD-FEC limit for a received power of −4.5 dBm (with an error floor appearing due to bandwidth limitations). Fig. 5-27 (inset) clearly shows the open PAM-4 eye patterns.

Third, a demonstrator setup, as shown in Fig. 5-28, has been built in our laboratory in order to provide a proof-of-concept of our AWGR-based beam-steered OWC system. This system enables the simultaneous wireless delivery of high-speed video streams. We previously demonstrated a bidirectional system setup featuring: 35 Gbit/s NRZ-OOK downstream transmission with an 80-ports AWGR (without defocusing, i.e., $p = 0$) and a large $f = 40$ cm condenser lens with 20 cm diameter; and 5 Gbit/s ASK US transmission via 60 GHz UWB radio technology [5-29]. We also showed real-time 10G Ethernet transmission in TCP/IP sessions before [5-31]. The setup, shown in Fig. 5-28, demonstrates downstream

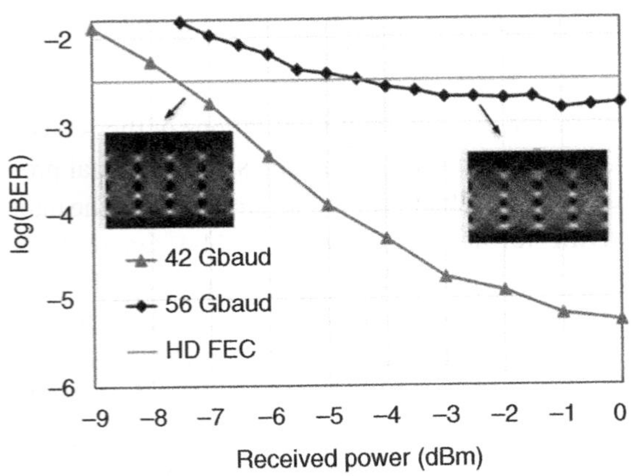

Fig. 5-27 BER performance of the received PAM-4 signals at 42 and 56 GBaud. *Source:* [5-30]/IEEE

delivery of two steerable IR beams, each transporting a 10 Gbit/s OOK signal, and US communication using ultra-wideband (UWB) 60 GHz radio channel for user localization and signaling. The IR beam-steering PRA was built based on the design equations presented in Section 5.5.1. The design parameters include a 14 × 14 2D fiber array, a ceiling-to-table distance $b_0 = 2.5$ meters, and a spot diameter at the table of 12 cm. By choosing a defocusing parameter $p = 0.21$, these design constraints yielded: lens focal length $f = 51.2$ mm; lens diameter $D_{lens} = 41.2$ mm (with $f/D = 1.24$); fiber spacing $\Delta y = 2.51$ mm. This system deploys a composite AWGR, consisting of a C-band AWGR with 96 ports and a lower-L-band (i.e., L^-) AWGR with 48 ports, where all output ports follow an ITU grid channel spacing of 50 GHz, and have a −3 dB bandwidth of 35 and 24 GHz, for the C-band AWGR and the lower-L-band AWGR, respectively. The output fibers were assembled into a honeycomb matrix array of 129 fiber ferrules spaced at a $\Delta y = 2.6$ mm pitch which fits into the physical aperture of a commercial high-speed $f = 50$ mm camera lens with $f/D = 0.95$ [5-28]. The table area covered by the PRA when doing the 2D beam steering amounts to 1.6 × 1.6 m^2. Two optical beam receivers were employed, each equipped with a lensed fiber collimator ($f = 37.1$ mm, NA = 0.24) feeding the received signal with a power of about −16 dBm via an SMF to an SFP+ transceiver capable of 10G Ethernet transmission. The SFP+ transceivers were hosted in network interface cards (NICs) and streamed high-definition video from two servers. One transceiver emitted at a fixed wavelength $\lambda = 1559.92$ nm, thereby providing a beam from port #65 of the 2D fiber array,

Fig. 5-28 Beam-steered OWC laboratory demonstrator. *Source:* [5-32]/Amir Masood Khalid

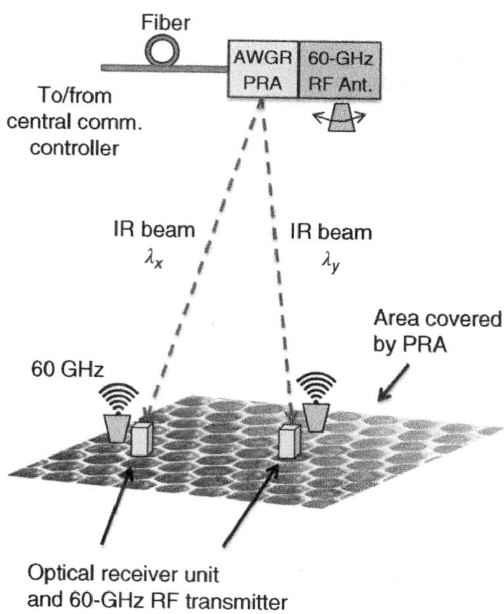

whereas the other transceiver was tunable across the C-band, enabling it to provide a second steerable beam.

For the US 60 GHz UWB link, the user's terminal is equipped with a horn antenna with 30 dBi gain, which emits toward a steerable similar horn antenna of a 60 GHz RF receiver mounted next to the PRA on the ceiling (see Fig. 5-28). At this receiver, half of the received power is monitored by an RF power meter, which serves as a sensor for user localization; more details are given in Section 6.1. The horn antenna is steered using a mechanical pan/tilt system, which is controlled by a Raspberry Pi 3 board. This board determines the user's coordinates and communicates them to the central communication controller (emulated on a laptop), which controls the tunable transceiver, thereby closing the loop for controlling the 2D beam steering.

A photograph of this laboratory demonstrator setup is shown in Fig. 5-29, highlighting the PRA, the steerable 60 GHz horn antenna near the PRA, and the cells which receive the optical beams on the table's surface. Fig. 5-30 shows the realized 2D array of 129 fibers, and the assembly of the fiber array with the $f = 50$ mm wide aperture camera lens. This commercially available multi-element camera lens has a high speed of $f/0.95$ [5-28]. This lens was primarily designed for portrait

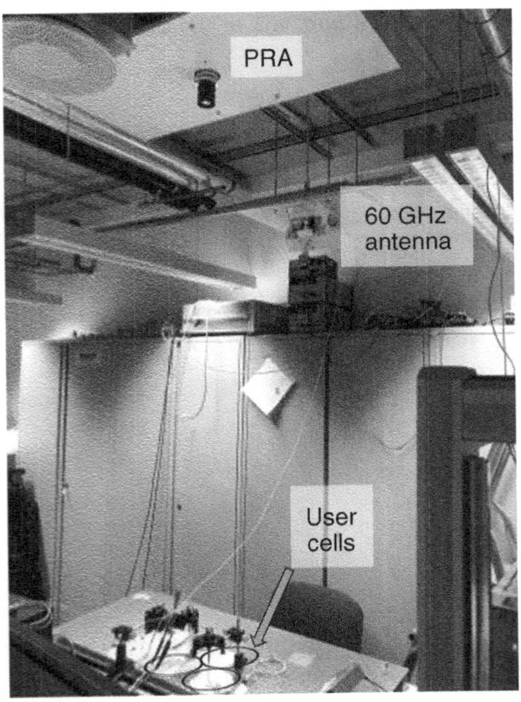

Fig. 5-29 Photo of the laboratory demonstrator

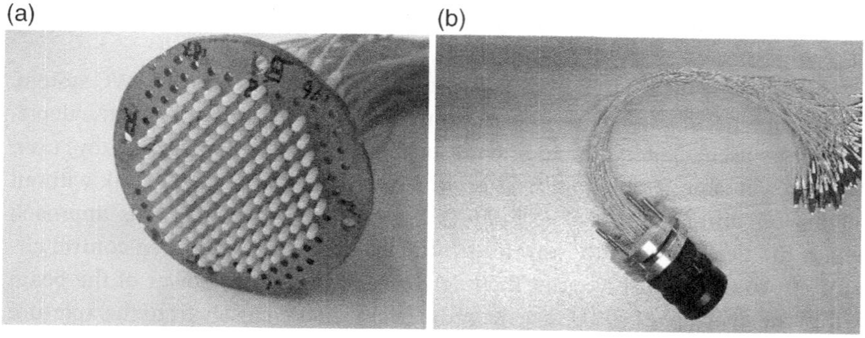

Fig. 5-30 Photos of (a) 2D fiber array and (b) assembly of fiber array and camera lens

photography, which ensures very low lens aberrations and retains the shape of the spots, even in the outer parts of the covered area. In the setup, simultaneous independent transmission by two IR beams of two high-definition video streams was successfully demonstrated. Each stream was embedded in a 10G Ethernet connection and ran at a video data rate of about 4.6 Gbit/s.

5.5.3 Concluding Remarks

Based on a high port-count AWGR with its output fibers arranged in a 2D array and a high-speed lens, individual wavelength-controlled 2D steering of narrow infrared beams can be realized. Thus, many ultrahigh-capacity wireless links for high-bandwidth wireless services can be created. Capacities of up to 35 Gbit/s NRZ-OOK and 112 Gbit/s PAM-4 per beam, with a reach beyond 2.5 meters, have been demonstrated. Using an 80-ports C-band AWGR, a total wireless system throughput beyond 8.9 Tbit/s can be achieved. By extending this setup to 192-ports (96 + 48 + 48) three-element composite AWGR operating in both the C- and the L-band, throughput volumes beyond 21 Tbit/s become feasible. The number of ports may be doubled by operating two AWGRs in a polarization diversity mode, wherein one AWGR is fed from the feeder fiber by light signals in one linear polarization state and the other by light signals in the orthogonal linear polarization state (cf. [5-32]); however, it must be noted that the polarization states in the feeder fiber need to be accurately controlled. By applying a defocusing technique, the size of the passive beam-steering module is decreased considerably without affecting the system's performance. Wireless delivery of multiple high-speed ultrahigh-definition video streams by means of multiple narrow infrared beams has been shown in a laboratory proof-of-concept demonstrator. Autonomic localization of the user's devices is aided by a 60 GHz UWB radio return path (see Chapter 6).

5.6 2D Beam Steering by Mechanical Translators[3]

For establishing the US link from each user device in a bidirectional OWC system, the use of a narrow low-power beam is preferred in order to minimize the depletion of the user's battery and to provide adequate eye safety for the (nearby) user. Moreover, it is preferable to use an arbitrary wavelength for the US link without relying on diffraction-based techniques for the beam steering. This approach avoids the need for a costly wavelength-tunable laser and associated control circuitry in the cost-sensitive user module. To capture the full power of the beam within the aperture of the US receiver, the beam should closely fit to the aperture and requires continuous accurate beam steering. Hence, for the US beam steering, we opted for a 2D translator stage which moves the output fiber of the US transmitter laterally over Δx and/or Δy with respect to the axis of the US lens 1, as shown in Fig. 5-31 ([5-33]. More details are provided in Chapter 8.

The (x, y) translator stage uses two stepper motors (in linear actuator stages, e.g., [5-34]), driven by a controller board. It may be noted that stepper motors do not need to stay activated after the US beam alignment has been completed; this minimizes power consumption at the (typically battery-operated) user site. The effective lens aperture of the lens D_{lens} limits the displacement Δx to $\Delta x_{\max} = \frac{1}{2} D_{\text{lens}} - f_1 (1 - p_1) \tan \alpha$ where $\tan \alpha = \lambda / (\pi\, w_0)$ with mode field radius w_0 of the single-mode fiber. The achievable maximum US steering angle $\varphi_{\max}$ is

$$\varphi_{\max} = \operatorname{atan}\left[\frac{D_{\text{lens}}}{2f_1} - (1 - p_1) \tan \alpha\right] \tag{5-29}$$

and the lateral steering resolution δx_c at the US receiver site at the ceiling is

$$\delta x_c = \varepsilon_x \left(\frac{d}{f_1} - 1\right) \tag{5-30}$$

where ε_x is the resolution of the (x, y) stepper motor stage, and d is the user-to-ceiling distance. For instance, in our laboratory setup we have $d = 2$ meters, use a triplet lens having a clear aperture with diameter $D_{\text{lens1}} = 11.5$ mm and effective focal length $f_1 = 20$ mm, apply a defocussing with $p_1 = 0.0271$ for a US beam diameter $D_{\text{beam2}} = 15$ mm, and use a 2D stepper motor constellation with resolution of $\varepsilon_x = 30\,\mu$m. We then achieve a steering resolution $\delta x_c = 2.97$ mm and

3 This section is largely based on T. Koonen, K.A. Mekonnen, F. Huijskens, and E. Tangdiongga, "Bi-directional all-optical wireless Gigabit Ethernet communication system using automatic self-aligned beam steering," *J. Lightwave Technol.*, vol. 41, no. 11, pp. 3446–3454, June 2023, https://doi.org/10.1109/JLT.2022.3231438.

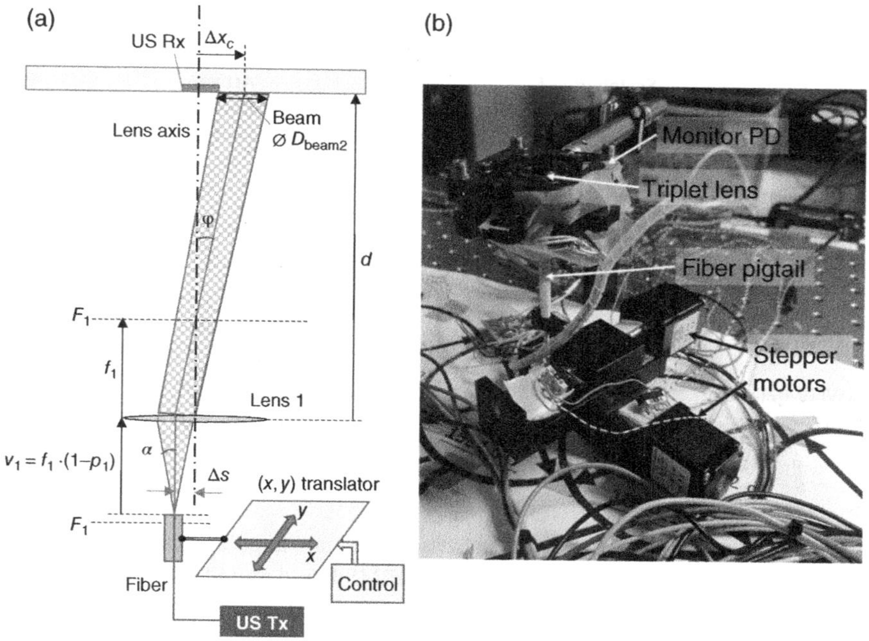

Fig. 5-31 2D angular steering of US beam by means of an (x, y) translator. (a) 2D US beam steerer. (b) Laboratory module for 2D US beam steering, with NEMA-11 stepper motors and triplet lens $f = 20$ mm

maximum US steering angle $\varphi_{max} = 10.8°$ which exceeds the maximum DS steering angle of 10° as is required (see Section 8.5).

5.6.1 Concluding Remarks

By 2D mechanical translation, continuous 2D beam steering can be achieved, which enables to position the US beam tightly within the US receiver's aperture. The US wavelength can be arbitrarily chosen, which allows the deployment of a cost-effective laser diode at the user's end.

5.7 Summarizing Conclusions

In Table 5-2, the beam-steering technologies discussed in this chapter are compared qualitatively.

When using MEMS-based mirrors or an SLM, the scalability toward many beams is limited as each beam needs a separate steering element. The FoV for covering the user area is relatively large when using MEMS mirrors, and it is small

Table 5-2 Comparison of beam-steering technologies

Technology	Scalability of number of beams	FoV coverage	Steering speed	Active/ passive module	Refs.
MEMS mirrors	Low	20° × 20°	Low	Active	[5-7]
SLM	Low	3° × 3°[a]	Medium	Active	[5-1, 5-2]
2D stepper motor translator stage	Not scalable	10.8° × 10.8°	Low	Active	[5-34]
2D grating	High	6° × 12°	High	Passive	[5-10, 5-16, 5-20]
AWGR+2D fiber array	Medium	17° × 17°	High	Passive	[5-23, 5-26, 5-29, 5-30]

[a] 18° with lens-based angle magnifier.

when using an SLM (but can be increased with an angular magnifier, at the expense of narrowing the beam). For both MEMS-based and SLM-based steering, active electronic control at the steering module is needed, which requires a separate control line. It supports beam steering at slow/medium speed (in the order of several milliseconds).

When using the diffractive 2D grating-based concept and remotely tuning the signal's wavelength, continuous steering of a beam is feasible in one direction and stepwise steering in the orthogonal direction. The number of beams is limited only by the resolving power of the gratings, which can be high. When the beams are carrying high-bandwidth signals, the dispersive nature of the gratings may noticeably stretch the beam spots, which may decrease the received power of a beam and cause some crosstalk with neighboring beams. Moreover, reflection gratings may exhibit a performance dependent on the polarization of the incident light, particularly at larger incidence angles and smaller groove spacings, which requires control of the polarization state of the multiple light signals fed by the fiber feeder network.

When using the AWGR + 2D fiber array based concept, by remotely tuning the wavelength, the beam is steered stepwise in both orthogonal directions. The number of beams is limited by the number of output ports of the AWGR, which can be beyond 100. The beam-steering speed is determined by the wavelength tuning speed of the laser diode. This tuning time may be no more than a few nanoseconds for the laser diode itself; the tuning control electronics may increase that to a few microseconds.

The modules based on the 2D grating concept as well as the AWGR + 2D fiber array concept are both fully passive, meaning they do not require local powering or maintenance. Their beam-steering control is done by tuning the wavelength of a remote laser diode and does not need a separate control line (as the wavelength also carries the data signal). The polarization dependence of the AWGR's performance is typically small (polarization-dependent loss PDL < 0.5 dB), hence no polarization state control in the fiber feeder network is needed.

For individual steering of the multiple downstream beams, we have preferred the wavelength-tuned diffractive beam-steering techniques using 2D gratings or an AWGR + 2D fiber array, given their high steering speed enabled by using fast tunable laser diodes, their easy scaling to many beams, and passive maintenance-free nature. Moreover, the AWGR + 2D fiber array technique is not noticeably impacted by polarization state fluctuations in the optical input signal, as typically the AWGR's PDL is <0.5 dB only; in contrast, these fluctuations can yield noticeable signal intensity variations in the 2D grating concept.

For steering the single US beam from each user device, diffractive beam steering is less attractive as it would need a relatively costly tunable laser diode and associated control circuitry, which is not suitable for the cost-sensitive user module. We have preferred a 2D mechanical translator stage which moves the fiber pigtail of a laser diode with arbitrary wavelength with respect to the optical axis of a lens, and which is able to steer the US beam at an angle exceeding the FoV of our DS receiver (ca. ±10° half-angle, see Section 7.5).

References

[5-1] A. Gomez, K. Shi, C. Quintana, et al., "Beyond 100-Gb/s indoor wide field-of-view optical wireless communications," *Photon. Technol. Lett.*, vol. 27, no. 4, pp. 367–370, Feb. 2015, https://doi.org/10.1109/LPT.2014.2374995.

[5-2] A. Gomez, C. Quintana, G. Faulkner, and D.C. O'Brien, "Point-to-multipoint holographic beamsteering techniques for indoor optical wireless communications," *Proc. SPIE 9772*, Bellingham, 2016, Paper no. 97720Q. https://doi.org/10.1117/12.2213252.

[5-3] Spatial light modulators [on-line]. Available: https://holoeye.com/products/spatial-light-modulators/

[5-4] N.A. Riza, "Reconfigurable optical wireless," *Proc. LEOS 12th Annual Meeting*, vol. 1, San Franscisco, Nov. 10–11, 1999, Paper MH4, pp. 70–71.

[5-5] J. Xu, M. Cua, E.H. Zhou, et al., "Wide-angular-range and high-resolution beam steering by a metasurface-coupled phased array," *Opt. Lett.*, vol. 43, no. 21, pp. 5255–5258, Feb. 2018.

[5-6] A. Gomez, C. Quintana, G. Faulkner, and D.C. O'Brien, "Point-to-multipoint holographic beamsteering techniques for indoor optical wireless communications," *Proc. SPIE vol. 9772*, Broadband Access Comm. Techn. X, 2016, 7 p.

[5-7] K. Wang, A. Nirmalathas, C. Lim, et al., "Full-duplex gigabit indoor optical wireless communication system with CAP modulation," *IEEE Photon. Technol. Lett.*, vol. 28, no. 7, pp. 790–793, Apr. 2016.

[5-8] M.D. Soltani, H. Kazemi, E. Sarbazi, et al., "Terabit Indoor Laser-Based Wireless Communications: LiFi 2.0 for 6G," *arXiv:2206.10532*, June 2022.

[5-9] M.Z. Chowdhury, M.K. Hasan, Md. Shahjalal, et al., "Optical wireless hybrid networks: Trends, opportunities, challenges, and research directions," *IEEE Commun. Surv. Tutor.*, vol. 22, no. 2, pp. 930–966, Jan. 2020.

[5-10] A.M.J. Koonen, C.W. Oh, K. Mekonnen, et al., "Ultra-high capacity indoor optical wireless communication using 2D-steered pencil beams," *J. Lightwave Technol.*, vol. 34, no. 20, p. 4802–4809, Oct. 2016, https://doi.org/10.1109/JLT.2016.2574855.

[5-11] T. Mullane, D. McDonald, T. Farrell, and J. Dunne, "Full Coverage microsecond wavelength switching of a monolithic widely tunable laser, from any channel to any channel, over the entire C band on the ITU G.682 ITU grid," *Proc. SPIE Opto-Ireland 2002: Optics and Photonics Technologies and Applications*, vol. 4876, pp. 260–265, 2003, https://www.spiedigitallibrary.org/conference-proceedings-of-spie/4876/.

[5-12] M. Born, and E. Wolf, *Principles of Optics: Electromagnetic Theory of Propagation, Interference and Diffraction of Light*, 7th ed., 1999, Cambridge University Press, ISBN 0521642221.

[5-13] P.M. van den Berg, "Diffraction theory of a reflection grating," *Appl. Sci. Res.*, vol. 24, pp. 261–293, Dec. 1971.

[5-14] Z. Yaqoob, A.A. Rizvi, and N.A. Riza, "Free-space wavelength-multiplexed optical scanner," *Appl. Opt.*, vol. 40, no. 35, pp. 6425–6438, Dec. 2001.

[5-15] C.W. Oh, E. Tangdiongga, and A.M.J. Koonen, "Steerable pencil beams for multi-Gbps indoor optical wireless communication," *Opt. Lett.*, vol. 39, no. 18, pp. 5427–5430, Sep. 2014.

[5-16] T. Chan, E. Myslivets, and J.E. Ford, "2-Dimensional beamsteering using dispersive deflectors and wavelength tuning," *Opt. Exp.*, vol. 16, no. 19, pp. 14617–14628, Sep. 2008.

[5-17] M. Shirasaki, "Large angular dispersion by a virtually imaged phased array and its application to a wavelength demultiplexer," *Opt. Lett.*, vol. 21, no. 5, pp. 366–368, Mar. 1996.

[5-18] Z. Yaqoob, M.A. Arain, and N.A. Riza, "High-speed two-dimensional laser scanner based on Bragg gratings stored in photothermorefractive glass," *Appl. Opt.*, vol. 42, no. 26, pp. 5251–5262, Sep. 2003.

[5-19] A.M.J. Koonen, C.W. Oh, and E. Tangdiongga, "Reconfigurable free-space optical indoor network using multiple pencil beam steering," *Proc. OECC/ACOFT 2014*, Melbourne, Jul. 6–10, 2014, Paper Tu3F-1.

[5-20] C.W. Oh, E. Tangdiongga, and A.M.J. Koonen, "42.8 Gbit/s indoor optical wireless communication with 2-dimensional optical beamsteering," *Proc. OFC 2015*, Los Angeles, Mar. 22–26, 2015, Paper M2F.3.

[5-21] Forward Error Correction (FEC): a primer on the essential element for optical transmission interoperability [on-line]. https://www.cablelabs.com/blog/forward-error-correction-fec-a-primer-on-the-essential-element-for-optical-transmission-interoperability.

[5-22] J.L. Wei, J.D. Ingham, D.G. Cunningham, et al., "Performance and power dissipation comparisons between 28 Gb/s NRZ, PAM, CAP and optical OFDM systems for data communication applications," *IEEE/OSA J. Lightwave Technol.*, vol. 30, no. 20, pp. 3273–3280, Sep. 2012.

[5-23] A.M.J. Koonen, C.W. Oh, A.M. Khalid, et al., "2D beam-steered high-capacity optical wireless communication," *Proc. IEEE Sum. Top. 2016*, Newport Beach, 2016, Paper TuC2.2.

[5-24] A.M. Khalid, A.M.J. Koonen, C.W. Oh, et al., "10 Gbps indoor optical wireless communication employing 2D passive beam steering based on arrayed waveguide gratings," *Proc. IEEE Sum. Top. 2016*, Newport Beach, Jul. 11–13, 2016, Paper TuC2.3.

[5-25] A.M.J. Koonen, A.M. Khalid, C.W. Oh, et al., "High-capacity optical wireless communication using 2-dimensional IR beam steering," *Proc. OECC 2017*, Singapore, Jul. 31–Aug. 4, 2017, Paper 1-4K-3.

[5-26] A.M.J. Koonen, F. Gomez-Agis, F.M. Huijskens, and K.A. Mekonnen, "High-capacity optical wireless communication using two-dimensional IR beam steering," *IEEE/OSA J. Lightwave Technol.*, vol. 36, no. 19, pp. 4486–4493, Oct. 2018, https://doi.org/10.1109/JLT.2018.2834374.

[5-27] Petzval field curvature [on-line]. Available: https://en.wikipedia.org/wiki/Petzval_field_curvature

[5-28] Fast f/0.95 camera lens with $f = 50$ mm [on-line]. Available: https://www.slrmagic.com/product-page/hyperprime-50mm-t0-95

[5-29] A.M. Khalid, P. Baltus, R. van Dommele, et al., "Bi-directional 35-Gbit/s 2D beam steered optical wireless downlink and 5-Gbit/s localized 60-GHz communication uplink for hybrid indoor wireless systems," *Proc. OFC 2017*, Los Angeles, Mar. 2017, Paper Th1E.6.

[5-30] F. Gomez-Agis, S.P. van der Heide, C.M. Okonkwo, et al., "112 Gbit/s transmission in a 2D beam steering AWG-based optical wireless communication system," *Proc. ECOC2017*, Göteborg, Sept. 17–21, 2017, Paper Th.2.B.1, https://doi.org/10.1109/ECOC.2017.8346059.

[5-31] A.M. Khalid, M. Torres Vega, K.A. Mekonnen, et al., "Real time 10Gb-ethernet transmission over 2D indoor passive beam steered optical wireless system based on high port arrayed waveguide gratings," *Proc. ECOC 2016*, Düsseldorf, Sep. 2016, Paper W.4.P1.SC4.40.

[5-32] X. Zhang, C. Li, Y. Jiao, et al., "Crosstalk-mitigated AWGR-based 2-D IR beam-steered indoor optical wireless communication system with a high spatial resolution," *J. Lightwave Technol.*, vol. 37, no. 15, pp. 3713–3722, Aug. 2019.

[5-33] T. Koonen, K.A. Mekonnen, F. Huijskens, and E. Tangdiongga, "Bi-directional all-optical wireless Gigabit Ethernet communication system using automatic self-aligned beam steering," *J. Lightwave Technol.*, vol. 41, no. 11, pp. 3446–3454, June 2023, https://doi.org/10.1109/JLT.2022.3231438.

[5-34] Linear actuators using stepper motors [on-line]. Available: https://www.helixlinear.com/Products/Linear-Actuators/PRA-Linear-Actuators/Profile-Rail-Linear-Actuator-PRA-11S-039~PRA-11S-039

6

User Localization

6.1 Introduction[1]

A powerful option for short-reach optical wireless communication (OWC) is to use narrow infrared optical beams, as introduced in Chapter 2. These beams need to be steered to the users and give each user an unshared personal wireless channel that can readily carry multiple Gbit/s [6-1]. For appropriately directing the beams, the location of the user device needs to be known accurately, and its movements need to be timely tracked. In addition, to establish the upstream path the user device needs to localize the ceiling unit accurately.

Various device localization techniques employing RF signals emitted or processed by the user have been reported, using triangular algorithms such as received signal strength (RSS), angle of arrival (AoA), and time difference of arrival (TDoA) [6-2]. Also, similar device localization techniques employing VLC signals have been reported, using multiple (existing) LED luminaires and signal processing at the user device [6-3]. VLC suffers less from multipath effects than RF and thus can achieve higher localization precision. It also is not affected by EMI.

For localizing and tracking a device in a hybrid system that uses beam-steered OWC techniques for DS and radio techniques for upstream (US), Wi-Fi positioning techniques may be deployed, or more accurately 60 GHz antenna pattern nulling techniques. In [6-4], at the device side, two high-gain horn antennas with 30 dBi gain are sending US data at 60 GHz, and RSS pattern nulling techniques at the central ceiling site are used to determine the device's location and direct the DS optical beam accordingly (see Fig. 6-1).

1 Sections 6.1–6.3 are largely based on A.M.J. Koonen, K.A. Mekonnen, F.M. Huijskens, et al., "Fully passive user localization for beam-steered high-capacity optical wireless communication system," *J. Lightwave Technol.*, vol. 38, no. 10, pp. 2842–2848, May 2020, https://doi.org/10.1109/JLT.2020.2980428.

Short-Reach Optical Wireless Communication: By Directed Narrow Beams, First Edition. Ton Koonen.

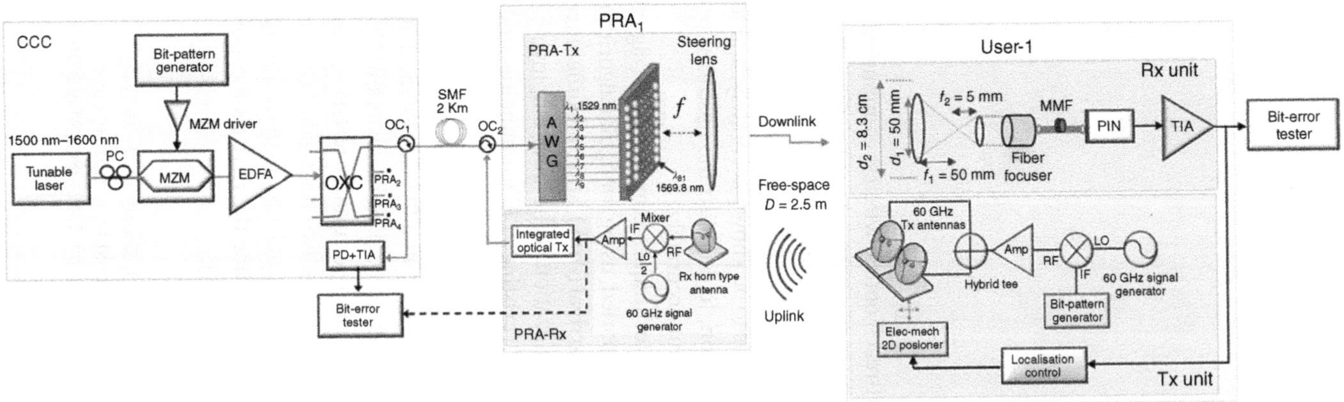

Fig. 6-1 Hybrid system with 2D beam-steered OWC downstream and 60 GHz-based radio communication upstream. *Source:* [6-4]/The Royal Society

In the hybrid OWC system demonstrator shown in Fig. 5-28 [6-5] with DS optical wireless links steered from an arrayed waveguide grating router (AWGR)-based pencil beam radiating antenna (PRA) and US RF wireless links, device localization was done by means of a 60 GHz RF beam which was sent by the mobile device and was found by a scanning horn antenna positioned at the ceiling next to the PRA. Reading the mechanical two-dimensional (2D) angular coordinates of this scanning antenna enabled us to discover the (x, y) position of the user device, and subsequently the IR beam's wavelength was tuned such that the beam was pinpointed accurately to the device. Obviously, the sending of the 60 GHz beam draws power from the device's battery. Moreover, parallax errors need to be compensated, as the scanning horn antenna is physically separate from the DS optical beam steerer.

Device localization can also be achieved by means of a camera mounted at the ceiling which can observe the room and localize multiple user devices simultaneously. When aided by vision analysis, this method enables quick localization and tracking of the devices. A device may be equipped with a beacon which is traced by the camera. In [6-6], a circular array of 12 near-IR LEDs emitting at $\lambda = 890$ nm is used, mounted around the aperture of the user's optical receiver (see Fig. 6-2). The LEDs have a large emission half angle of 50°, which fits within the FoV's half angle of 30° of the system. By means of a CMOS wide-FoV camera, multiple devices can be localized with an accuracy of 0.05° (i.e., 2.5 mm at 3 meters room height). In

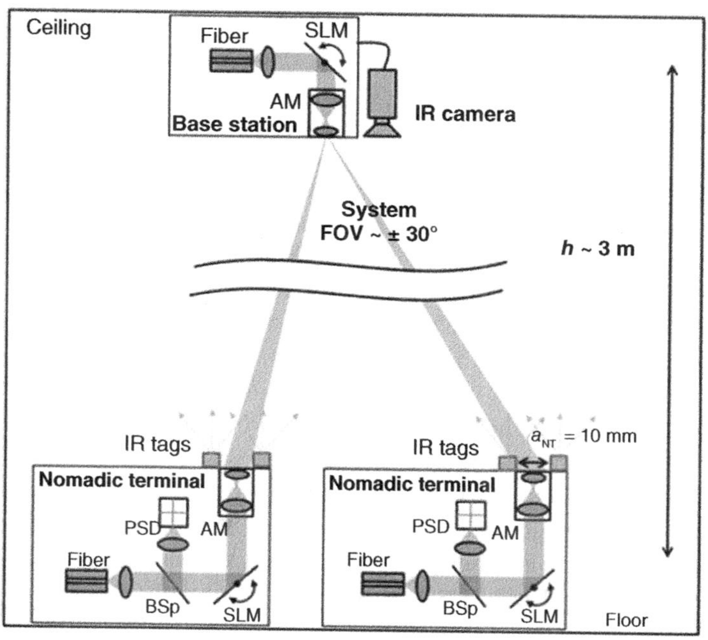

Fig. 6-2 Device localization using IR tags and a camera (AM: angular magnifier, BSp: beam splitter, and PSD: position-sensitive detector). *Source:* Gomez et al. [6-6]© [2016] IEEE

[6-7], a similar approach with four LED tags around the receiver's aperture has been realized. These tags blink in a specific sequence which represents the ID of that device. To achieve this, some power has to be drawn from the battery of the user's device. After processing the images captured by the camera and subsequently having determined the position coordinates of the devices while compensating for parallax errors, these positions have to be mapped and calibrated to the specific wavelengths needed by a diffractive beam steerer that has to direct the beams to those locations. A localization accuracy of less than 5 mm over a reach of 3 meters was achieved.

When using a 2D diffractive beam steerer (e.g., one based on a pair of crossed gratings, or on a high port-count AWGR, as discussed in Chapter 5 and summarized in Section 6.2), the localization may also be done by means of the wavelength-to-position mapping function of the steerer. The user's device may send light with a broad spectrum (such as the amplified spontaneous emission (ASE) light from an SOA, a semiconductor optical amplifier) US, and the 2D diffractive beam steerer will pass a narrow slice of this US light. By measuring the wavelength of this slice with a sensitive spectrum analyzer, the location of the device can be determined [6-8].

In each of the techniques aforementioned, active functions in the user device are needed, which draw extra power from the mobile device and thus compromise its battery lifetime. To counteract this battery depletion issue, we propose in Section 6.3 a novel concept for localization of the users' devices and accordingly for the alignment of the DS beams to them while building on our high-capacity IR beam-steered OWC system techniques [6-5–6-9]. This concept does not consume power at the user device and is based on optical retroreflectors (RR-s), which are passive means reflecting incident light in the direction it came from. The 2D DS beam steering is done by diffractive means in conjunction with wavelength tuning of each beam's signal, and the steering takes place in discrete steps (cf. the AWGR beam steerer, Section 5.5). The DS beams typically have a footprint exceeding considerably the aperture of the user's device, in order to enable the coverage of a large area with a limited number of discretely steerable beams. The concept only requires a simple fully passive function at each user's device (namely, a ring of miniature RR-s), thus not draining the device's battery. The position of a user device is found by scanning with a DS beam launched from the beam steerer at the ceiling, and monitoring whether light is returned by the device's RR-s with an optical power detector mounted next to the beam steerer.

Second, we propose in Section 6.4 a novel concept for localizing the US receiver in the ceiling unit in order to accurately align the US beams launched from the user's device. It also is based on applying passive RR means at the object to be localized, that is, the US receiver. The 2D US beam steering is done with (x, y) stepper motor stages which enable analog beam steering (see Section 5.6). These beams preferably have a footprint smaller than the US receiver's aperture, which allows low-power US beams in order to save the user's battery and provide adequate eye safety near the user.

The use of retroreflection provides calibration of the localization which is inherently aligned with the steering of the signal beam, as we use the beam for localization as well. This avoids bookkeeping of the localization information versus the steering information and avoids parallax errors which may occur when localization and steering are separated. We have validated both localization concepts in laboratory system experiments in which multiple high-definition video streams were carried by narrow infrared beams (see Sections 6.3.3 and 6.4.3, respectively).

6.2 Localization in an Indoor Beam-Steered System

Our BROWSE indoor beam-steered OWC system concept has been introduced in Chapter 4 and is shown again in Fig. 6-3. The fiber-to-the-home (FttH) access network is terminated at the entry of the house in a central communication controller (CCC) unit, and from there the services are routed to the individual rooms by a fiber network. At the ceiling of each room, PRAs launch the narrow IR beams to the respective mobile devices. A PRA is fully passive and contains optical diffractive elements that two-dimensionally direct each beam into a direction determined by the signal's wavelength. We have explored two options for implementing such a PRA, as introduced in Section 3.6 [6-1]: by means of a pair of orthogonally crossed diffraction gratings (see Fig. 4-2a), and alternatively by means of an AWGR with a high number of output fiber ports arranged in a 2D fiber array which is put in front of a lens (see Fig. 4-2b). Both options use diffraction techniques to map the one-dimensional (1D) tuning of the beams' wavelengths to the 2D positions in the area where the user devices are located. This 1D-to-2D mapping implies discrete

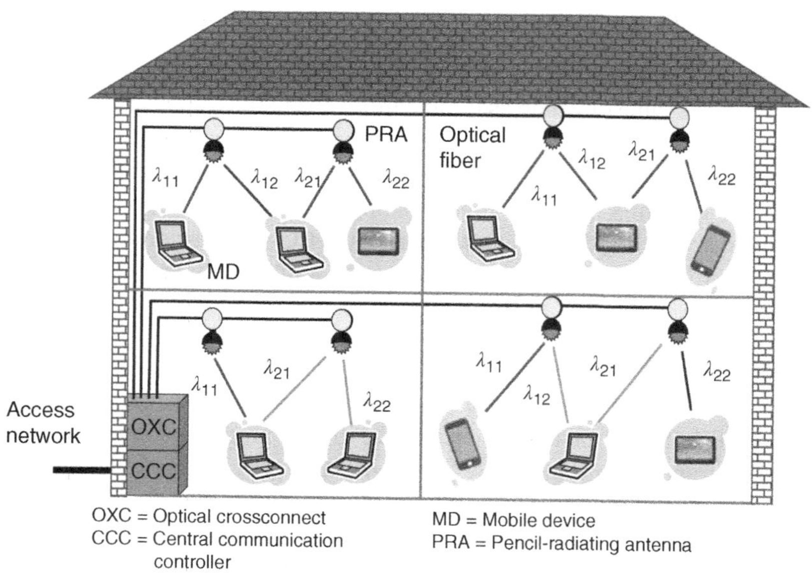

Fig. 6-3 Beam-steered indoor OWC system

steps in the beam steering. In the crossed-gratings beam steerer, there is continuous beam steering along a line by tuning the wavelength within a period of the small-FSR grating, and the steering makes discrete steps to the next lines when tuning to the next one of the FSR periods (see Section 5.4). In the AWGR-based beam steerer, each incoming wavelength is by the AWGR routed to a specific output fiber of the 2D fiber array, with its specific position with respect to the lens, and therefore after the lens the signal at that wavelength is emitted as a beam into a specific direction (see Section 5.5, and [6-5]). As shown in Fig. 5-18, the 2D fiber array is preferably positioned closer to the lens than the lens' focal plane, which yields slightly diverging beams. This defocusing yields a more compact PRA module and a better coverage of the user plane, which is less dependent on the user's distance to the lens. From the design condition that neighboring beam spots in the image plane should touch each other for optimum coverage of the user area, and from a desired spot diameter required for adequate received power by the user receiver's optical aperture, the lens' focal length f and the fiber pitch Δy in the 2D fiber array can be calculated as discussed in Section 5.5.

6.3 Localization When Using Discrete Diffractive Beam Steering

To enable localization of the user's device without drawing power from it, we have explored how to equip the device with passive means for that, in particular a passive retroreflector (RR) based on a matrix of optical corner cubes (CC-s) [6-9].

6.3.1 Corner Cube Retroreflectors

In a 2D case, a CC reflects an incoming light ray twice inside the CC, and the ray exits the CC in the same direction as it came from, due to the orthogonality of the two CC facets, as shown in Fig. 6-4. However, there is a displacement between the incoming and the retroreflected ray. The magnitude of this displacement Δ is dependent on the entry point of the incoming ray into the CC, and on the incident angle α with respect to the main axis of the CC. The maximum displacement Δ_{max} is proportional to the entry aperture D_{CC} of the CC; from Fig. 6-4, it follows that

$$\Delta = 2h = 2L\ \sin(\pi/4 - \alpha) \le D_{CC}\,\sqrt{2}\ \sin(\pi/4 - \alpha) = \Delta_{max} \tag{6-1}$$

When a ray enters the CC too close to an edge of it, it may not find a second reflection opportunity after the first reflection. Hence, the ray propagates in a direction very different from the incoming ray; see Fig. 6-5. Such non-retroreflected ray results when the incoming ray is outside the effective aperture of the CC $D_{CC,eff}$, which from Fig. 6-5 is found to be

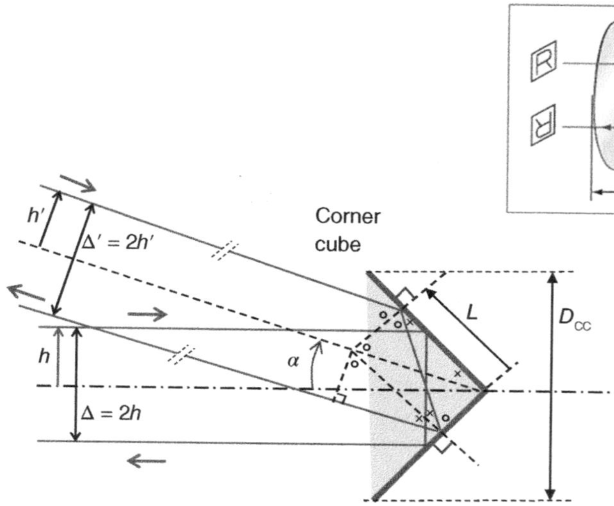

Fig. 6-4 Ray paths in an optical corner cube retroreflector in 2D view (inset: 3D view; *source:* from [6-10]/Edmund Optics Inc.)

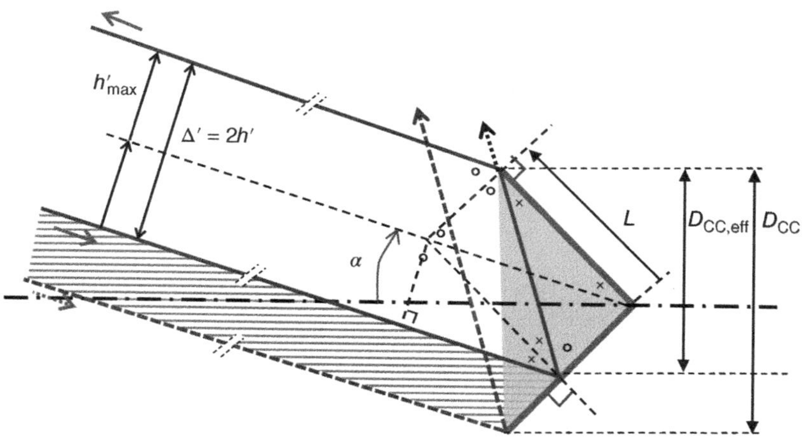

Fig. 6-5 Limited aperture of a corner cube retroreflector (in 2D view)

$$D_{\mathrm{CC,eff}} = \frac{1}{2} D_{\mathrm{CC}} \left[1 + \tan\left(\frac{\pi}{4} - |\alpha|\right)\right] \tag{6-2}$$

Thus, the aperture of CC to achieve retroreflection $D_{\mathrm{CC,eff}}$ is reduced when the incidence angle α increases, and equals half of its full aperture D_{CC} at $\alpha = 45°$.

For the analysis of the three-dimensional (3D) case, the CC model shown in Fig. 6-6 has been adopted. By using the ray tracing tools developed in Appendix

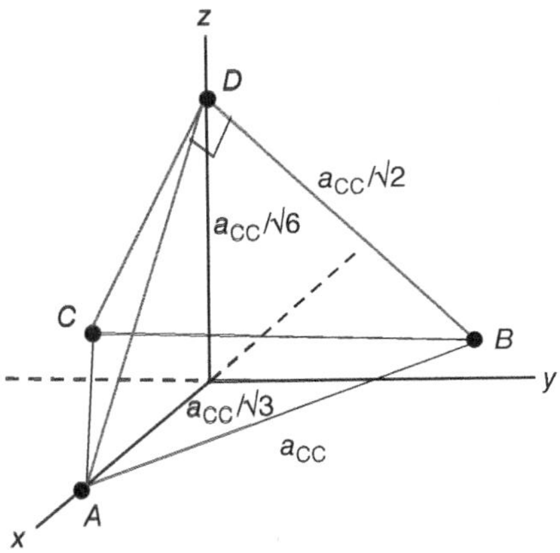

Fig. 6-6 3D corner cube model (entry plane ABC; orthogonal reflecting facets ABD, BDC, and CDA)

Section A.4 and implemented in MATLAB, a numerical analysis of the propagation of rays incident on a matrix of 3D CCs can be done. As an example, Fig. 6-7 shows the tracing of seven rays, of which four are retroreflected in the same direction as their corresponding incoming rays, but with a displacement. Two rays fall outside the CC's effective aperture and follow paths that diverge strongly from the incoming rays. Similar to the 2D case, it is found that this displacement depends on the ray's entry point, and its magnitude scales linearly with the side length a_{CC} of the CC.

A comprehensive 3D ray tracing analysis assuming a collimated incident beam shows that these non-retroreflected rays are caused by rays that are incident in

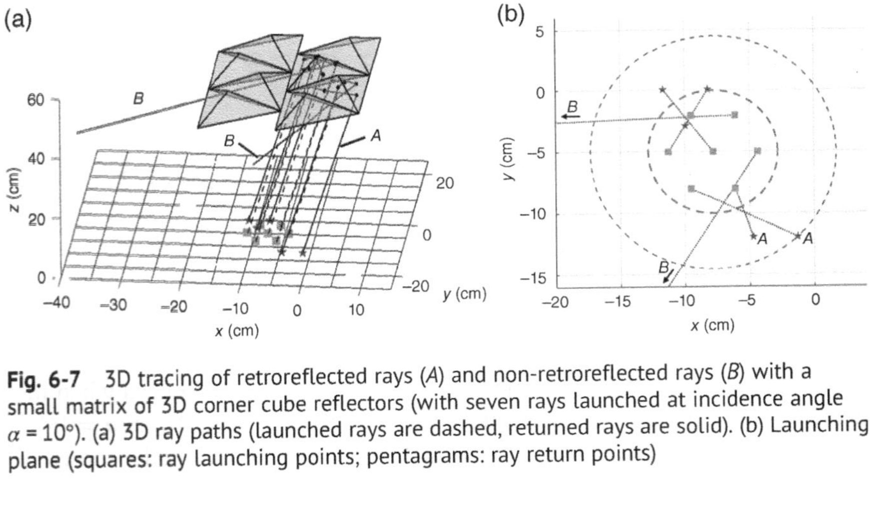

Fig. 6-7 3D tracing of retroreflected rays (*A*) and non-retroreflected rays (*B*) with a small matrix of 3D corner cube reflectors (with seven rays launched at incidence angle $\alpha = 10°$). (a) 3D ray paths (launched rays are dashed, returned rays are solid). (b) Launching plane (squares: ray launching points; pentagrams: ray return points)

areas close to the corners of the triangular entry facets of the CC-s. These areas, called "dead zones", are represented by the black spots shown in Fig. 6-8 for a CC matrix with CC element size $a_{CC} = 3.75$ cm. This figure is obtained by tracing 2611 rays of a collimated Gaussian beam which has an incidence angle α on the entry plane facets of the CC matrix. Comparing the cases of $\alpha = 0°$ (normal incidence) and $\alpha = 10°$, it can be observed that the dead zones become larger when α is increased, which is in line with the 2D CC case of Fig. 6-5 (see also [6-11]). Moreover, the fraction η_{CC} of the power of the incident beam which is retroreflected decreases when α increases: from $\eta_{CC} = 66\%$ at $\alpha = 0°$ to 62% at $\alpha = 10°$; this is

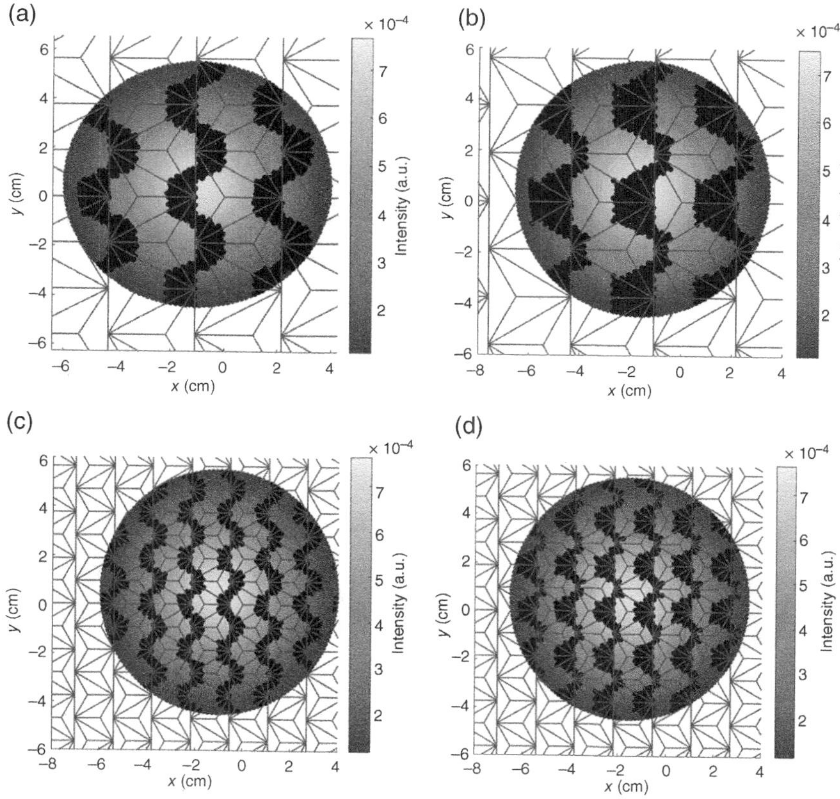

Fig. 6-8 Incidence of Gaussian beam on entry plane of a CC matrix with CC element size a_{CC}, showing retroreflected power fraction η_{CC} at beam incidence angle α (beam diameter D_{beam} = Ø10 cm; K = 2611 rays traced, black spots are non-retroreflected rays).
(a) CC matrix with a_{CC} = 3.75 cm; $\alpha = 0°$, η_{CC} = 66.1%.
(b) Matrix with a_{CC} = 3.75 cm; $\alpha = 10°$, η_{CC} = 61.9%.
(c) Matrix with a_{CC} = 1.875 cm; $\alpha = 0°$, η_{CC} = 66.7%.
(d) CC matrix with a_{CC} = 1.875 cm; $\alpha = 10°$, η_{CC} = 62.9%

also in line with the 2D CC case. The remaining 34–38% of this power is contained in the diverging non-retroreflected rays. The retroreflected fraction η_{CC} does not noticeably change when the size of the CCs is decreased, as shown in Figs. 6-8c and d where the CC's side length a_{CC} is halved compared to that in Figs. 6-8a and b.

The shape of the retroreflected beam arriving back in the plane from where the incident beam was launched has also been calculated by means of 3D ray tracing. Fig. 6-9 shows that this shape deviates from the launched Gaussian beam profile due to the ray offsets incurred in each CC element and due to the dead zones in the entry plane of the CC matrix. The distortion of the shape is worse when the size of

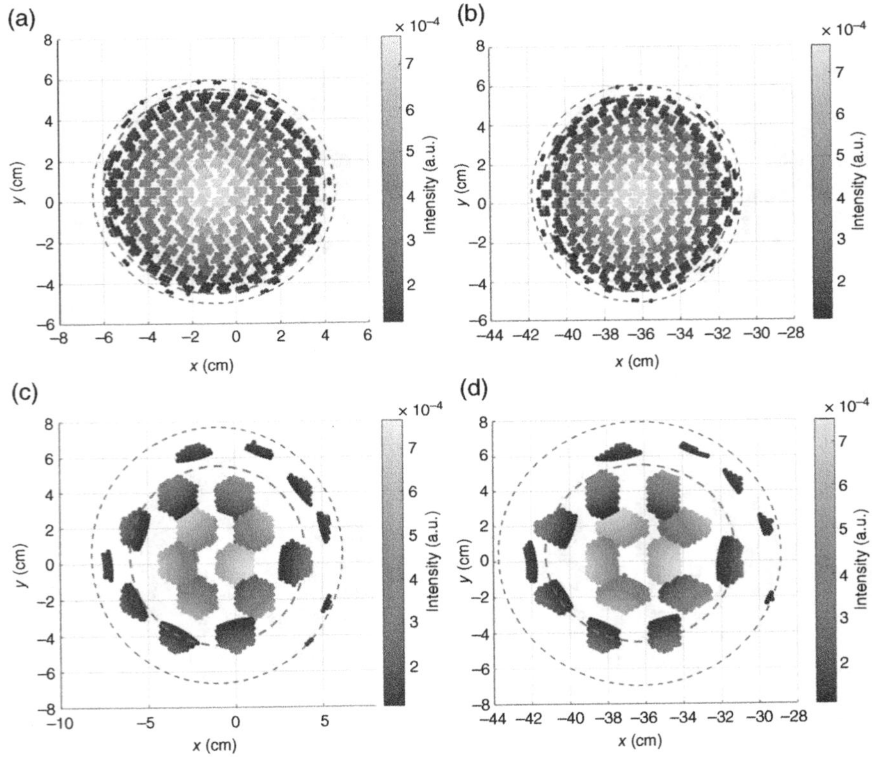

Fig. 6-9 Retroreflected rays returning from a CC matrix with CC element size a_{CC} = 0.938 cm and a_{CC} = 3.75 cm to the ray launching plane (incidence angle α, retroreflected fraction η_{CC} of the power of collimated Gaussian input beam with diameter D_{beam} = Ø10 cm, CC side length a_{CC}; K = 2611 rays traced).
(a) CC matrix, a_{CC} = 0.938 cm; α = 0°, η_{CC} = 66.4%.
(b) CC matrix, a_{CC} = 0.938 cm; α = 10°, η_{CC} = 63.2%.
(c) CC matrix, a_{CC} = 3.75 cm; α = 0°, η_{CC} = 66.1%.
(d) CC matrix, a_{CC} = 3.75 cm; α = 10°, η_{CC} = 61.9%

the CC elements is increased and thus the offsets and the dead zones increase, as shown in Fig. 6-8 and Fig. 6-9 (where a_{CC} is quadrupled from 0.938 to 3.75 cm).

The non-retroreflected rays, which typically comprise about one-third of the power of the incident rays, widely deviate from the retroreflected rays, as illustrated in Figs. 6-10a and c for normal beam incidence ($\alpha = 0°$) and oblique incidence ($\alpha = 10°$), respectively. Figs. 6-10b and c show that the landing points of the non-retroreflected rays B in the ray launching plane are far away from those of the retroreflected rays A. The landing points B are typically grouped in six sets, representing the six different facet reflection combinations possible as a non-retroreflecting ray has only two facet reflections inside the CC. Comparing Figs. 6-10b and d, it can be seen that the distance between landing points A and B is smaller at larger incidence angles α, but the distance typically remains

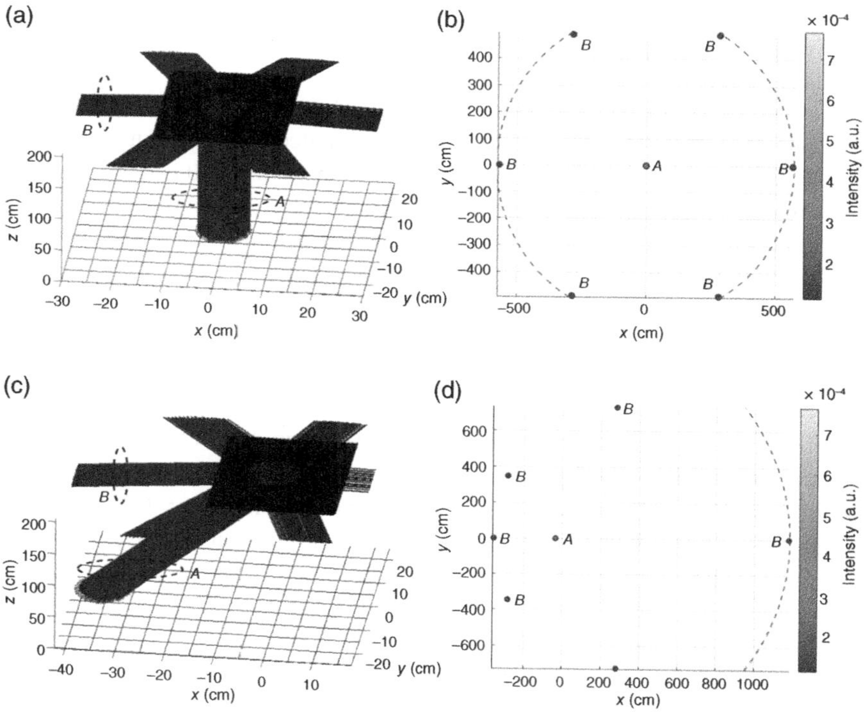

Fig. 6-10 Retroreflected (A) and non-retroreflected (B) rays returning from a CC matrix with CC element size $a_{CC} = 0.938$ cm for a collimated Gaussian input beam incident at angle $\alpha = 0°$ and $\alpha = 10°$ (beam diameter $D_{beam} = \varnothing 10$ cm, distance from ray launching plane to CC matrix entry $z_{CC} = 200$ cm, $K = 2611$ rays traced; retroreflected fraction $\eta_{CC} = 66.4\%$ of beam's power).
(a) Propagation of retroreflected and non-retroreflected rays for $\alpha = 0°$.
(b) Retroreflected and non-retroreflected rays ($\alpha = 0°$).
(c) Propagation of retroreflected and non-retroreflected rays for $\alpha = 10°$.
(d) Retroreflected (A) and non-retroreflected (B) rays ($\alpha = 10°$)

Table 6-1 Performance of retroreflector using a matrix of corner cubes with CC element size a_{CC} for a collimated Gaussian beam with diameter D_{beam} = Ø10 cm (calculated by ray tracing with K = 2611 rays; distance from the launching plane to the CC matrix entry z_{CC} = 200 cm)

a_{CC} (cm)	7.5		3.75		1.875		0.938	
α (deg)	0	10	0	10	0	10	0	10
D_{RR} (cm)	18.8	19.4	14.3	14.9	12.0	12.1	10.9	11.1
Δ_{RR} (μm)	2973	4201	374	381	65.2	167	213	127
η_{CC} (%)	64.8	59.6	66.1	61.9	66.7	62.9	66.4	63.2

Notes: Beam incidence angle α, diameter of retroreflected beam D_{RR}, offset of retroreflected beam with respect to launched beam Δ_{RR}, fraction of power in retroreflected beam with respect to power in launched beam η_{CC}.

large enough to avoid that the non-retroreflected rays are noticed by the OWC's localization system.

The various performance characteristics of CC matrices with different sizes obtained using the 3D ray tracing tools for a collimated incident Gaussian beam with a fixed diameter (of Ø10 cm) are summarized in Table 6-1. It can be seen that the diameter D_{RR} of the retroreflected beam and its offset Δ_{RR} with respect to the incident beam decrease when the CC side length a_{CC} is decreased, which is in line with the 2D CC analysis. The retroreflected power fraction η_{CC} decreases at larger beam incidence angle α, but is relatively independent of the CC side length a_{CC}.

The performance of a CC matrix has also been analyzed by 3D ray tracing for the case that the incident beam is not collimated, but divergent. As discussed in Section 5.5, in order to reduce the size of the AWGR-based beam steerer its 2D fiber array is positioned slightly closer to the lens, beyond its focal plane. This defocusing yields diverging beams which subsequently enter the CC matrix. Fig. 6-11 shows the 3D ray tracing results when a defocusing of $p = 0.2$ is applied, which yields a beam with diameter D_{beam} = Ø10 cm at the user's receiver having a divergence half-angle β given by (see Section 5.5.1)

$$\tan\beta = p\frac{\lambda}{\pi w_0} \tag{6-3}$$

and thus $\beta = 1.28°$ at a wavelength $\lambda = 1.5\,\mu m$ and SMF mode field radius $w_0 = 4.5\,\mu m$. Fig. 6-11a shows the 3D-traced retroreflected rays as well as the non-retroreflected rays which clearly deviate from the retroreflected ones. Fig. 6-11b shows the beam incident on the CC matrix with a diameter D_{beam} = Ø10 cm and also indicates the non-retroreflected rays in the CC's dead zones. The retroreflected beam returning to the ray launching plane is visualized in Fig. 6-11c, in which it can be seen that its diameter has shrunk to 6.1 cm upon

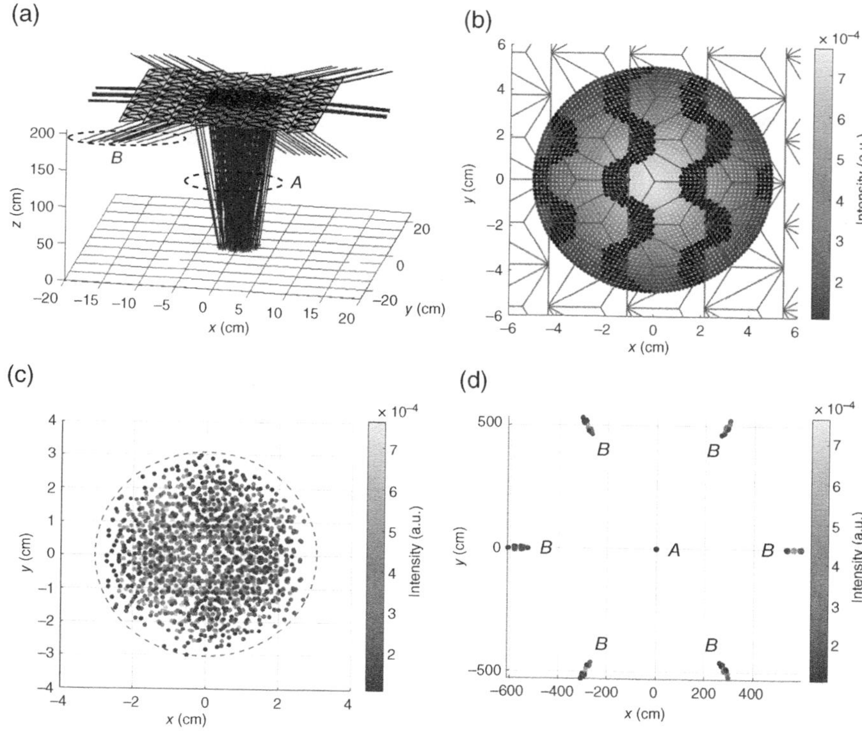

Fig. 6-11 Retroreflected (*A*) and non-retroreflected (*B*) rays returning from an 8 × 8 CC matrix with CC element size a_{CC} = 3.75 cm for a slightly divergent Gaussian input beam (defocusing $p = 0.2$, incidence angle $\alpha = 0°$, D_{beam} = Ø10 cm, z_{CC} = 200 cm, *K* rays traced; retroreflected fraction η_{CC} = 66.3% of beam's power).
(a) Retroreflected and non-retroreflected rays (*K* = 217).
(b) Incidence on the entry plane of CC matrix (black spots are non-retroreflected rays; *K* = 2611).
(c) Retroreflected rays at ray launching plane.
(d) Retroreflected and non-retroreflected rays at ray launching plane

its return from the CC matrix, as a result of the direction-preserving retroreflection process. Moreover, it can be seen that the profile of this retroreflected beam is no longer Gaussian, but distorted due to the various ray offsets occurring in the retroreflection process in the CC matrix. Fig. 6-11d illustrates that the non-retroreflected rays land in the launching plane far from the retroreflected rays, so will typically not interfere with the OWC system's localization process.

Similar results are obtained when the incidence angle of the slightly divergent beam ($p = 0.2$) is set at $\alpha = 10°$, as shown by Figs. 6-12a to d. Also in this oblique incidence case the retroreflected rays are nicely converging in their return to their launching site. Like for a collimated beam, the dead zones become slightly larger than at normal incidence, and the retroreflected power fraction decreases slightly

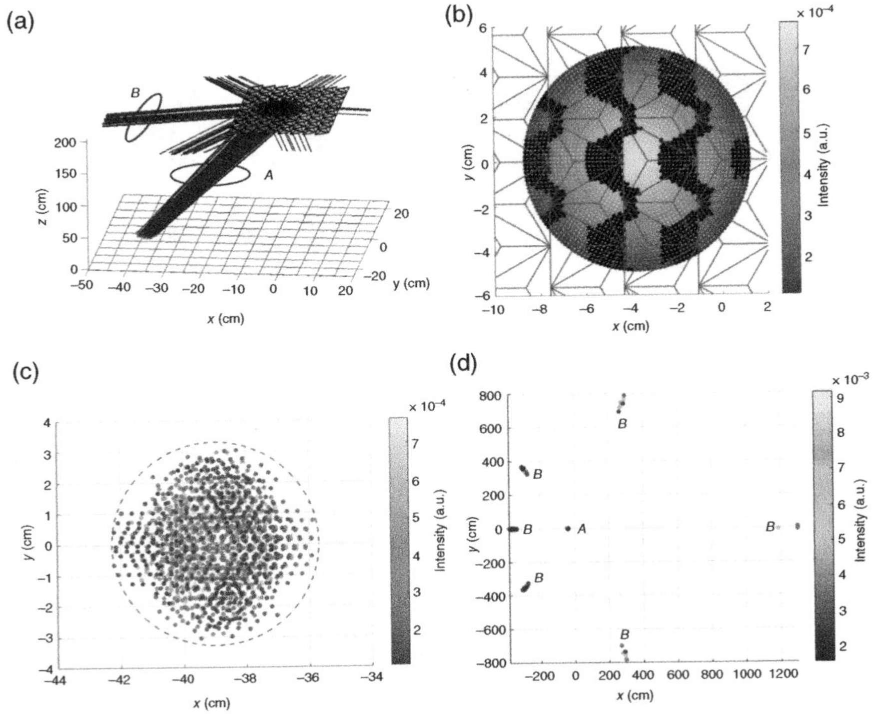

Fig. 6-12 Retroreflected and non-retroreflected rays returning from a CC matrix with CC element size a_{CC} = 3.75 cm, for a slightly divergent Gaussian beam (defocusing p = 0.2) with incidence angle α = 10° (D_{beam_CC} = Ø10 cm at CC matrix entry plane, z_{CC} = 200 cm, K rays traced; retroreflected fraction η_{CC} = 61.3% of beam's power).
(a) Retroreflected (A) and non-retroreflected (B) rays (K = 217).
(b) Incidence on entry plane of CC matrix (black spots are non-retroreflected rays; K = 2611).
(c) Retroreflected rays returning at ray launching plane (K = 2611).
(d) Retroreflected (A) and non-retroreflected (B) rays returning at ray launching plane (K = 2611)

to η_{CC} = 61.3%. The non-retroreflected rays are again landing far away from the retroreflected ones, although not all equally far, and thus will not be perceived by the OWC's localization system.

The results of the ray tracing when using a slightly divergent beam are summarized in Table 6-2. Like with a collimated beam, it can be observed that the retroreflected power fraction η_{CC} does not remarkably change when the size of the CC elements a_{CC} is decreased, and decreases a bit when the incidence angle α increases. As the retroreflected beam is converging when returning from the CC matrix to the ray launching plane, its diameter D_{RR} is substantially smaller than the beam diameter D_{beam_CC} at the entry of the CC matrix (note that the

Table 6-2 Performance of retroreflector using a matrix of corner cubes with size a_{CC} for a diverging Gaussian beam ($p = 0.2$) with beam incidence angles $\alpha = 0°$ and $\alpha = 10°$ ($D_{beam\_CC} = Ø10$ cm at CC matrix entry plane, $z_{CC} = 200$ cm; calculated with $K = 2611$ rays)

a_{CC} (cm)	3.75		1.875		0.938		0.469	
α (deg)	0	10	0	10	0	10	0	10
D_{RR} (cm)	6.26	6.68	3.86	4.16	2.58	2.74	2.08	2.14
Δ_{RR} (μm)	286	224	105	95	11.9	46.5	4.85	47.8
η_{CC} (%)	66.9	61.8	67.2	63.2	66.7	61.9	66.8	61.9

Notes: Diameter of retroreflected beam D_{RR}, offset of retroreflected beam with respect to launched beam Δ_{RR}, fraction of power in retroreflected beam with respect to power in launched beam η_{CC}.

defocusing p has been set such that $D_{beam\_CC} = Ø10$ cm). It may be noted that the beam convergence becomes stronger when the CC size a_{CC} is decreased; this trend of decreasing D_{RR} agrees with the 2D analysis results reported in Fig. 6-17. The diameter D_{RR} increases only slightly when the incidence angle α is increased.

Small imperfections in the geometry of the CC elements have an impact on the shape of the retroreflected beam. As illustrated in Figs. 6-6 and 6-7, in a perfect CC the top angle γ_{CC} between each of the reflecting facets ABD, BCD and CAD is exactly 90°, and thus a retroreflected ray runs exactly parallel to the ray entering the CC. When γ_{CC} deviates from 90°, however, this is no longer the case and the shape of retroreflected beam may be considerably altered. By means of ray tracing the impact of an imperfect top angle γ_{CC} has been evaluated. Fig. 6-13 shows how the beam widens and even breaks up into several parts when γ_{CC} deviates from 90° by an amount $\Delta\gamma_{CC}$ of only several minutes. The dependence of the width of the retroreflected beam $2R_{beam\_RR}$ on $\Delta\gamma_{CC}$ is shown in Fig. 6-13d; the width is only slightly larger than the diameter of the launching beam $2R_{beam}$ when $\Delta\gamma_{CC}$ is close to zero, but quickly grows when the deviation $\Delta\gamma_{CC}$ increases. The width also increases when the size a_{CC} of a CC element is increased, as shown by the graphs for $a_{CC} = 0.469$ cm to $a_{CC} = 1.875$ cm (this trend is in line with the results reported in Table 6-2).

6.3.2 Localizing the User Devices

Ideally, an RR mounted on the user device's OWC receiver will direct part of the incoming narrow IR beam back to the PRA into the AWGR output port it originated from, see Fig. 6-14. From the AWGR's input port, the reflected signal is subsequently returned to the central site, where it is monitored in the localization processor via an optical circulator (OC). This processor controls the wavelength of the tunable laser diode and thus the 2D beam steering (see Section 5.5). The

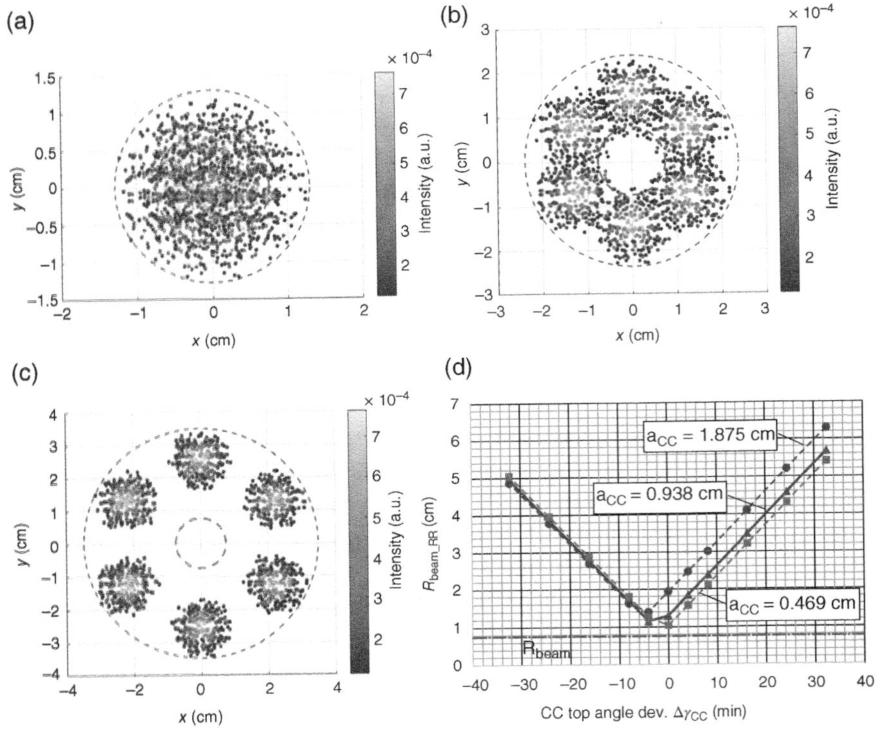

Fig. 6-13 Impact which a deviation $\Delta\gamma_{CC}$ (in minutes) from the ideal top angle $\gamma_{CC} = 90°$ of a CC element has on the shape of the retroreflected beam returning in the ray launching plane (CC element size a_{CC}, divergent beam ($p = 0.2$) with diameter Ø10 cm incident at $\alpha = 0°$ at CC matrix entry, $z_{CC} = 200$ cm; calculated by ray tracing with $K = 2611$ rays, outer dashed circle indicates the retroreflected beam diameter, inner dashed circle the diameter of the launched beam).
(a) $\Delta\gamma_{CC} = 0$; diameter of retroreflected beam $2R_{beam\_RR} = 2.58$ cm, $a_{CC} = 0.938$ cm.
(b) $\Delta\gamma_{CC} = 8.1$ min.; $2R_{beam\_RR} = 4.76$ cm, $a_{CC} = 0.938$ cm.
(c) $\Delta\gamma_{CC} = 16.3$ min.; $2R_{beam\_RR} = 6.96$ cm, $a_{CC} = 0.938$ cm.
(d) Widening of the retroreflected beam's radius R_{beam_RR} by deviations in the CC's top angle γ_{CC} (radius of divergent launching beam $R_{beam} = 0.756$ cm)

device localization process begins with a command from the localization processor to the tunable laser diode to start scanning the room with an IR beam by sweeping its wavelength. As soon as the processor detects a returning reflected signal, it halts and stores the actual wavelength of the tunable laser, and in this way has found and stored the location of the user's device. The laser is then set to this wavelength, and by using the correspondingly steered IR beam, the system has established a high-speed wireless communication link to the user. In this way, the localization is self-calibrating: the device position is directly mapped to the wavelength needed for directing the beam, and no separate calibration is needed. Also no parallax errors are incurred, as localization beam becomes the signal transmission beam.

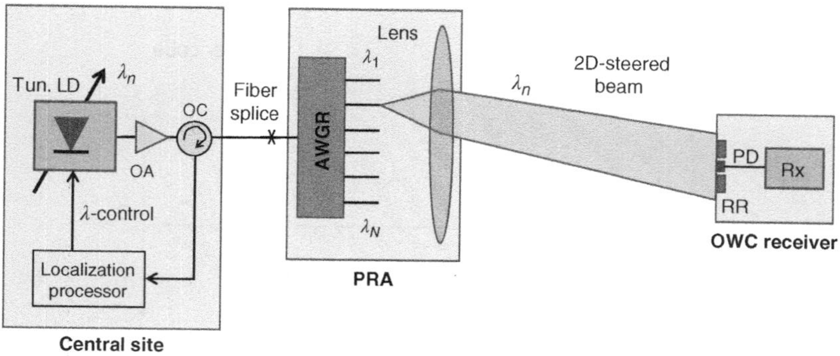

Fig. 6-14 Localizing the user's OWC receiver by means of an optical retroreflector (RR) (OA: optical amplifier; OC: optical circulator)

CC reflectors and cat's eye reflectors are well-known passive optical modules that can perform the retroreflecting function [6-12]. We have focused on matrices of miniature CC-s (and analyzed those in Section 6.3.1), which offer a relatively large effective aperture and are readily available commercially.

As reported in Section 5.5.1 and in [6-5]), in order to achieve a compact design of the PRA, it is required to position the 2D fiber array of the AWGR outputs not in the focal plane of the lens, but a bit closer to it; that is, to introduce an amount of defocusing $p = 1 - v/f > 0$ (see Fig. 5-18). Next to a reduction of the PRA's size, this defocusing yields a slight divergence of the beams, which improves the coverage of the user plane. It also makes the coverage less dependent on the actual distance to the lens, which relaxes the device's mobility requirements. However, the defocusing means that the displaced beam returning from the CC may not end in the same fiber port as where it came from.

In a 2D CC analysis according to Fig. 6-15 and assuming an ideal thin lens (hence without lens aberrations such as field curvature), at the fiber plane, this lateral offset δ of the returning beam is

$$\delta = \frac{f-v}{f}\,\frac{\Delta}{\cos\alpha} \leq p\,D_{CC}\,(1-\tan|\alpha|\,) \tag{6-4}$$

To minimize this offset issue, it is preferred that the displacement Δ between the incoming and the retroreflected ray is minimized, and therefore the aperture D_{CC} of the CC is minimized. However, this implies that only a very small fraction of the beam can be retroreflected, which compromises the detection process. We therefore opted to apply not a single small CC, but a matrix of many miniature CCs. Such CC arrays are readily available in retroreflecting foils; these are commercially available and are widely used already, for example, for road signage. We acquired such a foil containing a fine pattern of miniature molded CCs (from Orafol [6-13]);

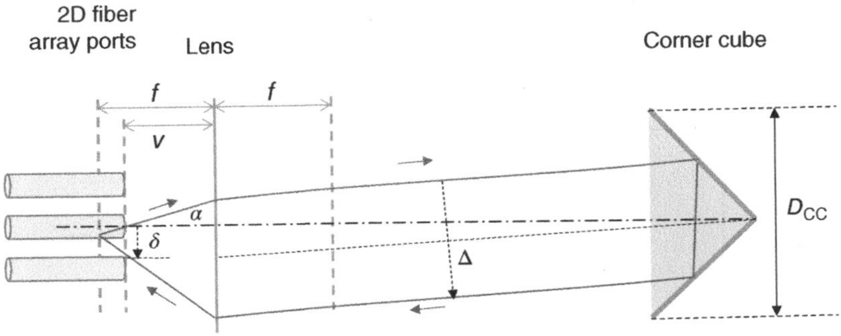

Fig. 6-15 Operating the retroreflecting CC with a defocused PRA

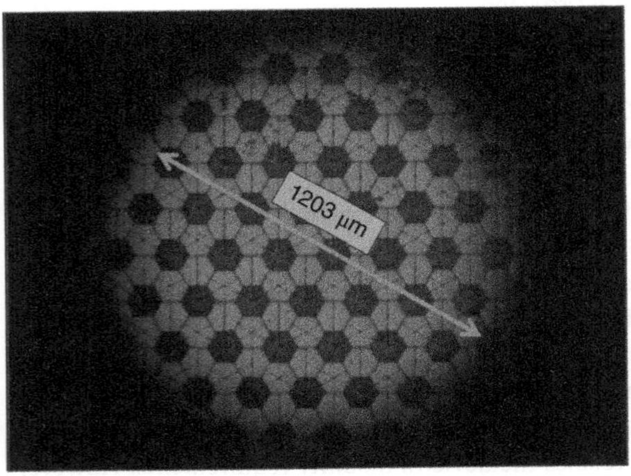

Fig. 6-16 Foil with many miniature corner cubes, each with a diameter of 100 μm. *Source:* Orafol [6-13]/ORAFOL Europe GmbH

Fig. 6-16 shows a microscopic view of its structure; the diameter of a CC in it is about 100 μm. It may be noted that the results obtained by 3D ray tracing shown in Fig. 6-8 are in accordance with this microscopic view and its dark dead zones.

With 2D ray tracing implemented in MATLAB, the impact of defocusing on the lateral offset δ when applying a matrix of CCs has been assessed. A lens with focal length $f = 50$ mm and negligible aberrations was assumed, and a CC matrix foil with a diameter of 35 mm. In the system design for indoor application, a distance lens-to-CC matrix (so PRA to user device) of $b = 2.5$ meters was assumed. With a fiber NA = 0.1, the beam diameter at the CC matrix was about 10 mm when the 2D fiber array was in focus (so $p = 0$), and 108 mm when the 2D fiber array was defocussed by 20% ($p = 0.2$). Fig. 6-17 shows the ray tracing results; the light rays are running toward the CC matrix, and the dark rays are returning from the CC matrix. The impact of the CC diameter has been analyzed by considering CCs with

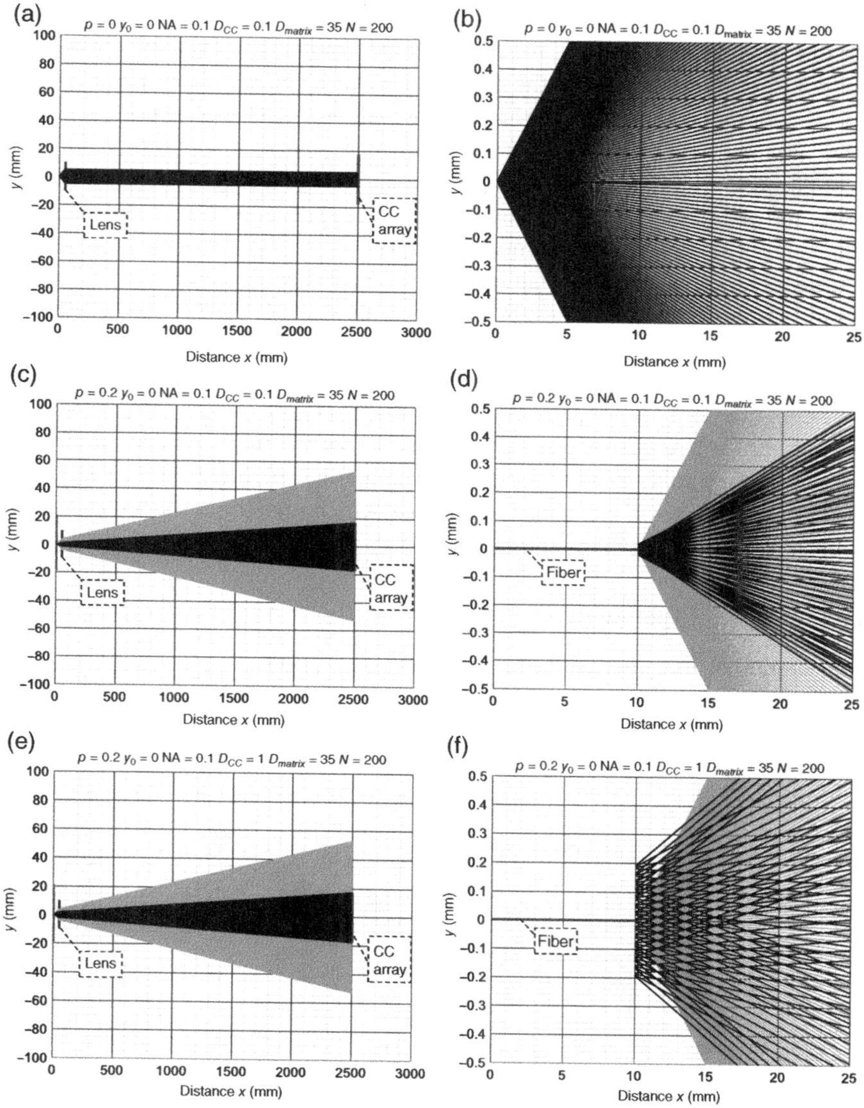

Fig. 6-17 Ray tracing with $N = 200$ rays when the 2D fiber array: in a) and b) is in focus; in c) and d) is 20% defocused w.r.t. lens and CC diameter D_{CC} = 100 μm; in e) and f) is 20% defocused w.r.t. lens and D_{CC} = 1 mm. *Light rays* are running toward the corner cube matrix; *dark rays* are returning from the CC matrix; lens $f = 50$ mm, CC matrix diameter D_{matrix} = 35 mm, distance lens-to-user area $b = 2.5$ m.
(a) Focused ($p = 0$), CC aperture D_{CC} = 100 μm.
(b) Focused ($p = 0$), D_{CC} = 100 μm; zoomed-in at fiber in array.
(c) Defocused ($p = 0.2$), CC aperture D_{CC} = 100 μm.
(d) Defocused ($p = 0.2$), D_{CC} = 100 μm; zoomed-in at fiber in array.
(e) Defocused ($p = 0.2$), CC aperture D_{CC} = 1 mm.
(f) Defocused ($p = 0.2$), D_{CC} = 1 mm; zoomed-in at fiber in array

diameter D_{CC} = 100 μm and = 1 mm. No lateral offset occurs when the 2D fiber array is in focus (p = 0); Fig. 6-17a shows the collimated beam for the whole lens-to-CC matrix link (and in the reverse direction), and Fig. 6-17b the situation at the endface of a fiber in the 2D array. However, when defocusing is introduced, for example, p = 0.2, a lateral offset δ occurs which is about 20% of the CC's diameter. Fig. 6-17c shows the diverging beam for the lens-to-CC matrix link for D_{CC} = 100 μm. Only a part of the downstream beam falls onto the CC matrix (with a diameter of 35 mm) and is returned (dark rays); the remainder of the beam is not returned. Fig. 6-17d shows the lateral offset of the returning rays at the fiber endface in a zoomed-in view. Figs. 6-17e and f show similar ray tracings, but for D_{CC} = 1 mm. These figures show the clear increase in the lateral offset at the fiber endface of the returning rays when the diameter of a CC is increased and thus underline the need for a matrix array of miniature CCs.

The ray tracing results reported in Section 6.3.1 for the 3D analysis of a matrix of CCs confirm a similar trend of decreasing lateral offset of the retroreflected beam when the size of the CC elements is decreased, which underlines the preference of using a foil containing many miniature CCs.

6.3.3 System Experiments

Our laboratory demonstrator setup is shown in Fig. 6-18.

In order to extend the area of coverage, and to circumvent possible line-of-sight blockings, the setup features two PRAs which can be selected by the optical

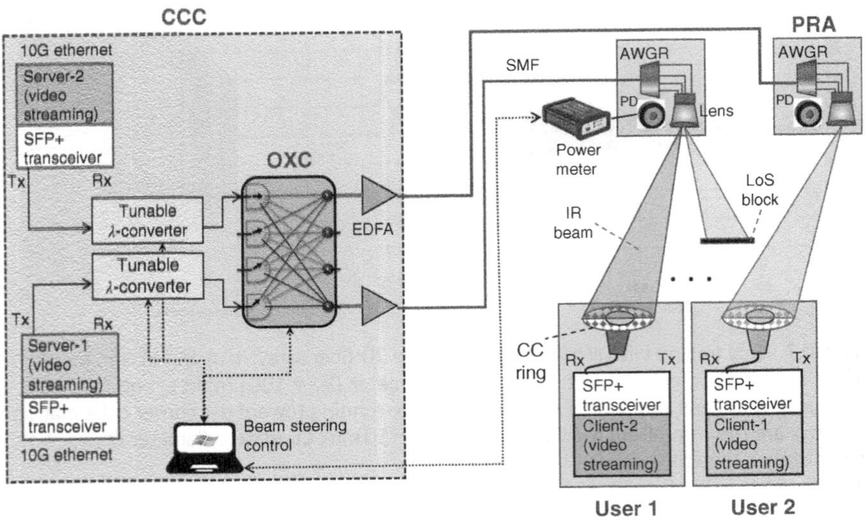

Fig. 6-18 Laboratory system demonstrator setup, with CC rings enabling user device localization

cross-connect (OXC) in the central communication controller site (CCC). Equipped with a $C + L^-$ band AWGR with channels spaced at 50 GHz and −3 dB bandwidths of 35 GHz and 24 GHz in the C-band and L^--band, respectively, a PRA can launch up to 129 beams, each slightly diverging upto a diameter of Ø10 cm in the user plane. It can cover a user area of about 1.3×1.3 m^2 with a PRA-to-user reach of 2.5 meters. The OXC is composed of a 4×4 MEMS switch, with insertion losses of 7.0 dB. In the demonstrator two user terminals are included, each equipped with a lens-based free-space optical receiver with a Ø3 cm aperture. A Ø4 cm CC foil with a central hole of Ø3 cm is mounted around this aperture. The physical details of the setup are shown in Fig. 6-19.

Two high-definition video streams are transmitted from the CCC, each embedded in a 10 Gbit/s Ethernet (GbE) stream. The video servers are equipped with SFP+ transceivers which are operated at a fixed wavelength. The tunable

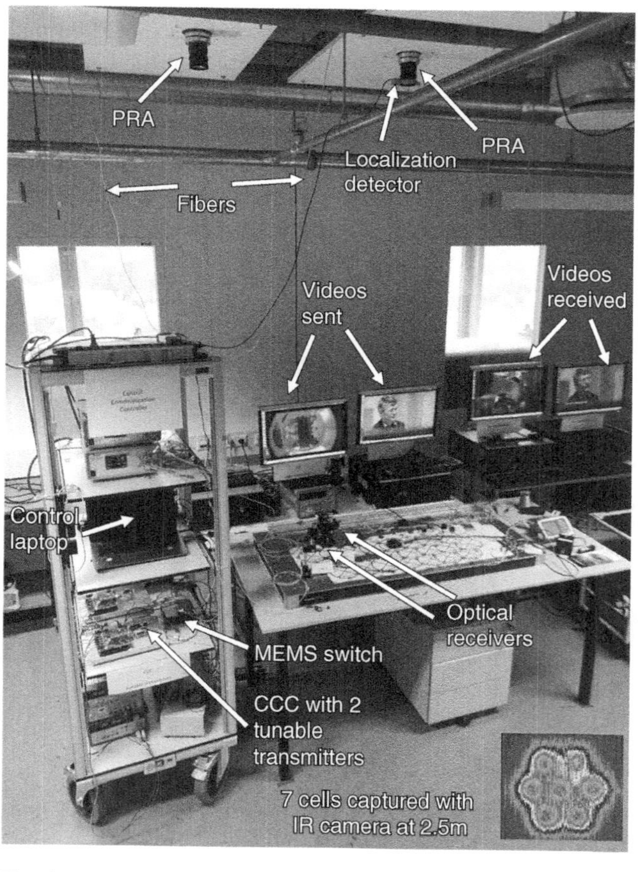

Fig. 6-19 Photo of laboratory demonstrator setup

wavelength converters translate these video streams to the wavelengths appropriate for the 2D beam steering to the respective user terminals by the PRA-s. Each user terminal is connected to a client for receiving the video stream and displaying it on a monitor. Good real-time transmission of both video streams was achieved, with $\text{BER} < 3 \times 10^{-12}$ at a transmitted beam power of 6.0 dBm and received power levels of −19 to −23 dBm.

We included the passive localization function by attaching a circular piece of Orafol foil containing the miniature CCs on an OWC receiver, as depicted in Fig. 6-18. Fig. 6-20 shows how this CC ring is retroreflecting light in the direction it came from, also when the CC ring is illuminated sideways. Due to the distortion of the profile of the retroreflected beam incurred in the retroreflection process in the CC matrix (see Fig. 6-9 and Fig. 6-11) and due to the opening in the center of the CC ring, this beam no longer has the Gaussian profile of the DS beam. Moreover, imperfections of the CC elements (i.e., deviations of the CC top angle from the ideal 90°) cause a broadening of the beam (see Fig. 6-13). This hampers the coupling of the retroreflected beam into the single-mode fiber (SMF) output ports of the PRA's AWGR [6-14] and thus the implementation of the basic RR-based localization concept outlined in Fig. 6-14. In addition, this basic concept is hampered by internal optical reflections in the PRA (we used a commercial 50 mm $F/0.9$ camera lens which had no antireflection coating for the $\lambda = 1.5\,\mu\text{m}$ window, and reflections at the connector joints between the 2D fiber array and the AWGR output fibers), and some beam scattering by the CC matrix foil.

Fig. 6-20 The CC ring mounted around the lens aperture of the user's receiver. (a) CC ring, photo made without flashlight. (b) CC ring, photo made with flash light which is retroreflected by the ring

Hence, we had to catch the returning rays with a small Ø1 cm localization power detector (PD) next to the lens. The optical power launched into the wireless channel was 6 dBm in all the measurements.

When the beam is scanning the user plane by tuning its wavelength, Fig. 6-21 shows the reflected power received by the localization detector for the case of two users, one equipped with a CC sheet with a diameter of 4 cm and the other of 5 cm. There is a 1-to-1 relation between the beam's wavelength and the device's position; hence, the power peaks in Fig. 6-21 indicate the pico-cells where the devices are located. The height of the power peaks is related to the size of the CC sheet, as shown in Fig. 6-22. An SNR > 5 dB adequate for the localization requires a CC foil diameter >4 cm. Background noise contributions are coming from the room

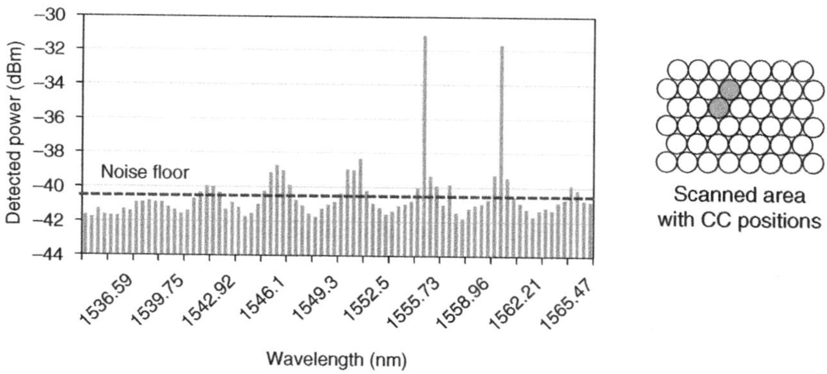

Fig. 6-21 Power received by localization detector when scanning across the system's pico-cells (for CC sheet diameter Ø4 cm)

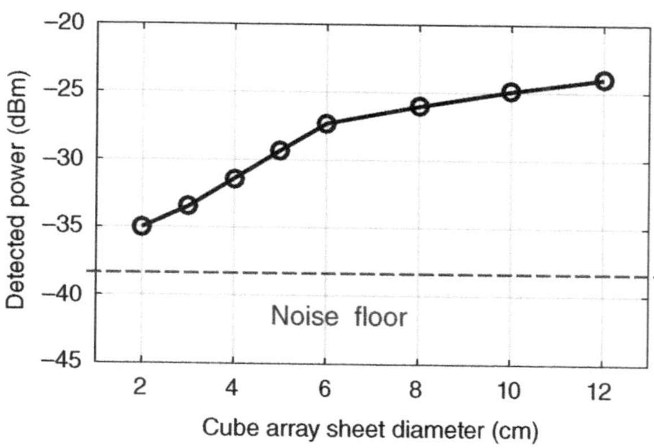

Fig. 6-22 Detected peak localization power versus diameter of circular CC sheet

illumination and from spurious reflections of the demonstrator table's surface. SNR values up to 15 dB have been achieved, which enabled localization clearly within the required resolution of a single pico-cell (diameter Ø10 cm). Scanning the whole user area takes about 15 seconds; the scan time is limited by the communication links between the power meter and the LabVIEW program running in the laptop which acts as a localization processor and performs the beam steering control. The localization time is intrinsically constrained by the time needed to tune the wavelength of a laser diode over the full range for scanning the PRA's whole area, which can be done within a few milliseconds. However, the scanning process involves additional steps: the stepwise tuning to each cell consecutively, deciding whether there is retroreflected power from that cell exceeding the background noise level, and subsequently moving to the next cell. Per cell, a total of 115 ms is needed for all this, of which only a few milliseconds are needed for the actual laser tuning. The majority of time is consumed by the LabVIEW software running in the laptop controller, the Arduino board which controls the laser tuning, and the readout time of the localization power meter. Scanning 129 cells in the area covered by the PRA thus takes about 15 seconds; this time may be reduced considerably by deploying more efficient algorithms implemented in embedded software.

The CC foil preferably surrounds the detector of the lens-based OWC receiver, which has an aperture of Ø3 cm. Hence, we also measured the returned localization power for a Ø4 cm CC foil with a central hole of Ø3 cm hosting the receiver's detector, and a Ø4 cm foil without a hole. As shown in Fig. 6-23, the central hole caused a reduction of returned power of about 3.2 dB (theoretically 3.6 dB when uniformly illuminated).

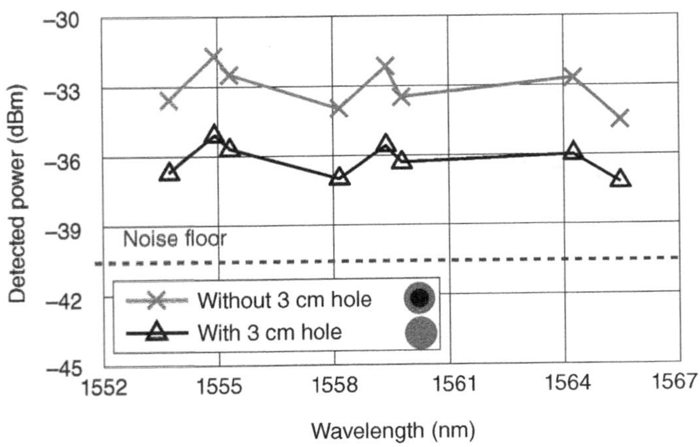

Fig. 6-23 Detected peak power versus wavelength, for Ø4 cm CC foil with Ø3 cm central hole, and Ø4 cm CC foil without the hole

6.3.4 Concluding Remarks

By means of narrow infrared optical beams on demand, wireless links to individual users offering high capacity, high energy efficiency as well as strongly protected privacy can be established. This technology may enable highly demanding broadband applications such as 8K ultrahigh-definition wireless video streaming, real-time virtual reality gaming or training without being hampered by wires, telepresence meetings, etc. Distinct advantages can be obtained in environments such as exhibition halls with frequently changing constellations of independent booths (which do not like capacity sharing), conference halls (where delegates want to download the proceedings instantly), electromagnetic radiation-sensitive areas (such as intensive care and surgery rooms in hospitals, inside airplanes, etc.), airport waiting areas near a gate (where people want to do fast downloading or other voluminous internet activities before getting on board), inside data centers between racks (where it provides minimum latency, as light travels 50% faster in the air than in silica fiber), business meeting rooms requiring high privacy and security, etc.

Efficient device localization is required in order to direct the beams on demand and instantly to the user. Applying passive retroreflection by means of a foil containing many miniature CCs has enabled energy-efficient device localization within the pico-cell resolution of our 2D beam-steered high-capacity optical wireless system setup. Real-time 10 GbE video streaming to mobile devices has been demonstrated.

6.4 Localization When Using Analog Beam Steering[2]

For US communication, that is, from the user device to the ceiling unit, 2D analog beam steering by, for example, 2D electro-mechanical transducer stages has been explored [6-15], see Section 5.6. The US beam power is restricted in order to save battery power consumption in the user's device and for eye safety of the user as well. For obtaining a sufficient power budget in the US link, it is therefore preferred to capture the full power of the US beam within the aperture of the US receiver at the ceiling. Alignment of the beam can then be achieved by means of a retroreflective ring surrounding the aperture and can be automated by scanning the US beam spot over this RR ring in the search for the aperture.

---

2 This section is largely based on T. Koonen, K.A. Mekonnen, F. Huijskens, and E. Tangdiongga, "Bi-directional all-optical wireless Gigabit Ethernet communication system using automatic self-aligned beam steering," *J. Lightwave Technol.*, vol. 41, no. 11, pp. 3446–3454, June 2023, https://doi.org/10.1109/JLT.2022.3231438, and on Dutch patent application N2032214 "Bi-directional all-optical wireless communication system with autonomous optical beam steering," inventor A.M.J. Koonen, filed June 17, 2022.

6.4.1 Center-of-Gravity Algorithm

Provided that the user is within the coverage area of the PRA, initially the relatively large RR ring surrounding the US receiver's aperture is found by taking coarse scanning steps of several cm·s (see also the spot scanning mechanism described in detail in Chapter 8). As soon as the RR ring has been found, the fine scanning in order to find the US Rx aperture itself is started. Fig. 6-24 illustrates the fine scanning in x- and y-directions.

When assuming a uniform power profile of the beam, at every position of the scanning beam, the power reflected by the RR ring can be analytically derived. It is proportional to the area of overlap between the beam and the RR ring, which is

$$A_{\text{overlap beam}-\text{RR}} = A_{\text{RR}}(R_1) - A_{\text{RR}}(R_2) \tag{6-5}$$

where if $R_n - R_{\text{beam}} < |D| < R_n + R_{\text{beam}}$

$$A_{\text{RR}}(R_n) = A\left(\frac{R_{\text{beam}}^2 - (R_n - D)^2}{2D}, R_n\right) + A\left(\frac{R_n^2 - (R_{\text{beam}} - D)^2}{2D}, R_{\text{beam}}\right) \tag{6-6}$$

with radius of beam spot R_{beam}, $n = 1$ or $= 2$, and wherein the area overlap function $A(d, R)$ is

$$A(d, R) = R^2\left\{\frac{\pi}{2} + \arcsin\left(\frac{d}{R} - 1\right) + \left(\frac{d}{R} - 1\right)\sqrt{1 - \left(\frac{d}{R} - 1\right)^2}\right\} \tag{6-7}$$

with distance D between the beam center and the center of the RR ring (which has an outer radius R_1 and inner radius R_2). The beam spot should preferably be not very small in order to get sufficient power reflected from the RR ring but should

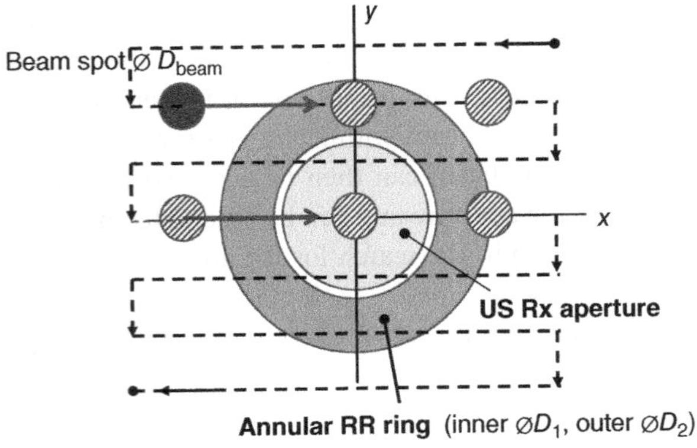

Fig. 6-24 Scanning the RR ring at the US receiver, in searching the receiver's aperture

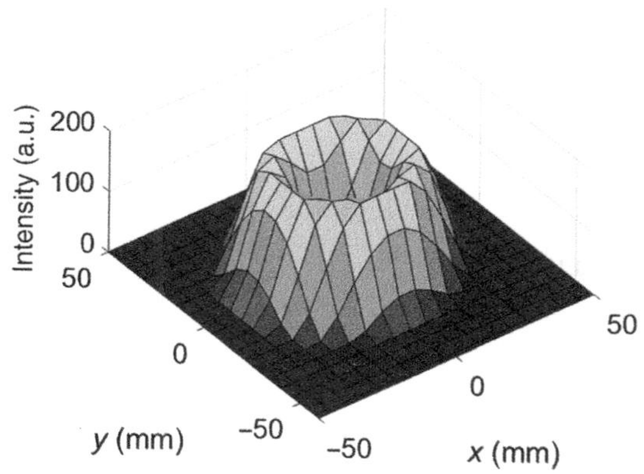

Fig. 6-25 Reflected power profile when scanning the RR ring with a uniform beam (calculated with scan step size 6 mm, beam spot diameter Ø15 mm, RR ring inner diameter Ø25 mm, outer diameter Ø62 mm; optical intensity in arbitrary units)

be smaller than the ring's width in order to obtain a good resolution, so $R_{beam} \leq \frac{1}{2}(R_1 - R_2)$. For $R_2 + R_{beam} < |D| < R_1 - R_{beam}$ the beam has full overlap with the RR ring, so then $A_{overlap\ beam\text{-}RR} = \pi R_{beam}^2$. For $|D| < R_2 - R_{beam}$ or $|D| > R_1 + R_{beam}$, there is no overlap, so $A_{overlap\ beam\text{-}RR} = 0$.

Based on this analysis, Fig. 6-25 shows the analytically calculated reflected power profile which can be monitored near the US transmitter at the user device, when the RR ring is scanned with a uniform beam. From the power results $\{p_i\}$ obtained at vector positions $\overline{r_i} = \begin{pmatrix} x_i \\ y_i \end{pmatrix}$, the Center-of-Gravity (CoG) vector position of the RR ring can be calculated as (cf. [6-16])

$$\overline{\mathrm{CoG}} = \begin{pmatrix} x_{\mathrm{CoG}} \\ y_{\mathrm{CoG}} \end{pmatrix} = \frac{1}{P_{tot}} \sum_{i=1}^{N} p_i \begin{pmatrix} x_i \\ y_i \end{pmatrix} \quad \text{with } P_{tot} = \sum_{i=1}^{N} p_i \tag{6-8}$$

As the US Rx aperture is centered inside the RR ring, localizing the CoG by scanning also yields the localization of this aperture, and thus the automatic alignment of the US beam into this aperture can be achieved.

6.4.2 Impact of Lens Aberrations on Beam Alignment Accuracy

The accuracy of the CoG algorithm depends on the diameter of the scanning beam, the start position of the scan, and the scan step size. For a uniform beam profile and beam diameters of Ø15 and Ø25 mm, Figs. 6-26a and b, respectively, show how much the estimated CoG location deviates from its exact value when the scan start position and the scan step size are varied. These graphs were calculated with the analytical approach introduced in Section 6.4.1. As expected, they have a

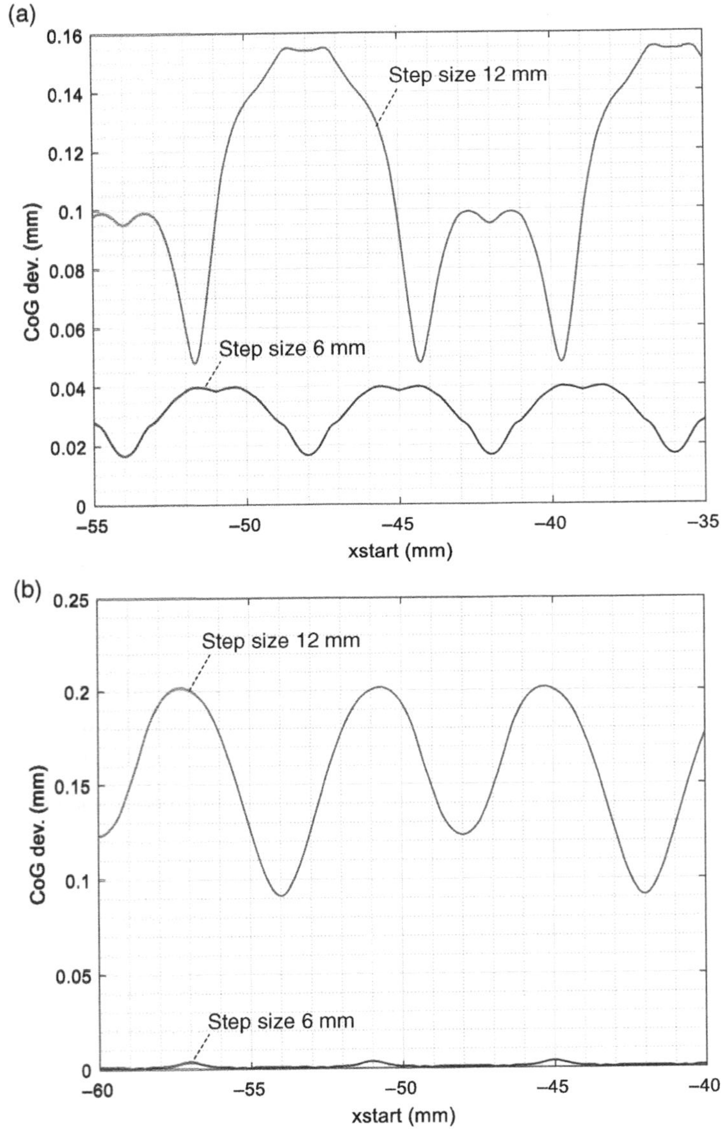

Fig. 6-26 Accuracy of the CoG algorithm versus start position of scan, for a uniform scanning beam, with (a) beam spot diameter Ø15 mm and (b) beam spot diameter Ø25 mm (RR ring inner diameter Ø25 mm, outer diameter Ø62 mm; curves for scan step size 6 and 12 mm)

periodicity equal to the scan step size. With the arbitrary start position of the scanning, a beam spot diameter of Ø15 mm, and a scan step size of 6 mm in the plane of the RR ring, the accuracy of the beam alignment with respect to the US Rx aperture of Ø25 mm is better than 40 μm, which is well within the required precision. With a larger beam spot diameter and the same scan step size, the accuracy even improves as the reflected power received during the scanning is larger; for a diameter of Ø25 mm (the maximum value which just fits the US Rx aperture) and step size of 6 mm, the accuracy is better than 4 μm. As expected, it can be seen that the accuracy clearly degrades for a larger step size. When doubling the step size from 6 to 12 mm, for the Ø15 mm beam diameter, the accuracy degrades from 40 to 150 μ m; for the Ø25 mm diameter, it degrades from 4 to 200 μm.

To achieve high localization precision, the CoG algorithm requires that the beam spot preserves its circular shape during the RR scanning. Hence, the US lens should have minimal off-axis aberrations. Ray tracing using a self-built MATLAB-based program (see Appendix) applied with several commercially available lens types shows that a Hastings triplet lens with $f = 20$ mm [6-17] gives much lower aberrations than a commonly used plano-convex lens with the same focal length [6-18], see Fig. 6-27.

Here η_{AP} is the fraction of power in the K launched rays which is captured by a Ø15 mm aperture (shown by the circle), and β is the ellipticity of the spot (cf. [6-19]), which indicates spot deformation and can be determined by the ray tracing as

$$\beta = 1 - \frac{\sigma_x}{\sigma_y} \tag{6-9}$$

where σ_x and σ_y are the RMS values of half the short and half the long axis of the deformed spot, respectively, which are calculated from the ray tracing results (assuming the spot is elongated in the y-direction) by

$$\sigma_x = \frac{1}{P_{\text{tot}}} \sqrt{\sum_{k=1}^{K} P_k^2 \left(x_k - x_{\text{CoG}}\right)^2} \tag{6-10}$$

$$\sigma_y = \frac{1}{P_{\text{tot}}} \sqrt{\sum_{k=1}^{K} P_k^2 \left(y_k - y_{\text{CoG}}\right)^2} \tag{6-11}$$

Here K rays are traced, P_k is the power in ray k, and P_{tot} is the total power of the K rays, so the power in the beam spot. For the triplet lens at a beam incident angle of $\varphi_x = 8.4°$ (see Fig. 6-27), we find $\beta = 12\%$, whereas for the plano-convex lens $\beta = 45\%$ already at a smaller $\varphi_x = 5.4°$. This indicates that the off-axis aberrations of the triplet lens are significantly smaller than those of the plano-convex lens. Note also that the ellipticity β must be low for efficient lens coupling of the US beam to the PD in the US Rx.

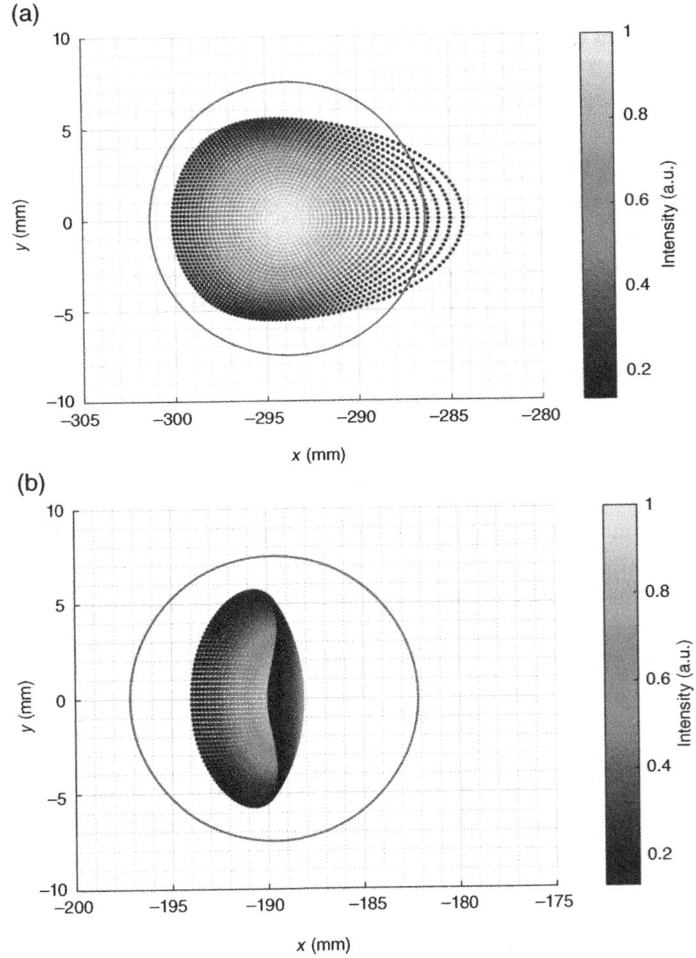

Fig. 6-27 US beam spot at the RR ring, using (a) triplet lens $f = 20$ mm: $\varphi_x = 8.4°$, $\eta_{AP} = 98\%$, $\beta = 12\%$, and (b) plano-convex lens $f = 20$ mm: $\varphi_x = 5.4°$, $\eta_{AP} = 100\%$, $\beta = 50\%$ (by ray tracing, with K=1027 rays of Gaussian beam from US transmitter at beam incidence angle φ_x; *circle*: aperture diameter Ø15 mm; *bars*: optical intensity, in arbitrary units)

Fig. 6-28 shows how due to off-axis lens aberrations the shape of the spot is changing when scanning across the RR ring. The changes with the plano-convex lens are clearly larger than with the triplet lens and do affect the accuracy of the CoG algorithm.

Fig. 6-29 illustrates the impact of these aberrations on the accuracy of the CoG algorithm. For both the triplet lens and the plano-convex lens, it shows the

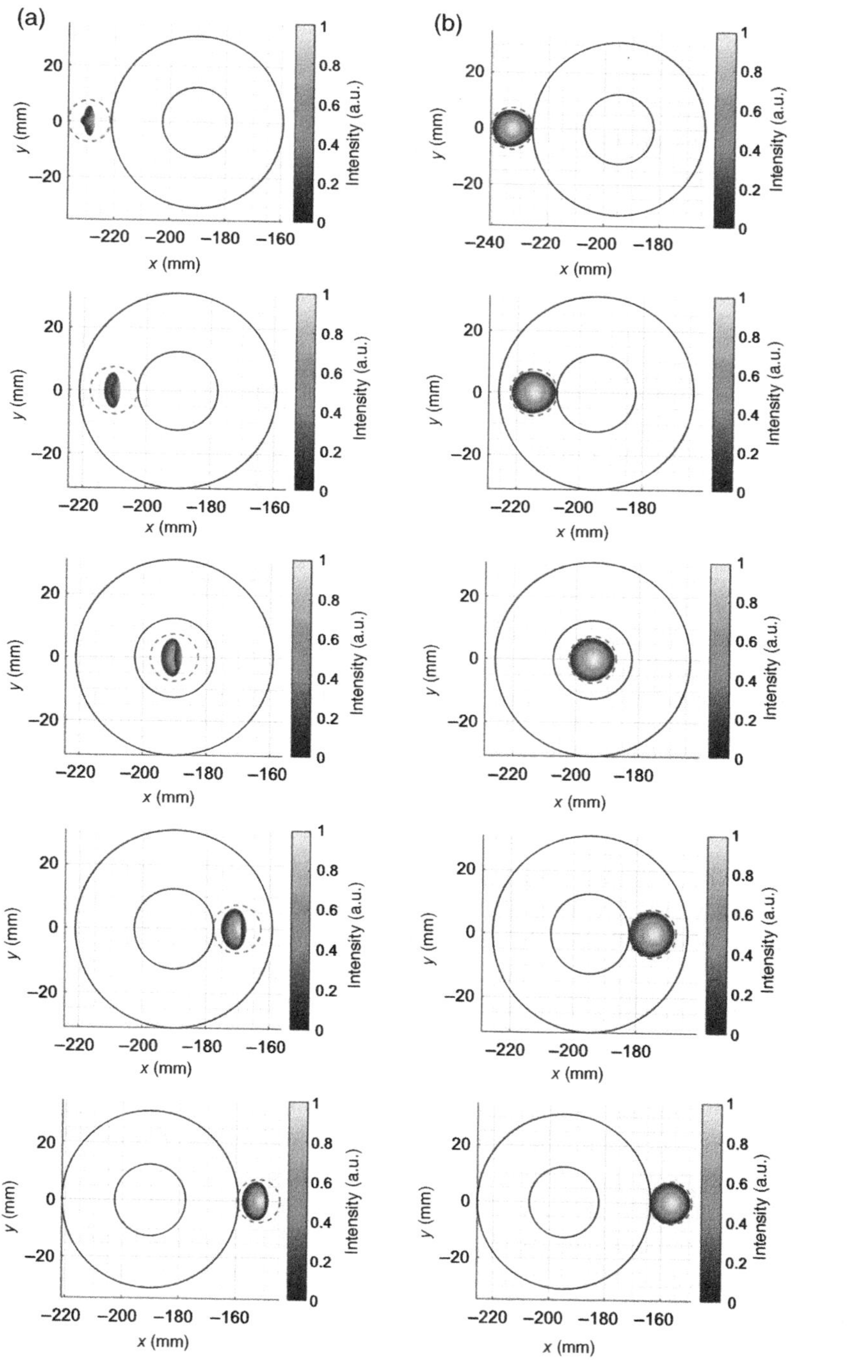

Fig. 6-28 Scanning with the US beam across the RR ring at the US receiver located at an FoV position $\varphi \approx 5.5°$, (a) using a plano-convex lens, $f = 20$ mm lens, (b) using a triplet lens, $f = 20$ mm; RR ring inner diameter Ø25 mm, outer diameter Ø62 mm; *dashed circle*: Ø15 mm, *bars*: optical intensity, in arbitrary units)

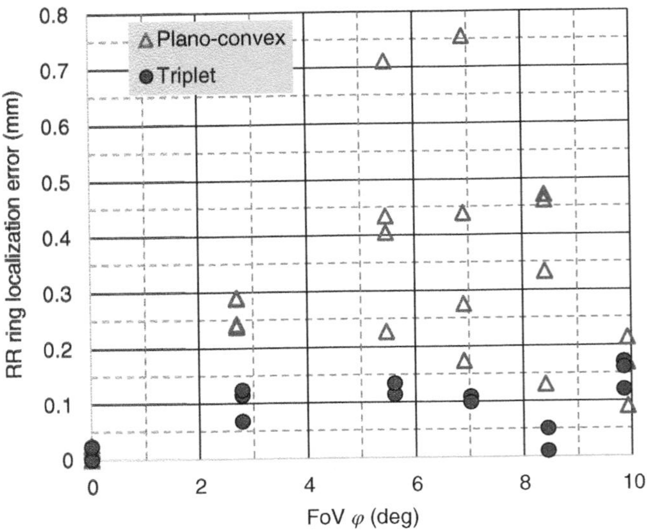

Fig. 6-29 Accuracy of the CoG algorithm versus the US receiver's FoV position φ when using a plano-convex lens $f = 20$ mm (*triangles*) and using a triplet lens $f = 20$ mm (*dots*) (system parameters: beam spot Ø15 mm at $\varphi = 0$, scan step size 6 mm, RR ring inner diameter Ø25 mm, outer diameter Ø62 mm)

deviations in the estimated CoG position from its exact position when the scan start position is varied. These deviations and their spreading are presented for various positions of the RR ring, that is, for various FoV positions φ. Each marker represents the results of scanning across the whole RR ring (outer diameter Ø62 mm) with a Gaussian beam spot of Ø15 mm waist and with a scan step size of 6 mm. Each marker involves ray tracing of $K = 469$ rays for 15×15 scan positions, so of more than 105,000 rays. As shown in Fig. 6-29, the localization errors made by the CoG algorithm depend somewhat on the position where the scanning starts. This trend agrees with the results shown in Fig. 6-26 obtained from the analytical calculations with a uniform beam. As shown in Fig. 6-29, the localization errors in general increase at larger FoV positions φ, which is due to the increasing off-axis lens aberrations. The localization errors and the spreading in the errors when the scan start position is varied are larger when the plano-convex lens is used than when the triplet lens is used. This underlines once more the preference for applying a low-aberration lens, such as a triplet lens. Over a FoV φ-range up to 10°, it can be seen in Fig. 6-29 that the localization accuracy of the CoG algorithm is better than 180 μm with the $f = 20$ mm triplet lens, whereas the accuracy is within 760 μm with the $f = 20$ mm plano-convex lens.

Based on these simulation results, for the US link in our system demonstrator a Hastings $f = 20$ mm triplet lens, a beam spot diameter of Ø15 mm, a scan step size of 6 mm, and an RR ring with an outer diameter of Ø62 mm and inner diameter of Ø25 mm have been selected.

6.4.3 System Experiments

The localization technique has been implemented in the bidirectional OWC laboratory setup described in full detail in Chapter 8 and shown in Fig. 8-6. The OWC receivers can handle GbE data speeds and have a wide FoV $\approx \pm 10°$, which eases beam alignment and provides some tolerance to tilt of the user's device. The DS beam alignment procedure uses the AWGR-based 2D steering of +10 dBm beams that are wider than the user's DS receiver aperture (Ø10 cm Gaussian beams versus Ø50 mm Fresnel lens). The steering is done by means of wavelength tuning controlled by LabVIEW running on a laptop computer, as discussed in Section 6.3.

The US beam alignment procedure uses 2D beam steering at the US transmitter by means of an (x, y) stepper motor stage controlled by the CoG algorithm which is implemented in an Arduino board. The input for the algorithm comes from a PD near the US transmitter which monitors the power reflected from the RR ring mounted around the Ø25 mm Fresnel lens of the US receiver at the ceiling. Fig. 6-30 shows the RR ring and how it retroreflects the light coming sideways. The US beams are narrower than the lens aperture (Ø15 mm versus Ø25 mm)

(a) (b)

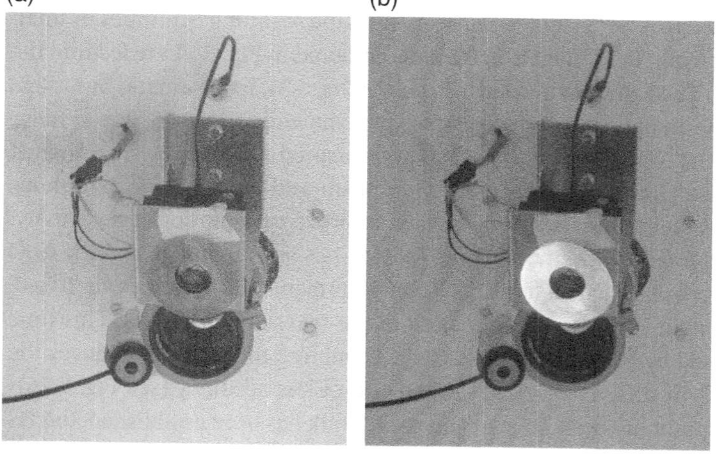

Fig. 6-30 The RR ring mounted around the lens aperture of the ceiling's upstream receiver. (a) RR ring, photo made without flashlight. (b) RR ring, photo made with flashlight which is retroreflected by the ring

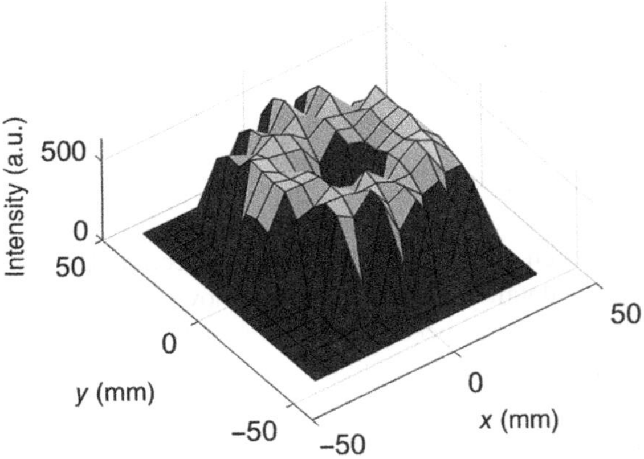

Fig. 6-31 Experimental results of monitoring the power reflected by the RR ring around the US receiver's aperture (upstream Gaussian beam spot Ø15 mm, RR ring with inner diameter Ø25 mm and outer diameter Ø62 mm; optical intensity, in arbitrary units)

as the US beam power has to be relatively low (only +2 dBm) in order to save battery power at the user's device and also for enhanced eye safety while being close to the user [6-20]. The results of monitoring the reflected power when scanning the narrow US beam across the RR ring are shown in Fig. 6-31. The shape of the 3D intensity graph is similar but a bit less smooth when compared with the theoretical one in Fig. 6-25, due to noise fluctuations in the monitored power. These measured results serve as the inputs for the CoG algorithm.

Upon the user's request, the bidirectional OWC connection is set up by first establishing the US link. The US receiver at the ceiling is localized by scanning from the user's device with the US beam in coarse steps until a first hit with the RR ring at the receiver is found. Subsequently, the scanning continues in smaller steps (about 6 mm each), and the results as gathered in Fig. 6-31 are fed into the CoG algorithm. The experimental surface plot of Fig. 6-31 approximates but is less smooth than the analytically derived surface plot shown in Fig. 6-25, due to noise in monitoring the reflected power levels. The measured plot may be smoothened by introducing a pilot tone on the scanning beam and associated synchronous detection techniques at the user device which improve the monitoring sensitivity. The scanning is finished when the whole RR ring has been scanned, and the CoG algorithm has converged. The total US beam alignment time takes about 10 seconds, which is mainly limited by the speed of the two stepper motors. This time may be shortened by increasing the step size (doubling the step size reduces the number of steps by a factor of 4), but at the cost of less accuracy (see Fig. 6-26).

After the US beam has been aligned and the US link has been established, the DS beam is aligned by scanning the user area using wavelength tuning and monitoring the power reflected by the RR ring at the user's DS receiver, as described in Section 6.3. The DS beam alignment takes less than 15 seconds, which is mainly limited by the LabVIEW control program running on a laptop at the headend site. Adequate alignment of both the US and DS beam is confirmed by good bidirectional TCP test results showing about 940 Mbit/s in DS and about 939 Mbit/s in the US direction, without packet loss, see Section 8.6.

6.4.4 Concluding Remarks

Automatic self-calibrated alignment of the US beam by means of a novel steering control protocol using the CoG algorithm as well as automatic self-calibrated alignment of the DS beam have been designed, realized, implemented, and validated in a laboratory testbed. Broadband GbE receivers with a wide FoV $\approx \pm 10°$ were employed, which eased the required beam alignment accuracy. With low-aberration triplet lenses, a good accuracy within 200 μm using the CoG algorithm and hence adequate US beam steering has been achieved within 10 seconds alignment time, mainly limited by the stepper motors. The automatic DS beam alignment takes less than 15 seconds, mainly limited by the wavelength-tuning control by means of LabVIEW. TCP measurements show 940 Mbit/s transfer speeds per user within the FoV $\approx \pm 10°$.

6.5 Summarizing Conclusions

Accurate self-calibrated steering of the DS and US beams in bidirectional indoor OWC systems without pre-establishing communication links between the ceiling unit and the user devices has been achieved by using passive RR techniques at the unit to be localized. By beam scanning and monitoring the reflected power, the beam is directly brought into the well-aligned position without needing calibration and without parallax errors, and the US and DS links are established without requiring active functions at the unit to be localized. Being able to do DS and US localization separately and independently enables setup of the OWC links upon an initiative from the user as well as upon an initiative from the central ceiling site, thus enlarging the flexibility to operate the system. The time consumed by the beam steering limits the system's suitability for mobile users. Its major benefits are for the delivery of dense high-capacity services to nomadic (and static) users.

References

[6-1] A.M.J. Koonen, "Indoor optical wireless systems: Technology, trends, and applications," *J. Lightwave Technol.*, vol. 36, no. 8, pp. 1459–1467, Apr. 2018, https://doi.org/10.1109/JLT.2017.2787614.

[6-2] H. Liu, H. Darabi, P. Banerjee, and J. Liu, "Survey of wireless indoor positioning techniques and systems," *IEEE Trans. Syst., Man, Cybern.*, vol. 37, no. 6, pp. 1067–1080, Nov. 2007.

[6-3] T.-H. Do, and M. Yoo, "An in-depth survey of visible light communication based positioning systems," *Sensors*, vol. 16, no. 5, Paper no. 678, 2016. On-line at http://www.mdpi.com/1424-8220/16/5/678.

[6-4] A.M. Khalid, P. Baltus, A.R. Dommele, et al., "Bi-directional 35-Gbit/s 2D beam steered optical wireless downlink and 5-Gbit/s localized 60-GHz communication uplink for hybrid indoor wireless systems," *Proc. OFC 2017*, Los Angeles, Mar. 21–23, 2017, Paper Th1E.6.

[6-5] A.M.J. Koonen, F. Gomez-Agis, F.M. Huijskens, and K.A. Mekonnen, "High-capacity optical wireless communication using two-dimensional IR beam steering," *J. Lightwave Technol.*, vol. 36, no. 19, pp. 4486–4493, Oct. 2018, https://doi.org/10.1109/JLT.2018.2834374.

[6-6] A. Gomez, K. Shi, C. Quintana, et al., "A 50 Gb/s transparent indoor optical wireless communications link with an integrated localization and tracking system," *J. Lightwave Technol.*, vol. 34, no. 10, pp. 2510–2517, May 2016.

[6-7] N.Q. Pham, K.A. Mekonnen, E. Tangdiongga, et al., "Accurate indoor localization for beam-steered OWC system using optical camera," *Proc. ECOC 2019*, Dublin, Sep. 22–26, 2019, Paper P45.

[6-8] K.A. Mekonnen, N. Calabretta, E. Tangdiongga, and A.M.J. Koonen, "High-capacity dynamic indoor network utilizing optical wireless and 60-GHz radio techniques," *Proc. MWP 2017*, Beijing, Oct. 23–26, 2017, Paper We.2.3.

[6-9] A.M.J. Koonen, K.A. Mekonnen, F.M. Huijskens, and E. Tangdiongga, "Fully passive user localisation for beam-steered high-capacity optical wireless communication system," *Proc. ECOC 2019*, Dublin, Sep. 22–26, 2019, Paper Tu.3.C.4.

[6-10] Corner cube retroreflectors [on-line]. Available: https://www.edmundoptics.com/f/n-bk7-corner-cube-retroreflectors/12447/

[6-11] M.K. Kilaru, J. Yang, and J. Heikenfeld, "Advanced characterization of electrowetting retroreflectors," *Opt. Exp.*, vol. 17, no. 20, pp. 17563–17569, Sep. 2009.

[6-12] Retroreflector [on-line]. Available: https://en.wikipedia.org/wiki/Retroreflector

[6-13] Photoelectric acrylic reflector material [on-line]. Available: https://www.orafol.com/en/europe/products/oralite-photoelectric-sheeting, https://update.orafol.com/en/americas/products/oralite-ac1000-photoelectric-acrylic-reflector-material.

[6-14] Private discussions with Dr. Peter Winzer, June 2024.

[6-15] A.M.J. Koonen, K.A. Mekonnen, F. Huijskens, and E. Tangdiongga, "Bi-directional all-optical wireless Gigabit Ethernet communication system using automatic self-aligned beam steering," *J. Lightwave Technol.*, vol. 41, no. 11, pp. 3446–3454, June 2023, https://doi.org/10.1109/JLT.2022.3231438.

[6-16] Center of gravity [on-line]. Available: https://www.sciencetopia.net/physics/center-gravity

[6-17] Triplet achromatic lenses [on-line]. Available: https://www.edmundoptics.com/f/hastings-triplet-achromatic-lenses/11968/

[6-18] Plano-convex lenses [on-line]. Available: https://www.edmundoptics.com/p/200mm-dia-x-200mm-fl-uncoated-plano-convex-lens/5719/

[6-19] Ellipticity [on-line]. Available: https://math.stackexchange.com/questions/537115/what-are-definitions-of-ellipticity

[6-20] Laser safety [on-line]. Available: https://en.wikipedia.org/wiki/Laser_safety.

7

Broadband OWC Receiver with Wide Field-of-View[1]

7.1 Introduction

A crucial element in a beam-steered OWC system, particularly in settings where devices may move, is a receiver which can capture a large part of the optical beam without requiring tedious alignment of the device regarding the position and the angle of incidence of the beam. The receiver hence should have a large aperture as well as a wide field-of-view (FoV, i.e., the range of angles in which the beam is captured). High-speed photodiodes typically have a small active area in order to minimize their bandwidth-limiting electrical junction capacitance, so a quite small aperture. The receiving aperture can be increased by a lens system, but as stated by the fundamental optics law of conservation of etendue [7-1], this is compromised by a decrease in the FoV.

Non-imaging optics such as a compound parabolic concentrator (CPC) can also increase the receiving aperture (see the OWC system shown in Fig. 5-3) but still need to obey the law of conservation of etendue [7-2, 7-3]. Hence, this comes at the expense of a reduced FoV, and the size of the CPC makes the receiver less compact.

The law of conservation of etendue may be circumvented by breaking the link between the etendue of the light collection area and that of the detector by means of wavelength conversion of the received light. For example, the light may be captured with a slab concentrator of which the slab waveguide is doped with fluorophore and is terminated with a photodetector [7-4]. Fig. 7-1 shows how the incoming light is converted to a longer wavelength by the fluorophore and guided

1 This chapter is an extended version of the chapter in paper "Beam-steered optical wireless communication for industry 4.0", by Ton Koonen et al., *IEEE J. Sel. Top. Quantum Electron.*, Dec. 2021, https://doi.org/10.1109/JSTQE.2021.3092837. It is also based on patent PEP 4052390B1, granted Oct. 23, 2024, priority date Oct. 31, 2019, "An optical wireless communication receiver with large photodetector surface area, large field of view and high bandwidth," by A.M.J. Koonen.

Short-Reach Optical Wireless Communication: By Directed Narrow Beams, First Edition. Ton Koonen.

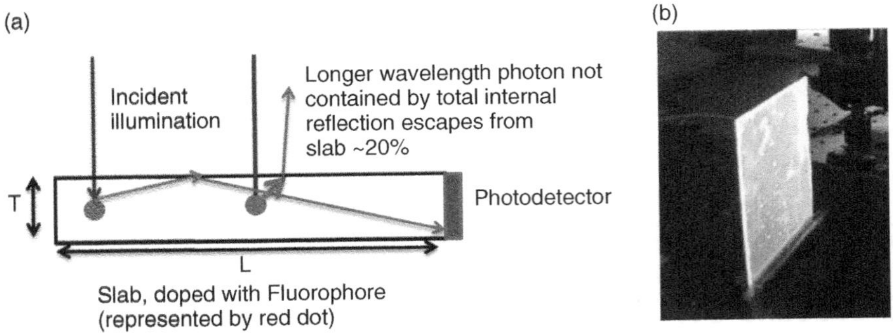

Fig. 7-1 Wide field-of-view receiver breaking the law of conservation of etendue ([7-4]/ Optical Society of America) [3-39]. (a) Slab fluorescent antenna. (b) Fluorescent concentrator

to the detector by reflections at the slab surfaces; thus, a wider FoV and higher gain are obtained. Using a quantum dot material as the fluorophore, a gain 50× that of an etendue-preserving concentrator with the same FoV can be reached. A similar concept has been reported using a doped fiber [7-5]. The long waveguide may lead to channel dispersion effects, thus reducing the bandwidth and making this solution less suitable for high data rates. Moreover, the conversion process requires a relatively high level of received optical power, and thus the receiver's sensitivity is compromised.

Another approach reported is to separate the functions of light collection and light detection and optimize them individually, for example, by capturing the light into a waveguide by means of surface grating couplers (SGCs) and subsequently using a fast waveguide-coupled photodiode for detection. With a $10 \times 10\,\mu m^2$ SGC and $3 \times 10\,\mu m^2$ 67 GHz uni-traveling-carrier (UTC) photodiode, 40 Gbit/s on-off-keyed transmission has been demonstrated [7-6]. As described in Sections 3.8 and 3.11.2, a larger aperture can be achieved by apodization of the SGC and by multiple SGCs of which the outputs are combined into a single waveguide (incurring power combining losses).

In an angular diversity receiver (ADR) [7-7, 7-8], multiple photodiodes are used in parallel to jointly increase the active area, but each one needs to be followed by a separate high-bandwidth preamplifier and their outputs need to be summed, which increases noise, requires careful phase synchronization among these amplifiers and complicates the receiver circuitry. As an example, Fig. 7-2 shows an ADR deploying a two-dimensional (2D) array of 7 metal-semiconductor-metal (MSM) hexagonal photodiodes, each having a side length of 50 μm and 0.12 pF capacitance, followed by a current-summing amplifier, yielding an effective aperture diameter of 260 μm and overall −3 dB bandwidth of 3.7 GHz.

In this chapter, we propose an alternative concept that extends the receiver's aperture and its FoV while it requires only a single preamplifier, thus avoiding high complexity and reducing power consumption which is particularly beneficial at the

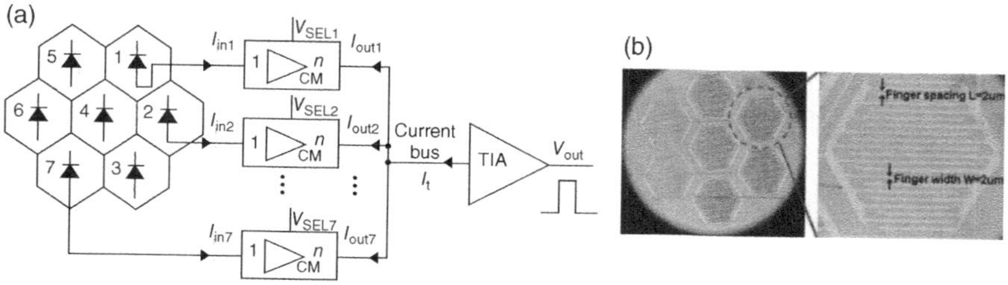

Fig. 7-2 Angular diversity receiver with 7 MSM photodiodes (Zeng et al. [7-7]/© [2009] IEEE). (a) Angular diversity receiver. (b) Hexagonal InGaAs MSM photodiode array

compact and (often) battery-operated user device. This concept was reported first in [7-9] and further elaborated in [7-10, 7-11]. Instead of a single photodiode, a 2D matrix of photodiodes is used to increase the active area while keeping its bandwidth equal to that of the single photodiode. This matrix constellation still has only a single signal output. The 2D photodiode matrix (PD matrix) concept is introduced and theoretically analyzed in Section 7.2, including its bandwidth characteristics, biasing conditions, and signal-to-noise ratio (SNR) performance of the PD matrix receiver in comparison with a single-PD receiver and a multi-PD ADR. In Section 7.3, it is studied both analytically and numerically by means of extensive ray tracing how the beam-to-photodiode coupling is improved by applying the PD matrix concept, including studies about the impact of the fill factor of the PD matrix, and about the impact on the receiver's performance when it is shifted within the footprint of the beam taking also into account the crosstalk power which may come in from neighboring beams. Section 7.4 discusses the assembly of the PD matrix and its implementation into a transimpedance preamplifier circuit board. Section 7.5 reports on the experimental validation of the receiver's performance employing a 4 × 4 PD matrix, including the assessment of its bit error ratio (BER) performance and FoV at 1 and 1.25 Gbit/s (Gigabit Ethernet, GbE) data rates. Section 7.6 analyzes the scaling of the PD matrix to larger sizes and its impact on the attainable FoV, and the chapter concludes with some summarizing remarks in Section 7.7.

7.2 Receiver Based on a Matrix of Photodiodes

A photodiode may electrically be modeled by the equivalent circuit shown in Fig. 7-3a, where R_s represents the (typically small) resistance of the bond wires, R_d represents the (large) parallel resistance representing the leakage current, C_d represents the capacitance of the reverse-biased photodiode's p-n junction (where incoming photons create electron-hole pairs), $i_d(t)$ represents the (small) dark current, and $i_s(t)$ represents the generated photocurrent where $i_s(t) = \mathcal{R}aP(t)$ in which $\mathcal{R}$ is the photodiode's responsivity, and a is the fraction of the optical

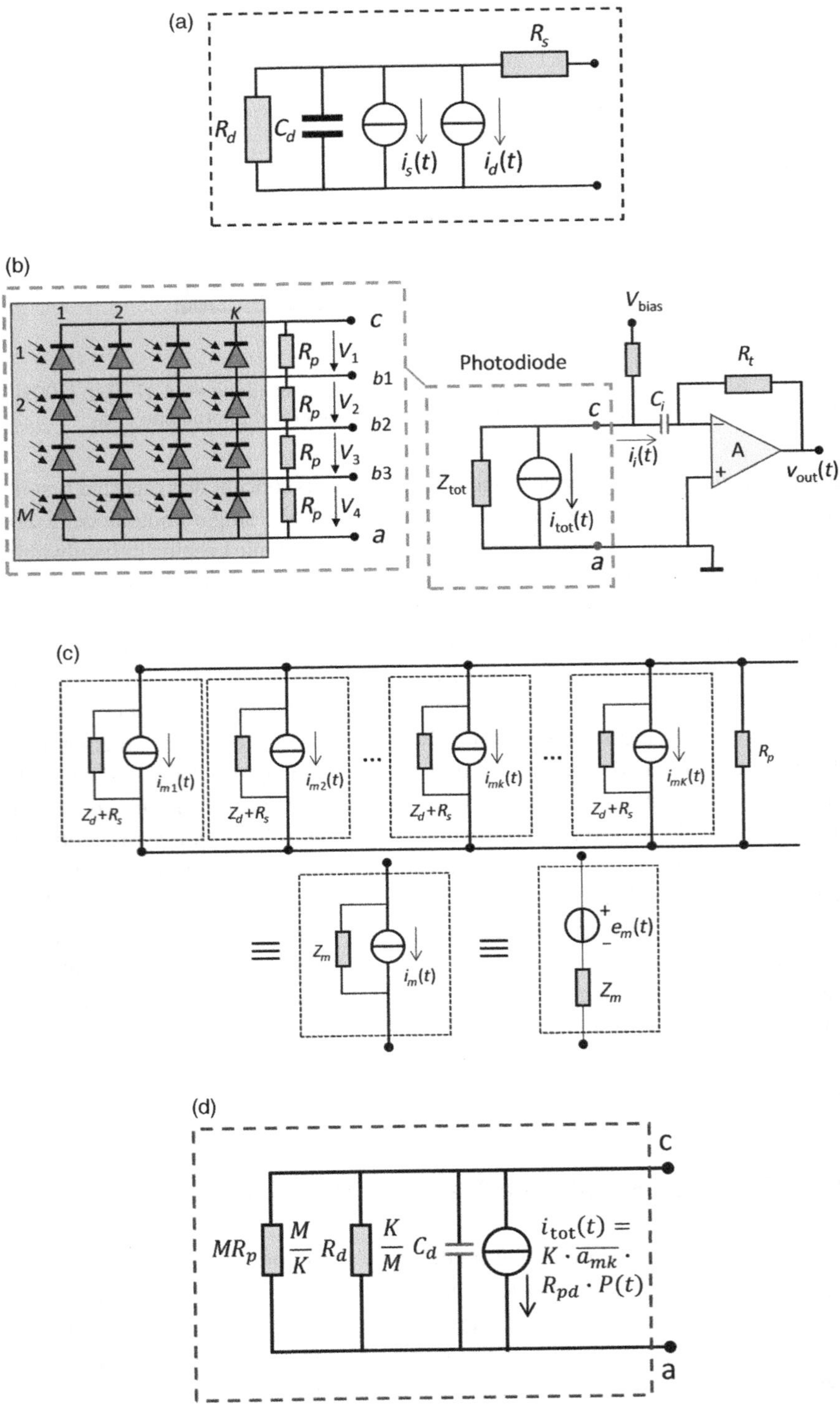

Fig. 7-3 OWC receiver employing a two-dimensional matrix of photodiodes. (a) Equivalent circuit of single photodiode. (b) 2D matrix of photodiodes applied in transimpedance amplifier. (c) Putting K photodiodes in parallel. (d) Equivalent circuit of 2D matrix of photodiodes (if $R_s << R_d$ and $i_d(t) << i_s(t)$)

beam's power $P(t)$ reaching the photodiode's active area (its p-n junction). By putting K photodiodes in parallel the total active area is increased K-fold, so the generated photocurrent as well, but also the capacitance increases K-fold and hence the bandwidth decreases K-fold. Putting photodiodes in series in order to increase the area is commonly not considered a viable option, as ideal current sources should not be cascaded. However, practical photodiodes are not ideal current sources. When a non-infinite parallel resistance R_d is assumed, M photodiodes may be cascaded, and thus the capacitance of this cascade is decreased M-fold and the bandwidth increases M-fold. The generated photocurrent does not increase by the cascading. We combined the putting of photodiodes in K parallel columns with putting them in series in M rows, leading to a two-dimensional $K \times M$ photodiodes matrix; using only two poles of it, this matrix can be applied in a transimpedance amplifier (TIA) as shown in Fig. 7-3b.

Following Thevenin's and Norton's theorems (illustrated in Fig. 7-3c), putting K identical photodiodes in parallel in the mth row may be represented by a single current source $i_m(t)$ with parallel impedance Z_m, or equivalently a single voltage source $e_m(t)$ with serial impedance Z_m, where a_{mk} is the fraction of the beam's power $P_0(t)$ which is incident on photodiode (m,k) in the 2D photodiode matrix. Thus follows

$$i_m(t) = \sum_{k=1}^{K} i_{mk}(t) = \frac{Z_d}{Z_d + R_s}\left\{K i_d(t) + \mathcal{R} P_0(t) \sum_{k=1}^{K} a_{mk}\right\} \tag{7-1}$$

with $Z_d = R_d \parallel (1/j\omega C_d)$, where $Z_1 \parallel Z_2$ denotes the connection of impedances Z_1 and Z_2 in parallel, and

$$Z_m = \left(\frac{Z_d + R_s}{K}\right) \parallel R_p \tag{7-2}$$

$$e_m(t) = i_m(t)\, Z_m = \left\{\frac{Z_d}{K} \,\middle\|\, \frac{R_p}{1 + R_s/Z_d}\right\}\left\{K\, i_d(t) + \mathcal{R}\, P_0(t) \sum_{k=1}^{K} a_{mk}\right\} \tag{7-3}$$

Furthermore, putting M rows in series each composed of K identical photodiodes yields the overall $K \times M$ photodiodes matrix represented by a single total output current source $i_{\mathrm{tot}}(t)$ and total parallel impedance Z_{tot} (or equivalent total output voltage source $e_{\mathrm{tot}}(t)$ with total serial impedance Z_{tot}) according to

$$e_{\mathrm{tot}}(t) = \sum_{m=1}^{M} e_m(t) = \left(\frac{Z_d + R_s}{K} \parallel R_p\right) \frac{Z_d}{Z_d + R_s}\left\{MK\, i_d(t) + \mathcal{R} P_0(t) \sum_{m=1}^{M}\sum_{k=1}^{K} a_{mk}\right\} \tag{7-4}$$

$$Z_{\mathrm{tot}} = \sum_{m=1}^{M} Z_m = M\left\{\left(\frac{Z_d + R_s}{K}\right) \parallel R_p\right\} \tag{7-5}$$

$$i_{\text{tot}}(t) = \frac{e_{\text{tot}}(t)}{Z_{\text{tot}}} = \frac{Z_d}{Z_d + R_s}\left\{Ki_d(t) + \frac{\mathcal{R}P_0(t)}{M}\sum_{m=1}^{M}\sum_{k=1}^{K} a_{mk}\right\} \tag{7-6}$$

$$i_{\text{tot}}(t) = \frac{K}{1 + R_s/Z_d}\left\{ i_d(t) + \mathcal{R}\, P_0(t)\, \overline{a_{mk}} \right\} \tag{7-7}$$

where $\overline{a_{mk}} = \frac{1}{MK}\sum_{m=1}^{M}\sum_{k=1}^{K} a_{mk}$

Hence, the photodiode matrix can be represented by the equivalent circuit shown in Fig. 7-3d composed of a current source $i_{\text{tot}}(t)$ and impedance Z_{tot} as

$$i_{\text{tot}}(t) = \frac{K\left[i_d + \mathcal{R}\,\overline{a_{mk}}\,P_0(t)\right]}{1 + \frac{R_s}{R_d}\left(1 + j\omega R_d C_d\right)} \tag{7-8}$$

$$Z_{\text{tot}} = \frac{M\,R_p\left[R_d + R_s\left(1 + j\omega R_d C_d\right)\right]}{R_d + \left(R_s + KR_p\right)\left(1 + j\omega R_d C_d\right)} \tag{7-9}$$

where $\overline{a_{mk}} = \frac{1}{M \cdot K}\sum_{m=1}^{M}\sum_{k=1}^{K} a_{mk}$ is the average fraction of the beam's power $P_0(t)$ which is incident on each photodiode, and Z_{tot} can be decomposed in a resistance $M(R_p \parallel (R_d + R_s)/K)$ in parallel with a capacitance $(K/M)C_d$.

7.2.1 Frequency Characteristics

The photodiode matrix can be readily applied in the simple transimpedance amplifier (TIA) configuration of Fig. 7-3b, of which the transimpedance $Z_T = v_{\text{out}}(t)/i_{\text{tot}}(t)$ near $\omega = 0$ (DC) and the -3 dB bandwidth $\omega_{-3\text{ dB}}$ for $R_s << R_d$ is

$$Z_T(\omega = 0) = \frac{A\,R_t}{1 + A + \frac{R_t}{MR_p} + \frac{K}{M}\frac{R_t}{R_d + R_s}} \tag{7-10}$$

$$\omega_{-3\text{ dB}} \approx \frac{M}{K}\frac{1}{C_d}\left[\frac{1 + A}{R_t} + \frac{1}{R_p}\right] + \frac{1}{R_d C_d} \tag{7-11}$$

Hence, the bandwidth for $R_p >> R_t/(1 + A)$ and $R_d >> (K/M)R_t/(1 + A)$ is given by

$$\omega_{-3\text{ dB}} \approx \frac{M}{K}\frac{1 + A}{C_d R_t} \tag{7-12}$$

that is, M/K times larger than with a single photodiode (for which $K = M = 1$). The transimpedance near DC referenced to the generated photocurrent is

$$Z_{T,\text{ref}}(\omega = 0) = \frac{v_o(t)}{\overline{a}\,\mathcal{R}P(t)} \approx K\frac{A}{1 + A}R_t \tag{7-13}$$

and is therefore K times larger than with a single photodiode.

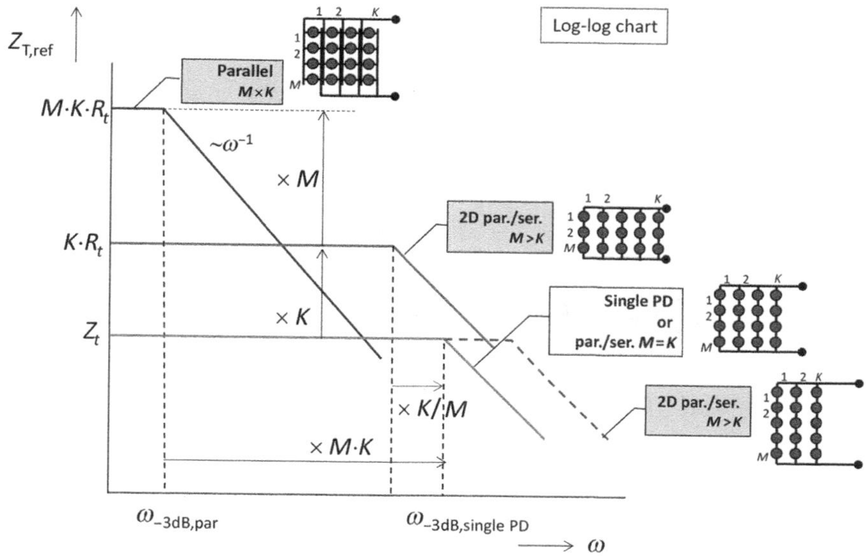

Fig. 7-4 Frequency characteristics of OWC receiver using a $K \times M$ matrix of photodiodes (with K columns and M rows)

Fig. 7-4 gives an overview of how the receiver characteristics will basically depend on the constellation of the photodiode matrix. When comparing a square $M \times M$ photodiode array to a single photodiode, it can be observed that the active detection area is M^2 times larger, the bandwidth is the same, and the output signal is M times larger.

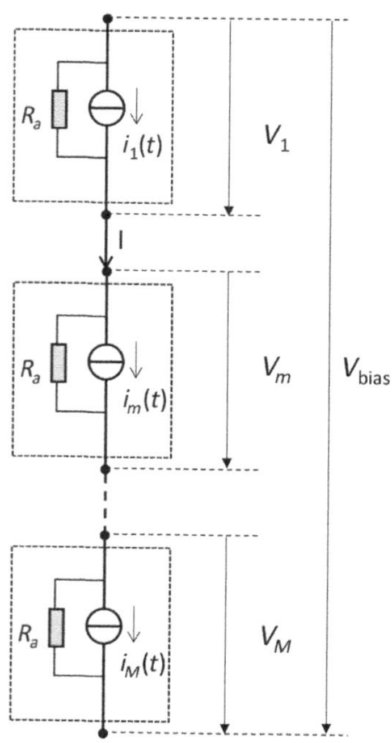

Fig. 7-5 DC bias conditions when cascading photodiodes

7.2.2 Bias Conditions in the Photodiode Matrix

The DC reverse-biasing voltages across the photodiodes in the matrix may get unbalanced when the photodiodes are unevenly illuminated, and this may impact the performance characteristics of the photodiodes. As shown in Fig. 7-3b, the grouping of the photodiodes in the matrix consists of K columns and M rows. Fig. 7-5 illustrates how the overall bias voltage V_{bias} across the total matrix is built up by the bias voltages V_m across the mth row (with $m = 1, 2, \ldots, M$). As each row transports the same current I, for the individual photocurrents i_m holds

$$i_1 + \frac{V_1}{R_a} = i_2 + \frac{V_2}{R_a} = \ldots = i_m + \frac{V_m}{R_a} = \ldots = i_M + \frac{V_M}{R_a} = I \tag{7-14}$$

where $R_a = \left(\frac{R_d + R_s}{K}\right) \| R_p$

Thus,

$$V_{\text{bias}} = \sum_{m=1}^{M} V_m = MR_a I - R_a \sum_{m=1}^{M} i_m. \tag{7-15}$$

$$V_m = R_a\,(I - i_m) = \frac{V_{\text{bias}}}{M} - R_a\left(i_m - \frac{1}{M}\sum_{m=1}^{M} i_m\right) = \frac{V_{\text{bias}}}{M} + R_a\left(\overline{i_m} - i_m\right) \tag{7-16}$$

where $\overline{i_m} = \frac{1}{M}\sum_{m=1}^{M} i_m$ is the average of the photocurrents in the m rows.

The photocurrent i_m is composed of the contributions of each of the K photodiodes in that row, so when assuming identical photodiodes:

$$i_m = \sum_{k=1}^{K} i_{mk} = \frac{R_d}{R_d + R_s}\left\{K i_d + \mathcal{R} P_0 \sum_{k=1}^{K} a_{mk}\right\} \tag{7-17}$$

$$V_m - \frac{V_{\text{bias}}}{M} \approx \left(R_d \,\|\, (K R_p)\right) \mathcal{R} P_0 \left(\overline{a_{mk}} - \frac{1}{K}\sum_{k=1}^{K} a_{mk}\right) \tag{7-18}$$

where $\overline{a_{mk}} = \frac{1}{MK}\sum_{m=1}^{M}\sum_{k=1}^{K} a_{mk}$ is the average of the fraction a_{mk} of the optical beam's power P_0 which is captured by the $(m, k)^{\text{th}}$ photodiode of the $M \times K$ matrix. From Eq. (7-18), it can be seen that the unbalance in the DC bias voltages V_m across the photodiode rows can be effectively reduced by applying a parallel resistance R_p which is considerably smaller than R_d/K (note that the leakage resistance R_d of a photodiode is typically tens of MΩ).

In order to analyze the impact of uneven illumination of the photodiode matrix, the fractions a_{mk} have been calculated by numerical integration of the intensity profile of the beam when the beam is being shifted diagonally across the photodiode matrix. The active area of each photodiode is Ø150 µm, and they are spaced by 17 µm (cf. Fig. 7-10).[2] The beam is captured and projected onto the photodiode matrix by a lens system as shown in Fig. 7-9 in Section 7.3.1.

2 These physical dimensions are typical data for 2.5 Gbit/s photodetectors (cf. Albis Optoelectronics [7-12])

First, the analysis has been done for a typical upstream transmission scenario, wherein an upstream Gaussian beam is assumed with beam spot waist diameter of 15 mm and a lens with $f = 5$ mm focal length and aperture diameter $D_{\text{lens}} = 25$ mm, see Chapter 8. Assuming an ideal thin lens that makes a defocused image of the beam onto the matrix (see Section 7.3), Figs. 7-6a and b illustrate the received power per photodiode for a beam incidence angle α of 15° and defocusing parameter $p = 0.2$ (where $p = x/f$, with $f - x$ being the PD matrix-to-lens distance; see Section 7.3.1). The finite lens diameter is cropping the beam image to a diameter of $p\,D_{\text{lens}}$, so to Ø5 mm. As Fig. 7-6c shows, the equivalent beam-to-matrix power coupling factor T_{eq} for $\alpha = 15°$ is about −21.3 dB[3]. The DC bias voltages across the mth row ($m = 1, 2, \ldots, 4$) calculated with Eq. (7-18) versus the incidence angle α are shown in Fig. 7-6d. When the photodiodes are evenly illuminated, their bias voltages are balanced at an equilibrium of $V_m = V_{\text{bias}}/M = 2.5$ V (for $m = 1, 2, \ldots, 4$). At uneven illumination when shifting the beam across the PD matrix, the bias voltages deviate from this equilibrium by +0.71 to −0.64 V for a resistance effectively in parallel of every photodiode $R_{\text{eff}} = KR_p//R_d = KR_a = 1$ MΩ. With $R_{\text{eff}} = 50$ kΩ, these deviations range only from +36 to −32 mV. Clearly, the bias voltage unbalances are effectively reduced by introducing a sufficiently small (external) parallel resistance R_p.

Second, a similar numerical analysis has been done for a downstream transmission scenario where typically the Gaussian incident beam spot has a waist that is larger than the lens' aperture. The results are shown in Fig. 7-7 when shifting the beam across the PD matrix, with a beam spot waist of 100 mm, and a thin lens with $f = 10$ mm and aperture diameter $D_{\text{lens}} = 50$ mm. The cropping of the beam's spot by the limited lens' aperture results in a flatter intensity profile of its image on the PD matrix than for the upstream case, yielding a slower decrease of T_{eq} versus α up to $\alpha \approx 7°$ (and a steep decline thereafter). The beam-to-matrix power coupling factor T_{eq} for $\alpha = 5°$ is about −22.5 dB, with a defocusing parameter $p = 0.05$ (i.e., cropped beam image of Ø2.5 mm). In a typical downstream scenario with beam power $P_0 = 10$ dBm, with this ideal thin lens, an α even up to 25° would allow GbE transmission (at a typical GbE receiver sensitivity of −20 dBm). As shown in Fig. 7-7d, when shifting the beam the bias voltage deviations from the $V_m = 2.5$ V equilibrium state range from +0.55 to −0.46 V for an effective parallel resistance $R_{\text{eff}} = KR_p//R_d = 1$ MΩ. With $R_{\text{eff}} = 50$ kΩ, the deviations range only from +27 to −23 mV. Hence, also for this downstream scenario, the effectiveness of reducing the parallel resistance R_p in order to reduce the voltage unbalances is clear.

On the other hand, the resistance R_p should remain large enough in order not to affect the frequency characteristics of the transimpedance receiver. If R_p is much larger than $R_t/(A + 1)$, the bandwidth nor the amplitude characteristics of the receiver are noticeably affected (see Eqs. (7-10) and (7-11)).

3 For upstream communication, a typical beam power $P_0 = 2$ dBm and receiver sensitivity of −20 dBm for GbE transmission is assumed.

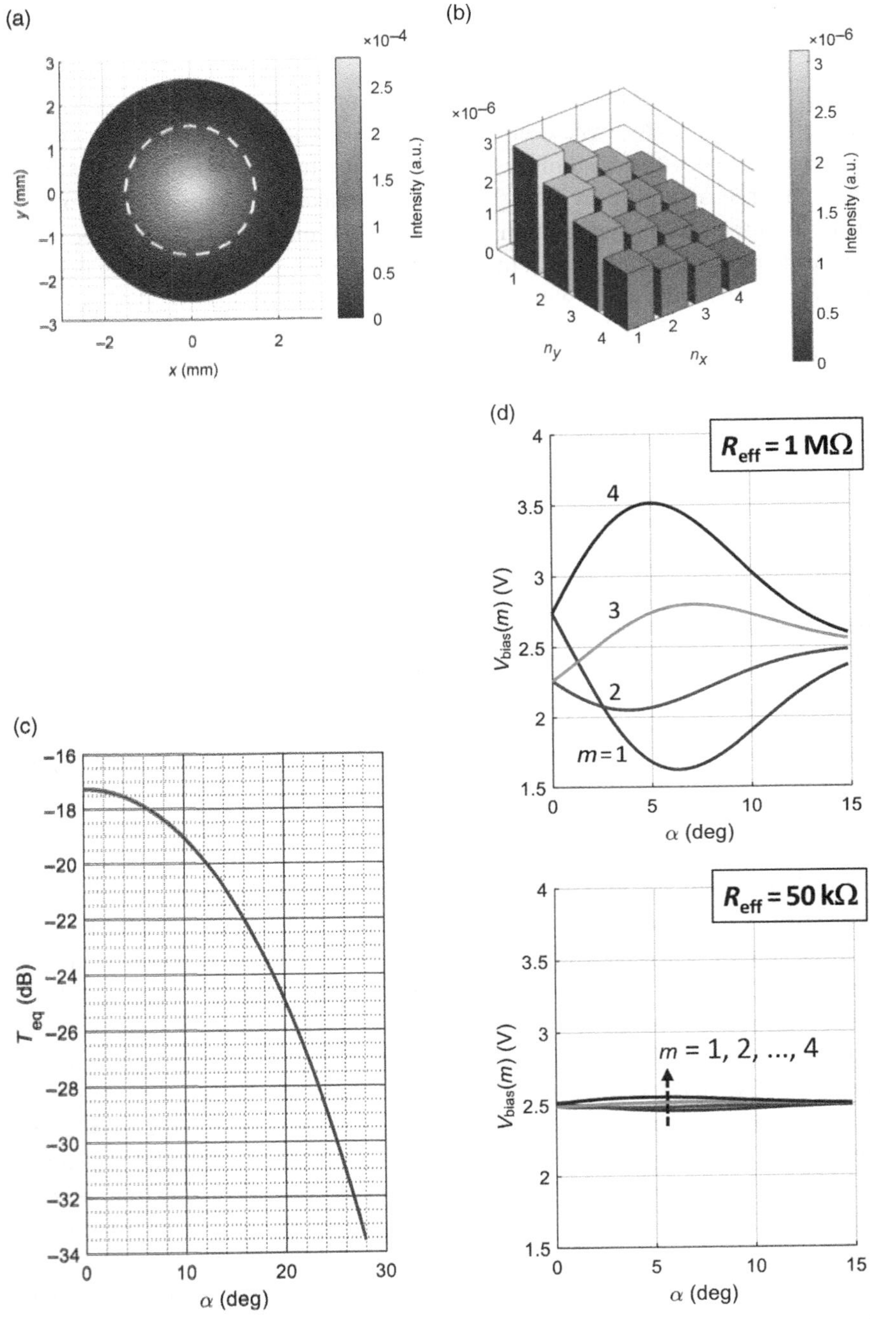

Fig. 7-6 DC bias conditions per photodiode in 4 × 4 PD matrix when beam incidence angle α is varied (for typical upstream scenario with Gaussian beam, spot waist D_{beam} = Ø15 mm, and ideal thin lens with aperture D_{lens} = Ø25 mm, f = 5 mm; bias voltage V_{bias} = 10 V). (a) Gaussian beam image onto PD matrix at α = 15° (defocusing p = 0.2; dashed circle: imaged beam waist). (b) Received power per PD at α = 15° (at p = 0.2). (c) Equivalent coupling loss T_{eq} of beam to PD matrix versus α. (d) DC bias voltage per row m of the PD matrix versus α, for R_{eff} = 1 MΩ and R_{eff} = 50 kΩ (at p = 0.2)

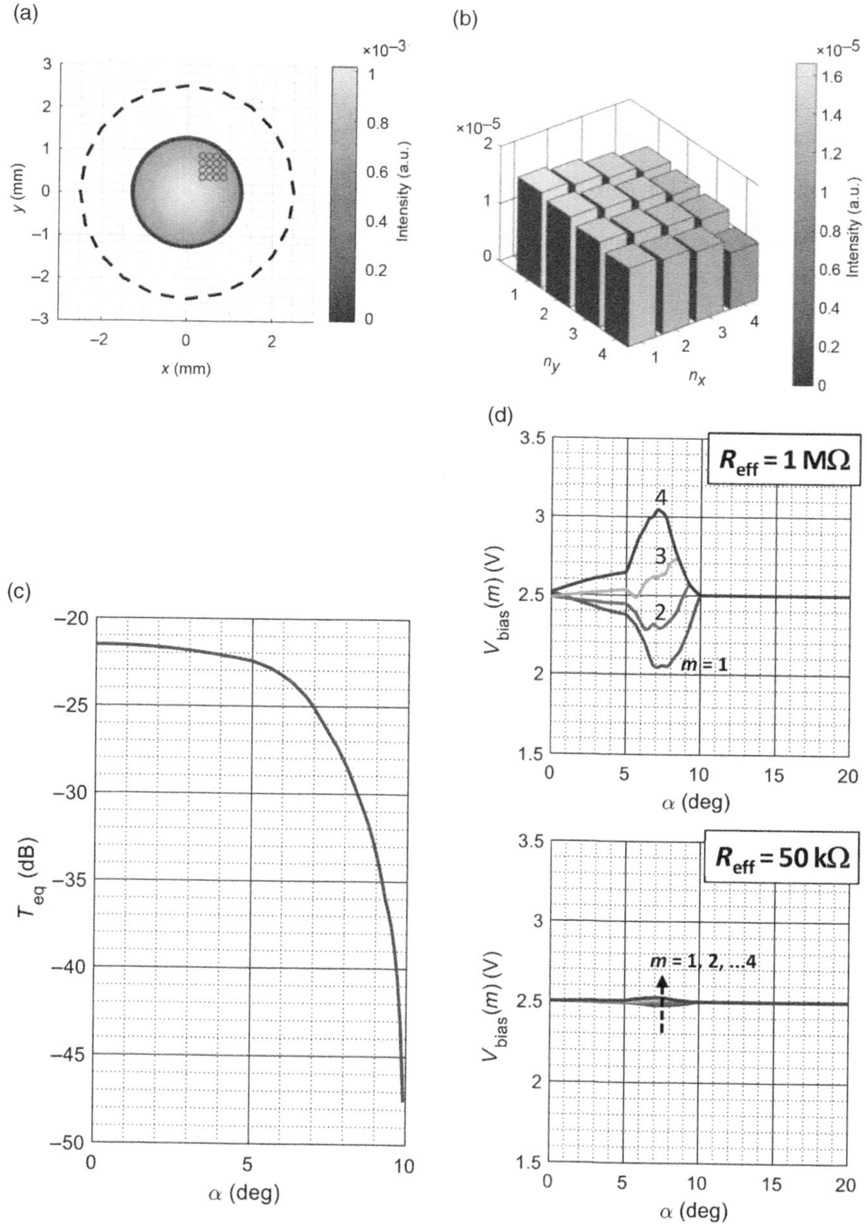

Fig. 7-7 DC bias conditions per photodiode in 4 × 4 PD matrix when beam incidence angle α is varied (for typical downstream scenario with Gaussian beam, spot waist D_{beam} = Ø100 mm; ideal thin lens with aperture D_{lens} = Ø50 mm, f = 10 mm; bias voltage V_{bias} = 10 V). (a) Gaussian beam image onto PD matrix at α = 5° (*dashed circle*: imaged beam waist; defocusing p = 0.05). (b) Received power per PD at α = 5° (at p = 0.05). (c) Equivalent coupling loss T_{eq} of beam to PD matrix versus α. (d) DC bias voltage per row m of the PD matrix versus α, for R_{eff} = 1 MΩ and R_{eff} = 50 kΩ (at p = 0.05)

7.2.3 SNR Performance

The signal-to-noise power ratio (SNR) performance of the OWC receiver using an $M \times K$ photodiode matrix has been analyzed in the following. Representing the beam's intensity-modulated instantaneous power $P_0(t)$ by

$$P_0(t) = \overline{P_0}\,(1 + m_0(t)) \text{ with modulation index } |m_0(t)| \leq 1 \tag{7-19}$$

the average generated signal power $\overline{i_{\text{tot}}^2}$ according to Eq. (7-7) when neglecting the dark current i_{d} is

$$\overline{i_{\text{tot}}^2(t)} = \overline{P^2(t)}\left(\frac{Z_d}{Z_d + R_s}\,K\mathcal{R}\overline{a_{mk}}\right)^2 = \left|\frac{Z_d}{Z_d + R_s}\right|^2 \overline{m_0^2}\left(\frac{A\mathcal{R}\overline{P_0}}{M}\right)^2 \tag{7-20}$$

where $A = KM\overline{a_{mk}}$ is the fraction of the average beam's power $\overline{P_0}$ which is captured by all $M \times K$ photodiodes jointly.

The shot noise variance of the photocurrent $i_{mk}(t)$ generated in photodiode (m, k) in a bandwidth B is

$$\overline{\sigma_{i,mk}^2} = 2qI_{mk}B = 2qB\mathcal{R}\overline{P_0}\,a_{mk} \tag{7-21}$$

where q is the charge of an electron, I_{mk} is the DC component of $i_{mk}(t)$, and B is the receiver's bandwidth. Hence, the shot noise variance of the photovoltage $e_{mk}(t)$ is

$$\overline{\sigma_{e,mk}^2} = |Z_d|^2\,\overline{\sigma_{i,mk}^2} = |Z_d|^2 2qB\mathcal{R}\overline{P_0}\,a_{mk} \tag{7-22}$$

As the shot noise voltage in each photodiode is statistically independent of the shot noise in the other photodiodes, the shot noise power generated in the kth column of the PD matrix is

$$\overline{\sigma_{i\_\text{tot},k}^2} = \frac{\overline{\sigma_{e\_\text{tot},k}^2}}{M^2\,|Z_d + R_s|^2} = \frac{|Z_d|^2\,2qB\mathcal{R}\overline{P_0}}{M^2\,|Z_d + R_s|^2}\sum_{m=1}^{M} a_{mk} \tag{7-23}$$

and the shot noise power in the output current signal of the PD matrix is

$$\sigma_{i,\text{shot}}^2 = \sum_{k=1}^{K}\overline{\sigma_{i_{\text{tot}},k}^2} = \left|\frac{Z_d}{Z_d + R_s}\right|^2 \frac{2qB\mathcal{R}\overline{P_0}}{M^2}\sum_{k=1}^{K}\sum_{m=1}^{M} a_{mk} = \left|\frac{Z_d}{Z_d + R_s}\right|^2 \frac{2qB\mathcal{R}\overline{P_0}}{M^2}A \tag{7-24}$$

The shot noise-limited SNR of a receiver using the PD matrix therefore is

$$\text{SNR}_{\text{shot, PD matrix}} = \frac{\langle i_{\text{tot}}^2\rangle}{\sigma_{i,\text{shot}}^2} = \frac{\overline{m_0^2}\,A\mathcal{R}\overline{P_0}}{2qB} \tag{7-25}$$

Similarly, when the receiver uses a single PD the shot noise power is

$$\overline{\sigma_{i,\text{shot}}^2} = 2qBI_{\text{PD}} = 2qB\mathcal{R}A\overline{P_0} \tag{7-26}$$

where I_{PD} is the DC component of the photodiode current $i_{PD}(t)$, and the output current signal power is

$$\overline{i_{out}^2} = \left(\mathcal{R} A \overline{P_0}\right)^2 \overline{m_0^2} \tag{7-27}$$

which yields a shot noise-limited SNR using a single PD

$$\text{SNR}_{\text{shot}} = \frac{\overline{i_{out}^2}}{\sigma_{i,\text{shot}}^2} = \frac{\overline{m_0^2}\, A\mathcal{R}\overline{P_0}}{2q\, B} \tag{7-28}$$

Hence, the shot noise-limited SNR of a receiver using the PD matrix is the same as when using a single PD.

Typically, the received optical power is low, and the thermal noise of the TIA electronics (with a white noise power density N_0 and bandwidth B) dominates the shot noise from the photodiodes. The thermal noise-limited SNR is, assuming $R_s << Z_d$,

$$\text{SNR}_{\text{therm}} = \frac{\overline{i_{tot}^2}}{\sigma_{\text{therm}}^2} \cong \frac{\overline{m_0^2}}{N_0 B} \left(K\overline{a_{mk}}\mathcal{R}\overline{P_0}\right)^2 \tag{7-29}$$

Thus, for the $M \times K$ PD matrix receiver, the SNR is a factor K^2 better than for the single-PD receiver (where $K = 1$) if the received optical power per photodiode is the same.

As to be discussed in Section 7.3.1, by introducing defocusing in the lens coupling of the optical beam to the PD matrix or to a single PD, the FoV of the receiver can be enlarged at the expense of reducing the beam's power captured by the PD matrix or the single PD. From Eq. (7-39), it can be derived that (theoretically) the single-PD receiver will have the same FoV as the PD matrix receiver if their defocusing factors p_{PD} for the single-PD receiver and p_{matrix} for the PD matrix receiver satisfy

$$p_{PD} = \frac{p_{\text{matrix}}\left(D_{\text{lens}} - D_{PD}\right) - \left(D_{\text{matrix}} - D_{PD}\right)}{D_{\text{lens}} - D_{\text{matrix}}} \tag{7-30}$$

For example, as considered in Section 7.3.1, for a lens with diameter $D_{\text{lens}} = 50$ mm and focal length $f = 10$ mm, a 4×4 PD matrix with diameter $D_{\text{matrix}} = 859$ µm and fill factor $\eta = 48.8\%$, and a single PD with diameter $D_{PD} = 150$ µm, an FoV $\alpha_{\max} = 11.3°$ is achieved with defocusing $p_{\text{matrix}} = 0.09$ for the PD matrix receiver and $p_{PD} = 0.0769$ for the single-PD receiver. According to Eq. (7-43), these defocusing factors yield an equivalent beam-to-photodode matrix (/photodiode) coupling factor $T_{eq,\max} = -29.5$ and -34.2 dB, respectively, which implies that the $\text{SNR}_{\text{therm}}$ of the PD matrix receiver is 9.3 dB better than of the single-PD receiver.

7.2.4 Comparison with Angular Diversity Receiver

An ADR also deploys multiple photodiodes, but each photodiode is followed by an individual TIA, whereafter the outputs of all TIAs are summed [7-7, 7-8]. When

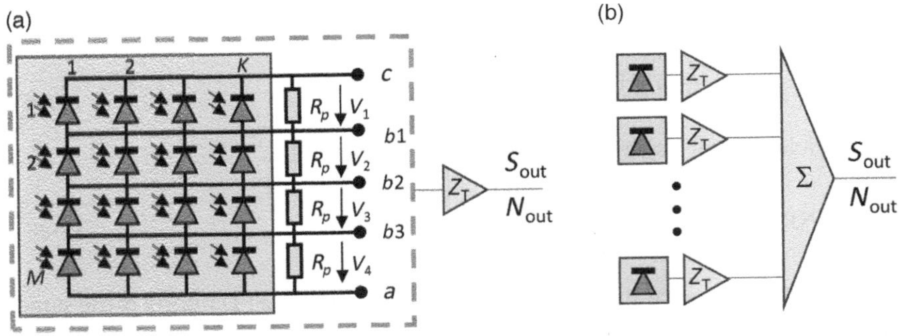

Fig. 7-8 Comparing the PD matrix receiver with an angular diversity receiver (ADR). (a) $N \times N$ PD matrix receiver. (b) ADR with N^2 PDs

compared to an ADR, the proposed PD matrix receiver has the same bandwidth and a better SNR as is shown in the following. Moreover, particularly at high frequencies, the TIAs of the ADR need to be phase-synchronous, which complicates their design.

As shown in Fig. 7-8, whereas an $N \times N$ PD matrix receiver needs only a single TIA, an ADR needs N^2 TIAs including a summation amplifier, and hence is less compact and consumes more power.

Assuming that thermal receiver noise dominates, we find the signal power at the output of the $N \times N$ PD matrix receiver

$$S_{\text{out}} = \overline{m_0^2}\left(N\overline{a_{mk}}\mathcal{R}P_o|Z_T|\right)^2 \tag{7-31}$$

where $\overline{a_{mk}}$ denotes the average fraction of the beam's power which is captured by a photodiode (m, k) in the matrix. For the noise power we find

$$N_{\text{out}} = F_1 N_0 B|Z_{\text{T}}|^2 \tag{7-32}$$

where the TIA has a transimpedance Z_{T}, input noise current power $\overline{i_{th}^2} = N_0 B$, and noise figure F_1. The SNR at the output of the PD matrix receiver is

$$\left(\frac{S_{\text{out}}}{N_{\text{out}}}\right)_{\text{PD matrix}} = \frac{\overline{m_0^2}\left(N\overline{a_{mk}}RP_o\right)^2}{F_1 N_0 B} \tag{7-33}$$

For the ADR, the signal power after the summation amplifier (with power gain G_2 and noise figure F_2) is

$$S_{\text{out}} = G_2\overline{m_0^2}\left(\sum_{n=1}^{N^2} a_n \mathcal{R}P_o|Z_{\text{T}}|\right)^2 = G_2\overline{m_0^2}\left(N^2\overline{a_n}RP_o|Z_{\text{T}}|\right)^2 \tag{7-34}$$

where $\overline{a_n}$ is the average fraction of the beam's power which is captured by photodiode n (with n = 1, 2, ..., N^2) in the ADR constellation. The thermal noise contributions of the TIAs are uncorrelated, and thus add incoherently, yielding a noise power after the summation amplifier

$$N_{\text{out}} = G_2F_2\left(\sum_{n=1}^{N^2} N_0B|Z_{\text{T}}|^2F_1\right) = F_2G_2N^2F_1N_0B|Z_{\text{T}}|^2 \tag{7-35}$$

The SNR at the output of the ADR is

$$\left(\frac{S_{\text{out}}}{N_{\text{out}}}\right)_{\text{ADR}} = \frac{G_2\overline{m_0^2}\left(N^2\overline{a_n}\mathcal{R}P_o|Z_T|\right)^2}{F_2G_2N^2F_1N_0B|Z_T|^2} = \frac{\overline{m_0^2}(N\overline{a_n}\mathcal{R}P_o)^2}{F_1F_2N_0B} \tag{7-36}$$

(note that the amplifier noise factors F_1 and F_2 are always larger than 1).

Thus, from Eqs. (7-33) and (7-36), it can be seen that the output SNR of the ADR is worse (by a factor F_2) than that of the PD matrix receiver.

7.3 Capturing the Optical Beam by the Photodiode Matrix

In order to create a large aperture and thus collect as much power of the incident optical beam onto the photodiode matrix as possible, an optical lens is used. To minimize the unbalance in the bias voltages of the photodiodes, the photodiodes should be illuminated as evenly as possible and therefore the spot projected onto the matrix should preferably cover the whole matrix. As shown in Fig. 7-9, this may be achieved by putting the matrix not in the focal plane of the lens, but slightly closer to it, so defocused over a distance x. This diminishes the captured power but increases the FoV angle (i.e., the maximum beam incidence angle α_{max} at which the spot still covers the whole matrix).

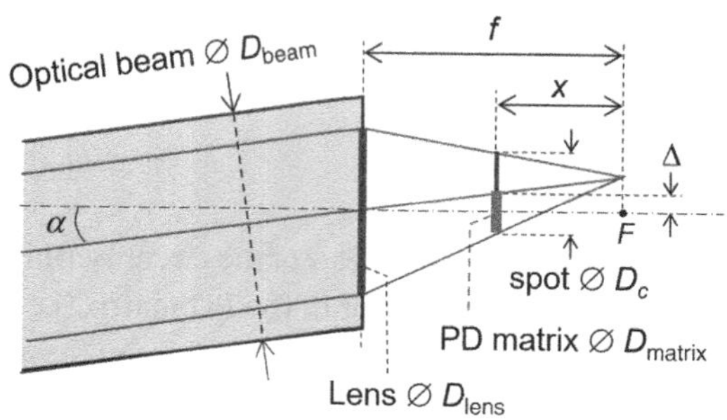

Fig. 7-9 Capturing the optical beam by the photodiode matrix in defocused position relative to the lens (x: defocusing w.r.t. focal plane, Δ: displacement of spot w.r.t. center of PD matrix)

7.3.1 Theoretical Analysis of the Beam-to-Photodiode Matrix Coupling

Fig. 7-9 illustrates how the optical beam is imaged by a lens system and captured by the photodiode matrix put in a defocused position with respect to the back focal plane of the lens (with focal point F). When assuming a beam with uniformly distributed intensity and diameter D_{beam}, and an aberration-free thin lens with diameter D_{lens} and focal length f, the diameter D_c of the beam's spot projected onto the matrix if the matrix is positioned with a defocusing factor $p = x/f$ (with $p < 1$) is

$$D_c = |p| \min(D_{\text{beam}}, D_{\text{lens}}) \tag{7-37}$$

When increasing the incidence angle α of the beam, the spot is moved away from the center of the photodiode matrix which has been put on the axis of the lens. The displacement of the spot Δ relative to the center of the PD matrix is

$$\Delta = (f - x) \tan \alpha = f(1 - p) \tan \alpha \tag{7-38}$$

The coupling of the beam to the PD matrix decreases as soon as the spot is no longer fully overlapping the PD matrix area. By defining the FoV $\alpha_{\max}$ as the maximum incidence angle of the beam onto the lens for which the spot still fully overlaps the PD matrix area, that is, for maintaining maximum coupling of the beam to the PD matrix, the FoV half angle $\alpha_{\max}$ is given by

$$\tan \alpha_{\max} = \frac{|D_c - D_{\text{matrix}}|}{2f(1-p)} = \frac{\big|\,|p| \min(D_{\text{lens}}, D_{\text{beam}}) - D_{\text{matrix}}\big|}{2f(1-p)} \tag{7-39}$$

Therefore, a lens with a small f is preferred in order to obtain a large FoV.

It may be noted that putting the matrix beyond the focal plane, further away from the lens, that is, at $p < 0$, will also increase D_c but will yield a smaller FoV. Hence, $0 \le p < 1$ is the preferred defocusing range.

Typically, $D_{\text{beam}} \ge D_{\text{lens}}$. The maximum coupling fraction $T_{\max}$ of the beam's power P_{beam} onto all photodiodes of the matrix when $\alpha \le \alpha_{\max}$ is then

$$\begin{aligned} T_{\max} &= \frac{P_{\text{matrix}}}{P_{\text{beam}}} \\ &= \eta_{\text{matrix}} \left(\frac{D_{\text{matrix}}}{D_c}\right)^2 \left(\frac{D_{\text{lens}}}{D_{\text{beam}}}\right)^2 \cos\alpha = \eta_{\text{matrix}} \left(\frac{D_{\text{matrix}}}{p D_{\text{beam}}}\right)^2 \cos\alpha \quad \text{for } D_{\text{matrix}}/D_{\text{lens}} < p < 1 \\ &= \eta_{\text{matrix}} \left(\frac{D_{\text{lens}}}{D_{\text{beam}}}\right)^2 \cos\alpha \qquad \text{for} \quad 0 \le p \le D_{\text{matrix}}/D_{\text{lens}} \end{aligned} \tag{7-40}$$

where D_{matrix} is the diameter of the circumscribing circle of the photodiode matrix, and η_{matrix} is the fill factor of the PD matrix (i.e., the sum of the active areas of all PDs divided by the total area of the circle). Assuming an $N \times N$ PD matrix where each photodiode has an active area with diameter D_0 and the spacing between the photodiodes is d, the diameter D_{matrix} of the matrix is (see Fig. 7-10)

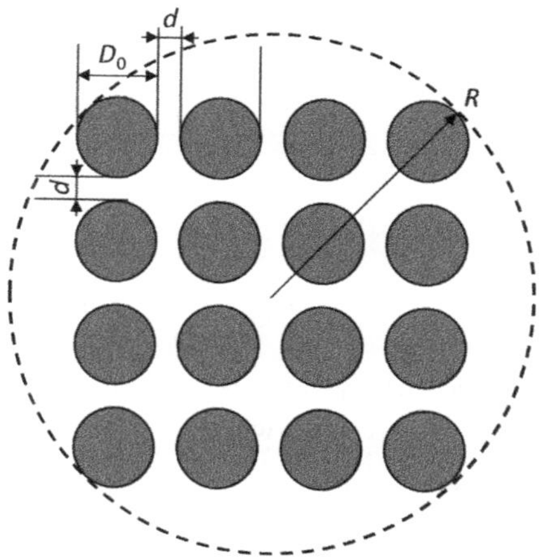

Fig. 7-10 $N \times N$ photodiode matrix (diameter PD active area D_0, distance between PDs d, diameter of circumscribing circle of matrix $D_{matrix} = 2R$)

$$D_{matrix} = 2R = \sqrt{2}(N-1)(D_0 + d) + D_0 \tag{7-41}$$

and the fill factor η_{matrix} is

$$\eta_{matrix} = \left(\frac{ND_0}{D_{matrix}}\right)^2 = \left(\frac{N}{1 + \sqrt{2}\,(N-1)\,(1 + d/D_0)}\right)^2 \tag{7-42}$$

Typical data for commercially available 1×4 arrays of 2.5 Gbit/s photodiodes [7-12] are $D_0 = 150$ µm and spacing which can be reduced to $d = 17$ µm, yielding a matrix diameter $D_{matrix} = 859$ µm and fill factor $\eta_{matrix} = 48.8\%$. This implies an additional contribution of 3.11 dB to the beam-to-PD matrix coupling loss when the matrix is uniformly illuminated.

By correcting for the current gain factor $1/N$ of the $N \times N$ photodiode matrix (see Eq. (7-6)), an equivalent beam-to-photodiode matrix maximum power coupling factor $T_{eq,max}$ if $D_{beam} \geq D_{lens}$ can be defined as

$$\begin{aligned} T_{eq,\,max} &= T_{max}/N = \frac{\eta_{matrix}}{N}\left(\frac{D_{matrix}}{p\,D_{beam}}\right)^2 \cos\alpha \quad \text{for } D_{matrix}/D_{lens} < p < 1 \\ &= \frac{\eta_{matrix}}{N}\left(\frac{D_{lens}}{D_{beam}}\right)^2 \cos\alpha \quad \text{for } 0 \leq p \leq D_{matrix}/D_{lens} \end{aligned} \tag{7-43}$$

If $D_{beam} < D_{lens}$, the maximum coupling fraction T_{max} when $\alpha \leq \alpha_{max}$ is

$$\begin{aligned} T_{max} &= \eta_{matrix}\left(\frac{D_{matrix}}{D_c}\right)^2 \cos\alpha = \eta\left(\frac{D_{matrix}}{pD_{beam}}\right)^2 \cos\alpha \quad \text{for } D_{matrix}/D_{beam} < p < 1 \\ &= \eta_{matrix} \quad \text{for } 0 \leq p \leq D_{matrix}/D_{beam} \end{aligned} \tag{7-44}$$

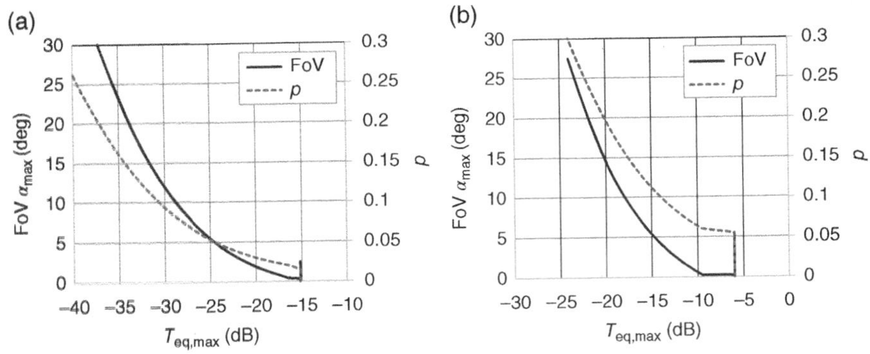

Fig. 7-11 Field-of-view α_{max} versus beam-to-photodiode matrix power coupling factor $T_{eq,max}$ with 4 × 4 PD matrix (single PD D_0 = 150 μm, PD pitch $D_0 + d$ = 167 μm), when defocusing p is varied from 0 to 30%, for (a) uniform beam D_{beam} = 100 mm, thin lens D_{lens} = 50 mm, f = 10 mm (*downstream*); (b) uniform beam D_{beam} = 15 mm, thin lens D_{lens} = 25 mm, f = 5 mm (*upstream*)

and the equivalent beam-to-photodiode matrix maximum power coupling factor $T_{eq,max}$ when corrected for the PD matrix current gain factor

$$\begin{aligned} T_{eq,max} &= \frac{\eta_{matrix}}{N}\left(\frac{D_{matrix}}{p\,D_{beam}}\right)^2 \cos\alpha \quad \text{for } D_{matrix}/D_{beam} < p < 1 \\ &= \frac{\eta_{matrix}}{N} \quad \text{for } 0 \le p \le D_{matrix}/D_{beam} \end{aligned} \tag{7-45}$$

Thus, the FoV of the receiver can be increased by defocusing the PD matrix, but at the expense of reducing the beam-to-PD matrix power coupling $T_{eq,max}$. As an example, for a uniform beam and an ideal thin lens, and a 4 × 4 matrix of Ø150 μm PIN photodiodes arranged at a pitch of 167 μm, by applying Eqs. (7-39), (7-43), and (7-45), Fig. 7-11 shows the compromise between FoV and $T_{eq,max}$ which can be set by adjusting the defocusing parameter p. In practice, the beam is not uniform and the lens is not ideal, hence for setting this compromise a numerical analysis is needed by means of ray tracing.

7.3.2 Numerical Analysis of the Beam-to-Photodiode Matrix Coupling

Typically, the beam diameter fills a significant part of (or is even considerably larger than) the lens diameter; hence, lens aberrations will affect the imaging of the beam onto the photodiode matrix. Also, the beam does not have an ideal uniform intensity profile but typically may be assumed to have a Gaussian profile. We have developed a 3D ray tracing program in MATLAB, based on vector analysis according to the modelling described in the Appendix. With this program, we have assessed for a non-ideal lens and a non-uniform collimated beam how the

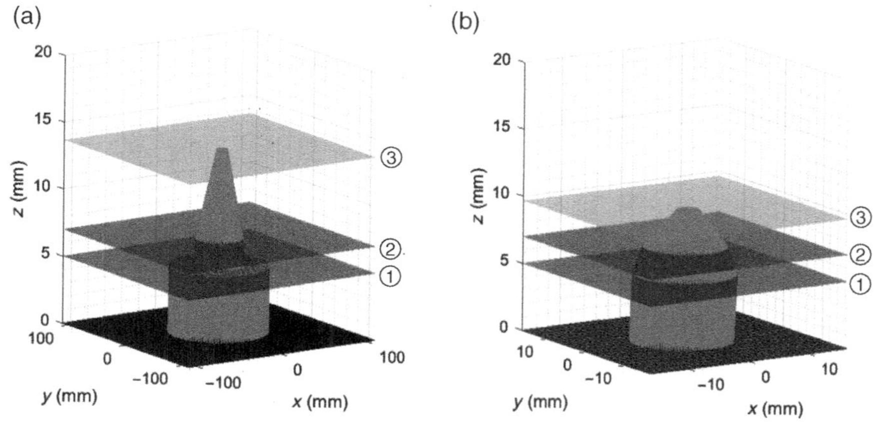

Fig. 7-12 3D ray tracing for analyzing beam-to-PD matrix coupling (with 1027 rays, at defocusing $p = 0.2$, and collimated beam with incidence angle $\alpha = 5°$; lens with Fresnel plane ① and flat plane ②; PD matrix plane ③). (a) Downstream: Gaussian beam D_{beam} = Ø100 mm, Fresnel lens D_{lens} = Ø50 mm, $f = 10$ mm, thickness 2 mm. (b) Upstream: Gaussian beam D_{beam} = Ø15 mm, Fresnel lens D_{lens} = Ø25 mm, $f = 5$ mm, thickness 2 mm

equivalent beam-to-PD matrix power coupling factor T_{eq} depends on the beam incidence angle α when the defocusing p is varied.

Figs. 7-12 and 7-13 (calculated by ray tracing with 1027 rays for the downstream case) show how the collimated beam is projected onto the photodiode matrix if the beam has a uniform profile or a Gaussian profile. With a Fresnel lens [7-13], a large light-collecting aperture D in combination with a short focal length f can be realized. Its high D/f number enlarges the light-collecting power of the OWC receiver. Its imaging properties are less good than those of well-corrected spherical lenses, but this is of less importance at the receiver site. When using an ideal thin lens (which has no lens aberrations) and a Fresnel lens, both with D_{lens} = 50 mm and $f = 10$ mm, Fig. 7-13 visualizes the impact of the lens aberrations introduced by the Fresnel lens for a defocusing $p = 20\%$.

By high-resolution ray tracing with 25,117 rays, we have analyzed the realistic case of a Gaussian beam with beam waist diameter D_{beam} = 100 mm and a Fresnel large aperture lens (optimized for minimum spherical aberration, and modeled for ray tracing according to the algorithms described in Appendix Section A.3) with effective focal length f = 10 mm, thickness d = 2 mm, and diameter D_{lens} = Ø50 mm, as used in our laboratory experimental setup. We compared it with the theoretical case of a uniform beam with a diameter D_{beam} = Ø100 mm and an ideal aberration-free thin lens with a diameter D_{lens} = Ø50 mm, as analyzed with Eqs. (7-39) and (7-43). Examples of the modeling with 3D ray tracing are shown in Fig. 7-14. We have varied p from 0 to 30%, and derived by the ray tracing how the equivalent beam-to-photodiode power coupling factor T_{eq} (including the

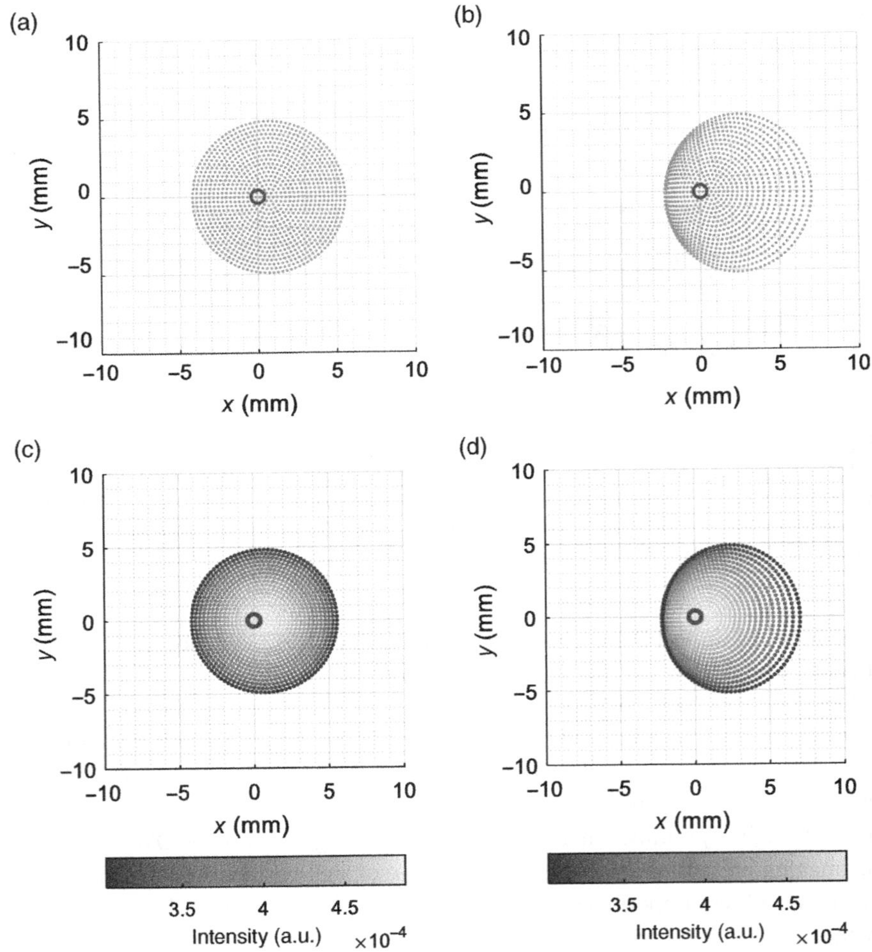

Fig. 7-13 Beam with diameter D_{beam} = Ø100 mm projected onto photodiode matrix with diameter D_{matrix} = Ø859 μm (thick circle) for $\alpha = 5°$ and defocusing $p = 20\%$ (both thin lens and Fresnel lens D_{lens} = Ø50 mm, f = 10 mm; 1027 rays traced). (a) Uniform beam, thin lens. (b) Uniform beam, Fresnel lens. (c) Gaussian beam, thin lens. (d) Gaussian beam, Fresnel lens

PD matrix fill factor η_{matrix}) depends on the beam incidence angle α and defocusing parameter p.

The results in Fig. 7-14 calculated for a 4 × 4 PD matrix show how at a desired coupling factor T_{eq} typically the FoV maximum angle α_{max} can be increased by increasing the defocusing p. Increasing p also brings a decrease of T_{eq} at a given incidence angle α. The peaks that are observed in the T_{eq} versus α curves of Fig. 7-14a for the Fresnel lens and Gaussian beam at larger α and p are caused

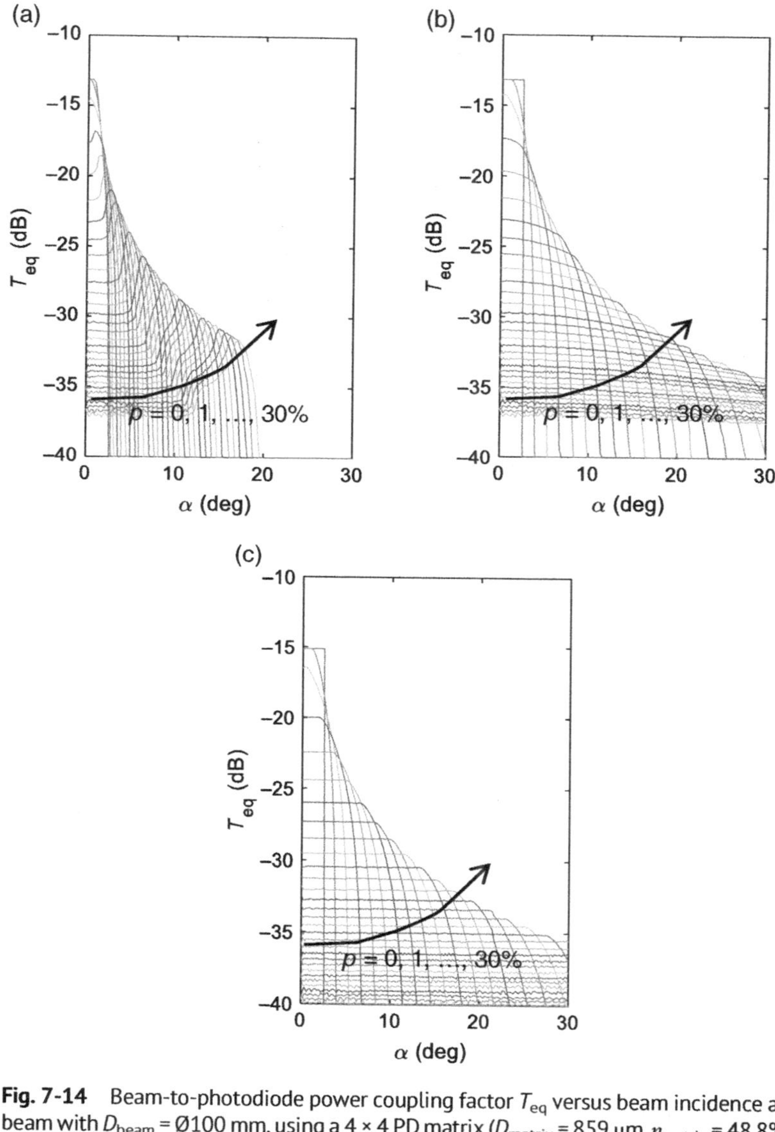

Fig. 7-14 Beam-to-photodiode power coupling factor T_{eq} versus beam incidence angle α of beam with $D_{beam} = Ø100$ mm, using a 4 × 4 PD matrix ($D_{matrix} = 859$ µm, $\eta_{matrix} = 48.8\% = -3.11$ dB, $D_0 = 150$ µm, $d = 17$ µm) and defocusing $p = 0, 1, \ldots, 30\%$. Calculated by ray tracing with 25,117 rays. (a) Gaussian beam $D_{beam} = Ø100$ mm, Fresnel lens $f = 10$ mm, $D_{lens} = Ø50$ mm. (b) Gaussian beam $D_{beam} = Ø100$ mm, ideal thin lens $f = 10$ mm, $D_{lens} = Ø50$ mm. (c) Uniform beam $D_{beam} = Ø100$ mm, ideal thin lens $f = 10$ mm $D_{lens} = Ø50$ mm

by lens aberrations. These peaks are absent for the aberration-free ideal thin lens with a Gaussian beam in Fig. 7-14b and with a uniform beam in Fig. 7-14c.

The FoV α_{max} decreases when the coupling T_{eq} has to be larger (and thus p smaller). In Fig. 7-15, the solid curves show the results from ray tracing (25,117 rays, $p = 0$ up to 30%; results are accurate up to $T_{eq} = -24$ dB) with the Gaussian beam and Fresnel lens for the 4 × 4 PD matrix and for the single PD. These have been calculated for both the typical downstream case (in Fig. 7-15a; beam size $D_{beam} = Ø100$ mm, lens $D_{lens} = Ø50$ mm, $f = 10$ mm) and the upstream case (in Fig. 7-15b; $D_{beam} = Ø15$ mm, $D_{lens} = Ø25$ mm, $f = 5$ mm), as applied in our laboratory experiments. The dashed curves are from the analytically derived Eqs. (7-39), (7-43), and (7-45) which hold for a uniform beam and an ideal thin lens. The results confirm that the FoV α_{max} obtainable with a 4 × 4 photodiode matrix ($D_{matrix} = 0.86$ mm, $\eta_{matrix} = 48.8\%$) is substantially higher than with a single photodiode ($D_0 = 150$ µm).

7.3.3 Impact of the Fill Factor of the Photodiode Matrix

For an $N \times N$ PD matrix, substituting from Eq. (7-42)

$$D_{matrix} = \frac{ND_0}{\sqrt{\eta_{matrix}}} \tag{7-46}$$

into Eq. (7-43) yields if $D_c > D_{matrix}$ (i.e., if the beam spot fully overlaps the PD matrix)

$$T_{eq} = \frac{\eta_{matrix}}{N}\left(\frac{D_{matrix}}{D_c}\right)^2 \cos\alpha = N\left(\frac{D_0}{D_c}\right)^2 \cos\alpha = N\left(\frac{D_0}{pD_a}\right)^2 \cos\alpha \tag{7-47}$$

with $D_a = \min(D_{beam}, D_{lens}) = D_c/p$. From Eq. (7-39) follows

$$\tan\alpha_{max} = \frac{D_0 N}{2f\left(\sqrt{\frac{T_{eq}}{N\cos\alpha}} - \frac{D_0}{D_a}\right)}\left(1 - \sqrt{\frac{T_{eq}}{\eta_{matrix} N\cos\alpha}}\right) \tag{7-48}$$

Hence, for the theoretical case of a uniform beam and ideal thin lens and if the beam spot fully overlaps the PD matrix, it may be concluded that the matrix fill factor η_{matrix} does not impact the equivalent beam-to-photodiode matrix power coupling factor T_{eq}, and hardly impacts the FoV α_{max} when T_{eq} is small.

In the case of a Gaussian beam and ideal thin lens, the small impact of the fill factor η_{matrix}, in particular at small T_{eq} and thus at larger FoV, is confirmed by the results of a numerical integration analysis shown in Fig. 7-16a when the beam is larger than the lens' aperture (the typical downstream case), and Fig. 7-16b when it is smaller (the typical upstream case).

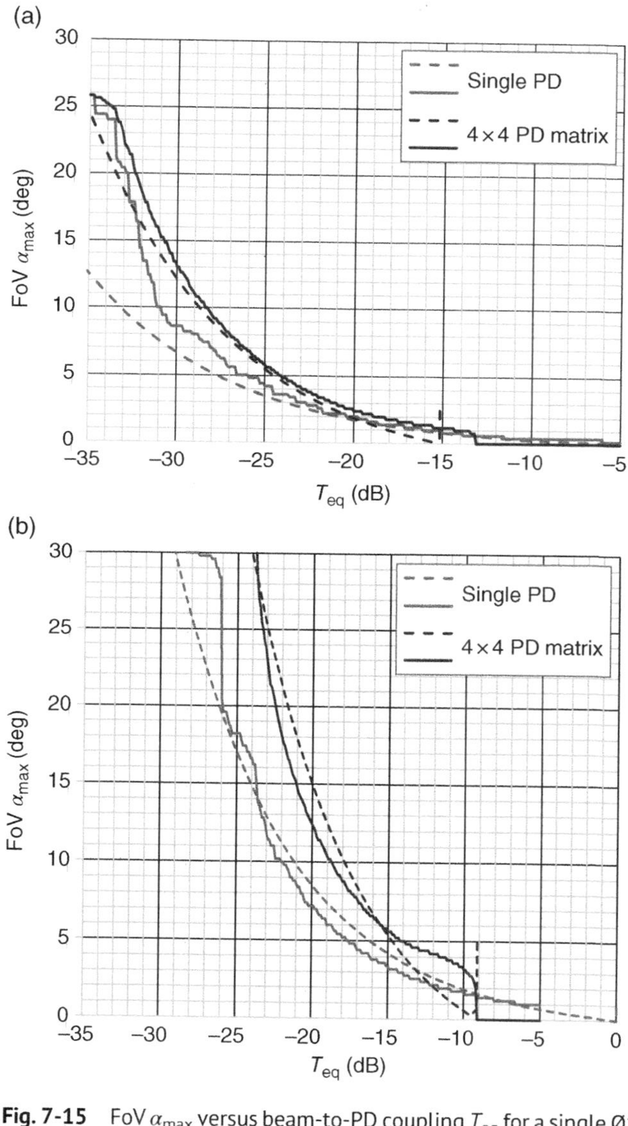

Fig. 7-15 FoV α_{max} versus beam-to-PD coupling T_{eq} for a single Ø150 μm PD and for a 4 × 4 matrix of Ø150 μm PDs spaced at 17 μm; *solid curves*: by ray tracing with 25,117 rays for Gaussian beam and Fresnel lens; *dashed curves*: by theoretical analysis with Eqs. (7-39), (7-43), and (7-45), for uniform beam and ideal thin lens. (a) *Downstream*: beam D_{beam} = Ø100 mm, lens D_{lens} = Ø50 mm with f = 10 mm. (b) *Upstream*: beam D_{beam} = Ø15 mm, lens D_{lens} = Ø25 mm with f = 5 mm

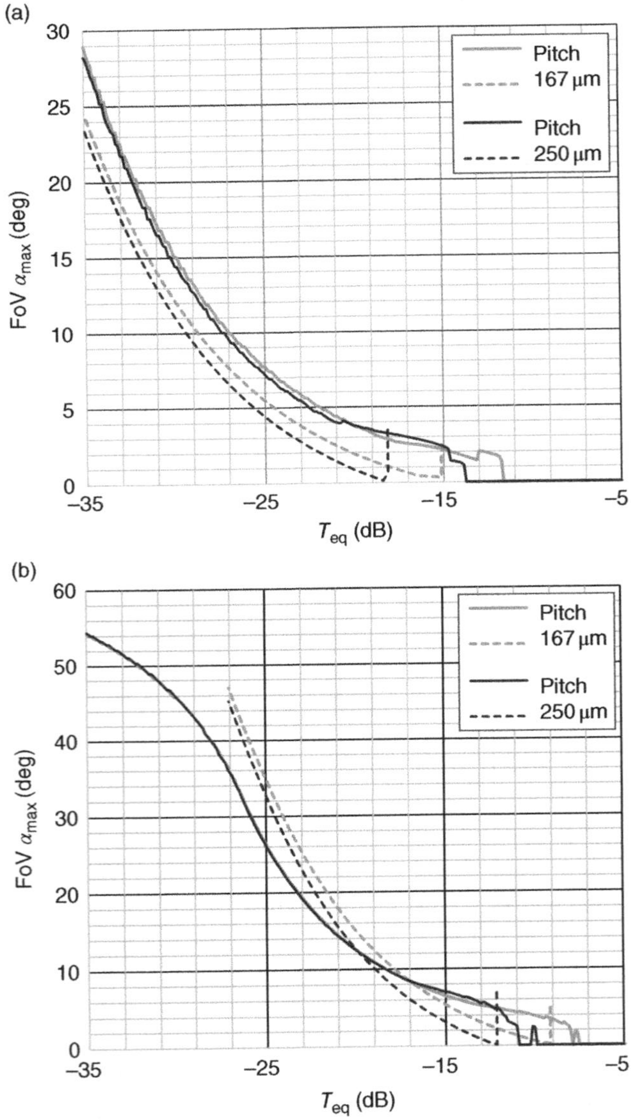

Fig. 7-16 Impact of fill factor η_{matrix} (in 4 × 4 PD matrix, with PD Ø150 μm, PD pitch 250 μm, η_{matrix} = 24.6%, PD pitch 167 μm, η_{matrix} = 48.8%; *solid curves*: by numerical integration, for Gaussian beam; *dashed curves*: by theoretical analysis, for uniform beam). (a) *Downstream*: with Gaussian beam D_{beam} = Ø100 mm; ideal thin lens Ø50 mm, f = 10 mm. (b) *Upstream*: with Gaussian beam D_{beam} = Ø15 mm; ideal thin lens Ø25 mm, f = 5 mm

7.3.4 Shifting the Receiver's Aperture over the Beam's Footprint

In the analysis so far, it has been assumed that the lens aperture of the receiver was positioned in the center of the beam's footprint, that is, the signal's cell. When the aperture is shifted, however, the power captured by the receiver will change, depending on the intensity profile of the beam and the diameter of the aperture. The power will decrease as part of the signal beam is lost. In addition, when the neighboring cells are illuminated by beams as well, crosstalk from those beams into the signal's cell may occur. Moreover, the shape of the beam's spot reaching the photodetector may have been deformed by lens aberrations and beam cropping by the limited lens aperture, which impacts the FoV of the receiver.

7.3.4.1 Impact of Crosstalk from Neighboring Beams

For a uniform beam profile with diameter $D_{\text{beam}} = 2R_{\text{beam}}$, the fraction of the beam's power captured by the receiver aperture with lens diameter $D_{\text{lens}} = 2R_{\text{lens}}$ can be analytically calculated, as illustrated in Fig. 7-17. When the lens center is displaced by x_{AP} from the beam's center, the position x_1 of the cord connecting the intersection points of the beam's circle and the lens' circle is for $R_{\text{beam}} - R_{\text{lens}} < x_{\text{AP}} < R_{\text{beam}} + R_{\text{lens}}$

$$x_1 = \frac{R_{\text{beam}}^2 - R_{\text{lens}}^2 + x_{\text{AP}}^2}{2x_{\text{AP}}} \tag{7-49}$$

and the areas A_1 and A_2 are

$$A_1 = R_{\text{beam}}^2 \left\{ \frac{\pi}{2} - \arcsin\left(\frac{x_1}{R_{\text{beam}}}\right) - \frac{x_1}{R_1}\sqrt{1 - \left(\frac{x_1}{R_{\text{beam}}}\right)^2} \right\} \tag{7-50}$$

$$A_2 = R_{\text{lens}}^2 \left\{ \frac{\pi}{2} - \arcsin\left(\frac{x_{\text{AP}} - x_1}{R_{\text{lens}}}\right) - \left(\frac{x_{\text{AP}} - x_1}{R_{\text{lens}}}\right)\sqrt{1 - \left(\frac{x_{\text{AP}} - x_1}{R_{\text{lens}}}\right)^2} \right\} \tag{7-51}$$

For $x_{\text{AP}} \leq R_{\text{beam}} - R_{\text{lens}}$, there is a full overlap of the lens aperture by the beam, and for $x_{\text{AP}} \geq R_{\text{beam}} + R_{\text{lens}}$, there is no overlap. Using these equations, the fraction of the beam's power captured by the lens, that is, the beam-to-aperture power coupling factor T_{AP} versus the receiver shift x_{AP} for a uniform beam is

$$\begin{aligned} T_{\text{AP}} &= \frac{P_{\text{captured}}}{P_{\text{beam}}} = \frac{A_1 + A_2}{\pi R_{\text{beam}}^2} && \text{for } R_{\text{beam}} - R_{\text{lens}} < x_{\text{AP}} < R_{\text{beam}} - R_{\text{lens}} \\ &= 1 && \text{for } 0 \leq x_{\text{AP}} \leq R_{\text{beam}} - R_{\text{lens}} \\ &= 0 && \text{for } x_{\text{AP}} \geq R_{\text{beam}} - R_{\text{lens}} \end{aligned} \tag{7-52}$$

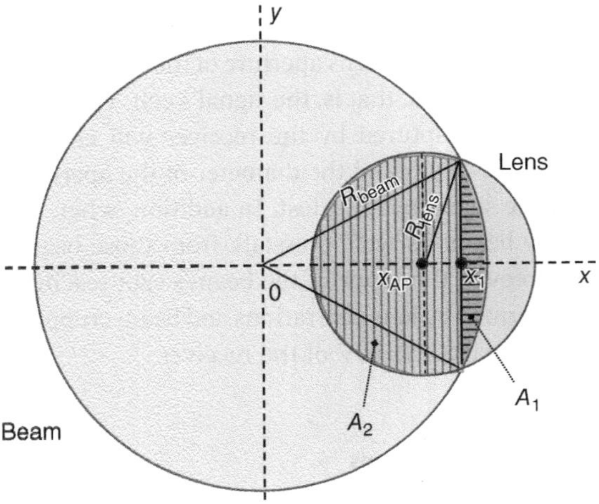

Fig. 7-17 Shifting the receiver's lens aperture with diameter $D_{lens} = 2R_{lens}$ by a distance x_{AP} with respect to the beam's footprint with diameter $D_{beam} = 2R_{beam}$

It may be noted that due to the (x, y) symmetry of the cell pattern, similar equations hold when the receiver moves in the $-x$ direction, or the $+y$ or $-y$ direction.

As illustrated in Fig. 7-18 (see also Fig. 5-22), the receiver may also capture power from neighboring beams if these beams are addressing neighboring cells. Thus, crosstalk from these beams into the signal cell may occur. Assuming that the cells are contiguous (in order to have optimum coverage of the user area) and receive identical power, the crosstalk coupling factor of the power from the

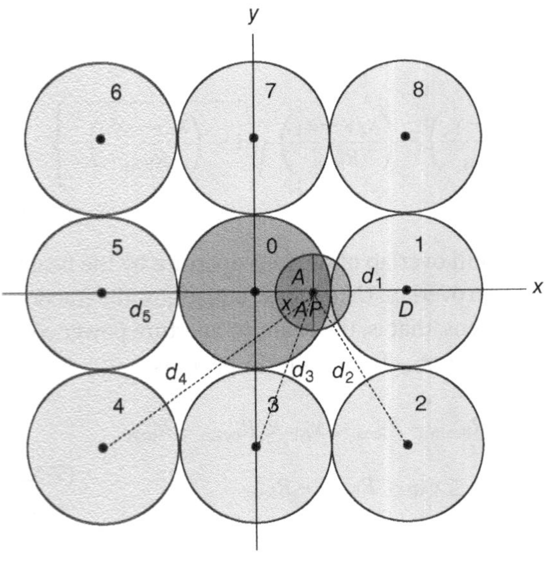

Fig. 7-18 Moving the receiver among multiple cells (cell 0: signal cell; cells 1–8: neighbor cells; circle A: receiver lens aperture)

closest adjacent neighbor beam (addressing cell 1) into the receiver's signal beam (addressing cell 0) is

$$XT_1(x_{AP}, D) = T_{AP}(d_1) = T_{AP}(D - x_{AP}) \tag{7-53}$$

where d_1 is the distance between the center of cell 1 and the center of the signal cell 0, and D is the spacing of the cells. Similarly, the crosstalk coupling factor from the other neighbor beams is given by

$$XT_n(x_{AP}, D) = T_{AP}(d_n) \quad \text{for } n = 2, .., 8 \tag{7-54}$$

where according to Fig. 7-18, the distances d_n (for $n = 2$–5) are

$$\begin{aligned} d_2 &= \sqrt{d_1^2 + D^2} = \sqrt{D^2 + (D - x_{AP})^2} \\ d_3 &= \sqrt{x_{AP}^2 + D^2} \\ d_4 &= \sqrt{(D + x_{AP})^2 + D^2} \\ d_5 &= D + x_{AP} \end{aligned} \tag{7-55}$$

As the cells are contiguous, $D = D_{\text{beam}}$. Due to the symmetry around the x-axis, the shifting of the receiver along the x-axis implies $XT_6 = XT_4$, $XT_7 = XT_6$, and $XT_8 = XT_2$. The optical signal-to-crosstalk power ratio $\text{SXR}_{1,\text{opt}}$ due to the crosstalk factor XT_1 from the closest beam addressing cell 1 is

$$\text{SXR}_{1,\text{opt}}(x_{AP}, D) = \frac{T_{AP}(x_{AP})}{XT_1(x_{AP}, D)} = \frac{T_{AP}(x_{AP})}{T_{AP}(D - x_{AP})} \tag{7-56}$$

The crosstalk from the beams addressing the neighboring cells can be regarded as incoherent interchannel crosstalk contributions into the signal cell, and typically the crosstalk from the beam addressing the closest cell 1 dominates the other crosstalk contributions from beams addressing the cells 2 to 8. The total crosstalk is the sum of all contributions, hence the signal-to-crosstalk power ratio $\text{SXR}_{\text{tot,opt}}$ is given by

$$\text{SXR}_{\text{tot,opt}}(x_{AP}, D) = \frac{T_{AP}(x_{AP})}{\sum_{n=1}^{8} XT_n(d_n)} \tag{7-57}$$

Following the analysis in [7-14], the bit error ratio (BER) when applying the optimum decision threshold in the receiver, assuming equally likely binary symbols "0" and "1" being sent, where P_0 is the power received for a "0" bit and P_1 for a "1" bit, is

$$\text{BER} = Q\left(\frac{\mathcal{R}(P_1 - P_0)}{\sigma_1 + \sigma_0}\right) \tag{7-58}$$

where $Q(.)$ is the tail distribution function of the standard normal probability distribution, $\mathcal{R}$ is the responsivity of the photodiode, and σ_0 and σ_1 are the standard noise deviation for the "0" and the "1" bit, respectively. If the received signal is impaired by crosstalk with a crosstalk coupling ratio ε, and assuming that the noises are independent of the signal power (so no avalanche photodiode (APD) or optical amplifier is used) and the thermal noise of the receiver dominates, the power penalty incurred by single-neighbor crosstalk can be deduced from

$$P_1 - P_0 = P_1' - P_0' \tag{7-59}$$

where P_0' and P_1' represent the received powers in the "0" and "1" bit, respectively, including the crosstalk. Without crosstalk and assuming a perfect extinction ratio, $P_1 = P_s$ and $P_0 = 0$, where P_s is the peak received power per bit. The worst-case crosstalk impact occurs when $P_1' = P_s'$ and $P_0' = \varepsilon P_s'$, where ε is the crosstalk-to-signal ratio (so equivalently for the nth crosstalk contributor $\varepsilon_n = XT_n/T_{\text{AP}}$). Hence, the power penalty PP due to the crosstalk from a single neighbor is (in dB)

$$PP = 10\log\left(\frac{P_s'}{P_s}\right) = -10\log(1-\varepsilon) \tag{7-60}$$

With multiple interferers, their crosstalk power contributions in the worst case may be added. Hence, for M interferers with crosstalk ratios ε_n ($n = 1, 2, \ldots, M$), the power penalty PP is (in dB)

$$PP = 10\log\left(\frac{P_s'}{P_s}\right) = -10\log\left(1 - \sum_{n=1}^{M} \varepsilon_n\right) \tag{7-61}$$

The power penalty incurred due to interchannel crosstalk from a single interferer and from multiple ones (with equal ε_i) is shown in Fig. 7-19. For a crosstalk ratio

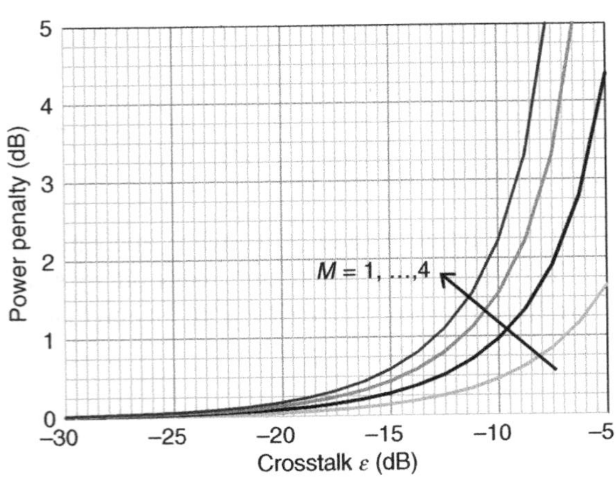

Fig. 7-19 Power penalty impact of interchannel crosstalk versus crosstalk ratio $\varepsilon = 1/\text{SXR}$ per interferer (for $M = 1, \ldots, 4$ interferers with same ε)

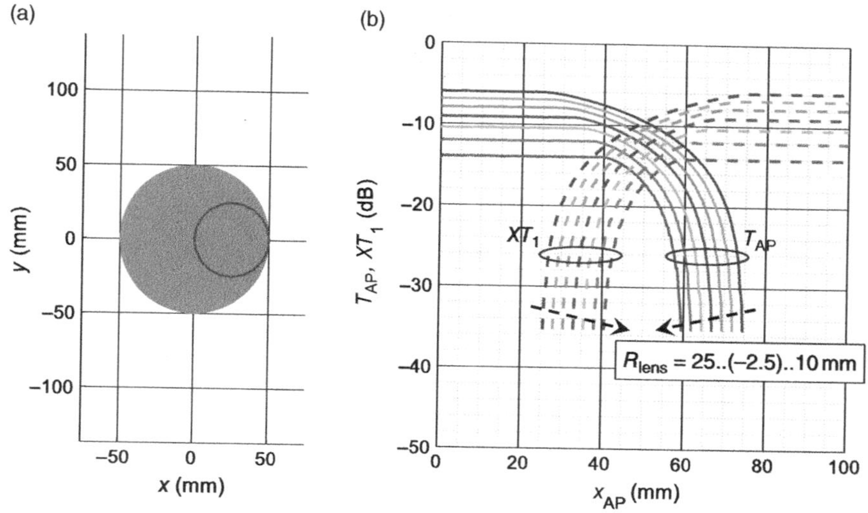

Fig. 7-20 Shifting the receiver's lens aperture with diameter $D_{lens} = 2R_{lens}$ over a distance x_{AP} with respect to the center of the footprint of a uniform beam with beam diameter D_{beam} = Ø100 mm (cell spacing D = 100 mm). (a) R_{lens} = 25 mm, x_{AP} = 25 mm (*circle*: lens aperture). (b) Aperture coupling factor T_{AP} (*solid curves*) and crosstalk coupling factor XT_1 from beam addressing the closest neighbor cell (*dashed curves*) versus shift x_{AP}

$\varepsilon < -10$ dB, the power penalty is less than 0.5 dB with a single interferer, while it increases with multiple interferers, for example, to 2.25 dB for $M = 4$ interferers.

For a uniform beam with a diameter of Ø100 mm, $T_{AP}(x_{AP})$ and XT_1 (x_{AP}, D) have been calculated as a function of the receiver shift x_{AP}. It may be noted that due to the strictly confined beam profile only the closest neighbor beam addressing cell 1 contributes crosstalk and the other beams addressing cells 2–8 do not. The results are shown in Fig. 7-20 for lens apertures varying from $D_{lens} = 20$ to $= 50$ mm. The aperture coupling factor T_{AP} will not change as long as the shift $|x_{AP}| \leq R_{beam} - R_{lens}$, but it will decrease for larger shifts. For smaller lens apertures, less power is captured and T_{AP} is smaller, but it will start to decrease only at larger shifts (i.e., for $|x_{AP}| > R_{beam} - R_{lens}$). The crosstalk factor XT_1 has been calculated for a cell spacing $D = 100$ mm, that is, when the cells are contiguous. As all cells are identical, T_{AP} and XT_1 have opposite behavior versus x_{AP}. The resulting optical signal-to-crosstalk ratio SXR_{opt} versus the shift x_{AP} is shown in Fig. 7-22a. When $x_{AP} = D_{beam}/2 = 50$ mm, that is, when the receiver is at the edge of the signal cell, the receiver gets as much signal as crosstalk from the neighboring interferer, hence, $SXR_{opt} = 0$ dB. For a power penalty <0.5 dB, using a lens diameter D_{lens} = Ø50 mm, a receiver shift $|x_{AP}|$ up to 32 mm is allowed, while for D_{lens} = Ø20 mm, the allowable shift $|x_{AP}|$ increases to 44 mm.

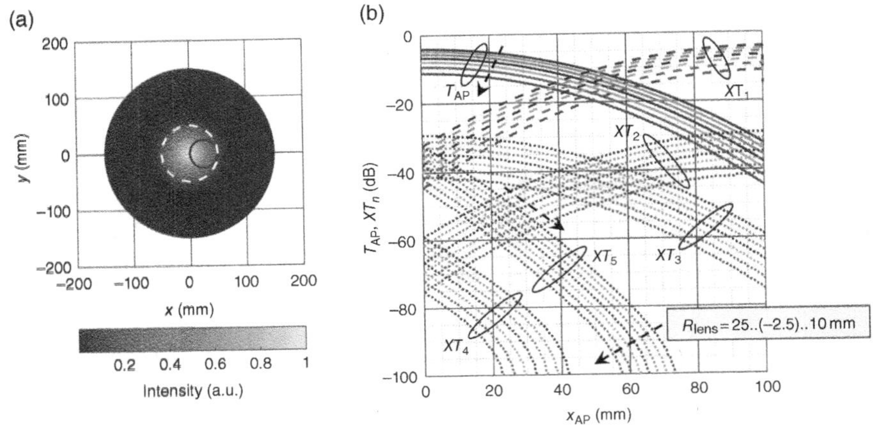

Fig. 7-21 Shifting the receiver's lens aperture with diameter $D_{lens} = 2R_{lens}$ over a distance x_{AP} with respect to the center of the signal cell (no. 0 in Fig. 7-18) when using a Gaussian beam with beam waist diameter D_{beam} = Ø100 mm. (a) Receiver aperture shifted by x_{AP} = 25 mm relative to the beam spot (*solid circle*: lens aperture, R_{lens} = 25 mm; white *dashed circle*: beam waist; ray tracing with 25,117 rays). (b) Aperture coupling factor T_{AP} (*solid curves*) and crosstalk coupling factors XT_n from beams addressing neighbor cells (*dashed curve* XT_1 for the closest neighbor; *dotted curves* XT_2 through XT_5) versus shift x_{AP}. Lens apertures D_{lens} = Ø20 ... Ø50 mm. Cell spacing D = 100 mm. Ray tracing with 50,311 rays per beam.

Similarly, for a Gaussian beam with a waist of Ø100 mm and multiple interfering beams addressing neighbor cells, the aperture coupling factor $T_{AP}(x_{AP})$ and crosstalk coupling factors XT_n (x_{AP}, D) versus the receiver shift x_{AP} have been calculated by means of ray tracing, as shown in Fig. 7-21. The ray tracing is done by launching the rays from an aperture with a diameter equal to 3× the beam waist, thus omitting only a fraction of 1.5×10^{-8} of the Gaussian beam's power (see Table A-1). It can be observed that T_{AP} starts to decrease as soon as the receiver is shifted, but it decreases slower when the lens aperture is larger. Hence, it is advantageous to maximize the lens aperture D_{lens} in order to reduce this decline.

It may be noted that crosstalk contributions XT_1 and XT_2 increase when T_{AP} decreases (so for increasing x_{AP}), but that XT_3, XT_4, and XT_5 decrease. Clearly, the dominating crosstalk is XT_1 which comes from the beam addressing the closest neighbor cell 1. It may be noted that the lens aperture $D_{lens} = 2R_{lens}$ should be minimized to minimize the crosstalk. The optical signal-to-crosstalk ratio $SXR_{1,opt}$ as well as the signal-to-crosstalk ratio $SXR_{tot,opt}$ taking the crosstalk contributions from all beams addressing neighboring cells into account are shown for a range of lens diameters in Fig. 7-22b. The difference between both sets of curves is small, in particular for a larger receiver shift x_{AP}, which shows the strong dominance of the crosstalk from the beam addressing the closest cell 1 above the crosstalk contributions

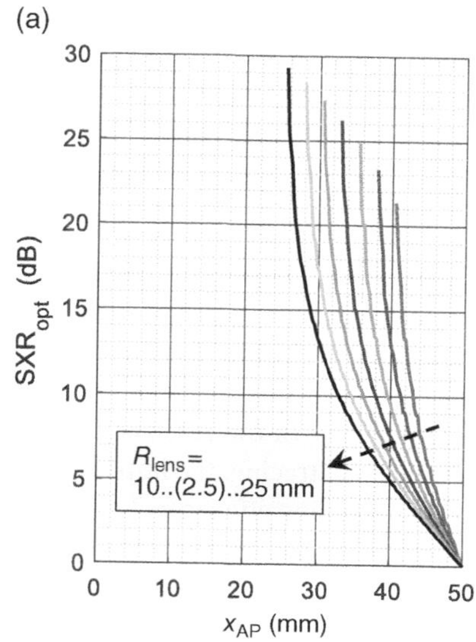

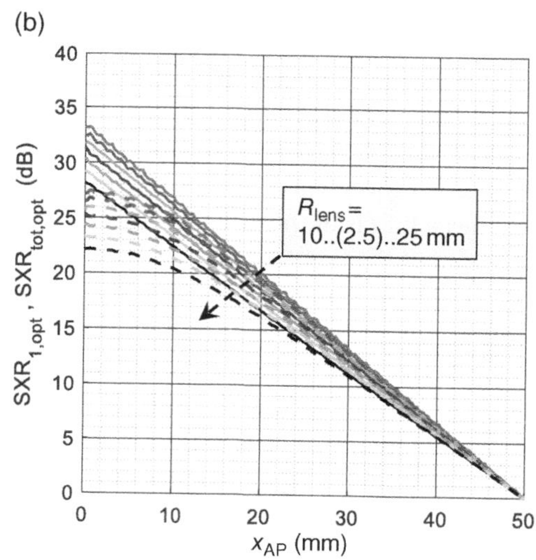

Fig. 7-22 Optical signal-to-crosstalk ratio SXR_{opt} versus receiver shift x_{AP}, for varying lens apertures (beam diameter D_{beam} = Ø100 mm). (a) For uniform beams (only crosstalk from beam addressing cell 1). (b) For Gaussian beams (*solid curves*: crosstalk from beam addressing cell 1; *dashed curves*: crosstalk from the beams addressing all neighboring cells)

from the beams addressing the other cells. When the receiver is shifting to the edge of the signal cell, so for x_{AP} approaching $D_{beam}/2 = 50$ mm, the SXR curves converge towards $SXR_{opt} = 0$ dB. For a power penalty <0.5 dB with a single interferer, with a Gaussian beam D_{beam} = Ø100 mm and a lens diameter D_{lens} = Ø50 mm a receiver shift $|x_{AP}|$ up to 32 mm is allowed, while for D_{lens} = Ø20 mm, the allowable shift $|x_{AP}|$ increases somewhat to 35 mm. The aperture coupling factor T_{AP}, however, is considerably reduced when D_{lens} decreases from Ø50 to Ø20 mm.

It may be noted that in the case of using the AWGR-based beam steerer (cf. Section 5.5), crosstalk into the signal cell may also stem from adjacent wavelength channels in the AWGR which leak into the signal's output port of the AWGR; typically, such crosstalk is less than −20 dB relative to the signal and does not noticeably impact the system's performance.

7.3.4.2 Impact of Cropping the Beam's Spot by the Lens' Aperture

Next to the decrease in captured power caused by the shift of the receiver's aperture, the change in the intersection of the beam with the lens' aperture also affects the shape and intensity profile of the spot projected on the photodiode and thus may affect the FoV of the receiver.

By means of ray tracing, the impact of the receiver shift on the equivalent beam-to-PD power coupling efficiency T_{eq} as well as on the FoV of the receiver have been assessed. Fig. 7-23 shows how the shape and intensity profile of the spot change due to the shift. The impact of the shift may be reduced by optimizing the defocusing parameter p at each new position of the receiver; this implies additional control complexity for adapting the distance of the PD matrix with respect to the lens. Even when applying this re-optimization of p, the impact of the shift on the T_{eq} and the FoV is significant. For the 4×4 PD matrix receiver, the results of ray tracing in Fig. 7-24a show that the FoV attainable for $T_{eq} = -30$ dB is reduced from 13° to 2° (half-angle) when the receiver is shifted away from the cell's center over 50 mm. For the single PD receiver, Fig. 7-24b shows that the attainable FoV then declines a bit faster, from 12° to less than 1°.

To limit the complexity in the cost-sensitive receiver, it may be preferred to optimize and fix p in order to obtain a maximum FoV at an acceptable T_{eq} when the receiver is positioned in the center of the beam, and not to alter this defocusing p when the receiver is moved. By ray tracing, the dependence of T_{eq} on the beam incidence angle α for various settings of p when changing the shift x_{AP} has been assessed, similarly as done in the ray tracing analysis reported in Section 7.3.2. For the 4×4 PD matrix receiver and shift $x_{AP} = 0$, Fig. 7-25a shows that a $T_{eq} > -30$ dB for α up to the FoV half-angle $\alpha_{max} = 8.2°$ is achieved when adopting a fixed $p =$ 13.5%. The table lists the reduction in T_{eq} and in the FoV half-angle α_{max} when the shift x_{AP} of the receiver with respect to the center of the Gaussian beam grows from 0 to 40 mm, as illustrated in Fig. 7-23. Similarly, Fig. 7-25b shows the impact of the shift for the single PD receiver, in which a fixed $p = 7.0\%$ is adopted to guarantee a $T_{eq} > -30$ dB for α up to $\alpha_{max} = 3.7°$.

It may be concluded that the impact of the receiver shift on the equivalent beam-to-PD power coupling T_{eq} is approximately the same for the PD matrix receiver and the single-PD receiver, but that the FoV performance of the PD matrix receiver is clearly better.

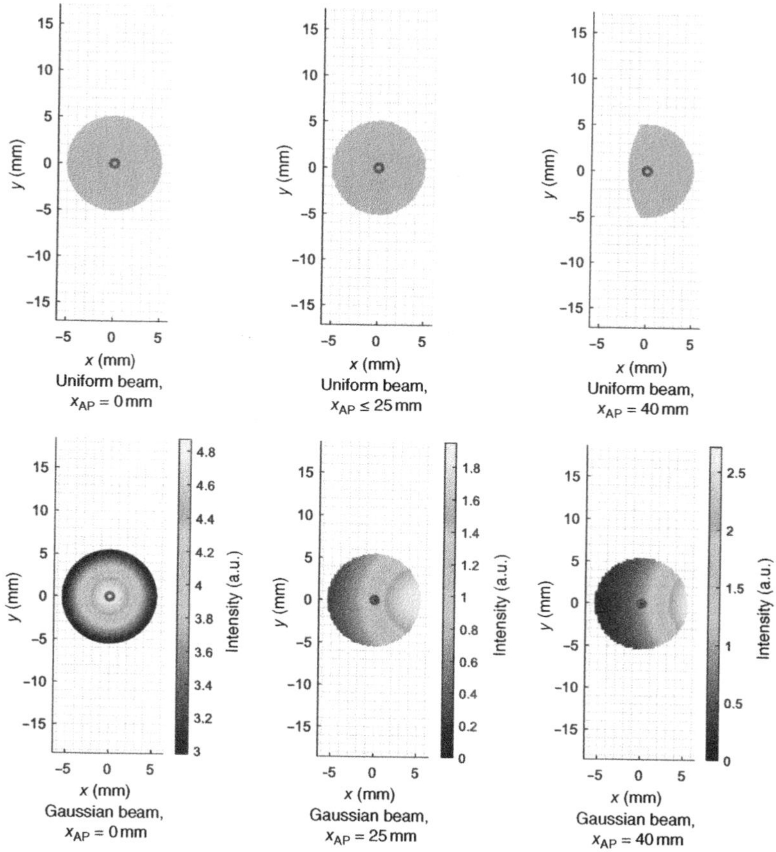

Fig. 7-23 Beam spot projected on PD matrix (located at $(x, y) = (0, 0)$) when the beam's center is shifted over a distance x_{AP} with respect to the receiver aperture. *Assumptions*: beam diameter D_{beam} = Ø100 mm, and Fresnel lens with aperture D_{lens} = Ø50 mm, f = 10 mm, with defocusing p = 20%.

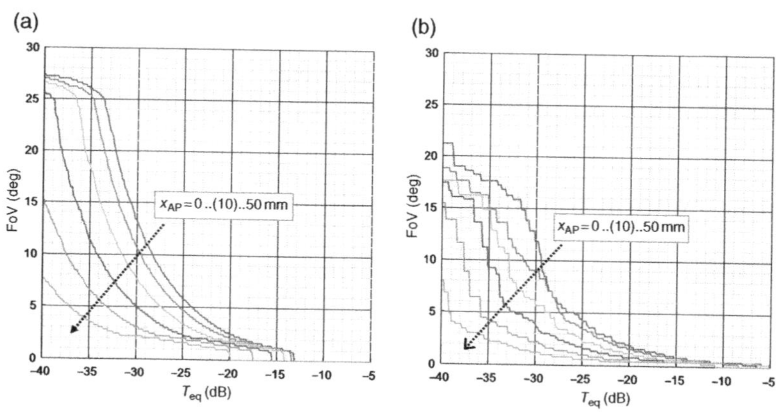

Fig. 7-24 Impact of shifting the receiver over a distance x_{AP} with respect to the center of the beam's footprint when applying a) the 4 × 4 PD matrix and b) a single PD. *Assumptions*: Gaussian beam, waist diameter Ø100 mm, and receiver with Fresnel lens, aperture Ø50 mm, f = 10 mm. (a) FoV versus T_{eq} when p is re-optimized at each x_{AP} for 4 × 4 PD matrix receiver (calculated by ray tracing for 25,117 rays). (b) FoV versus T_{eq} when p is re-optimized at each x_{AP} with single-PD receiver (calculated by ray tracing for 10,267 rays)

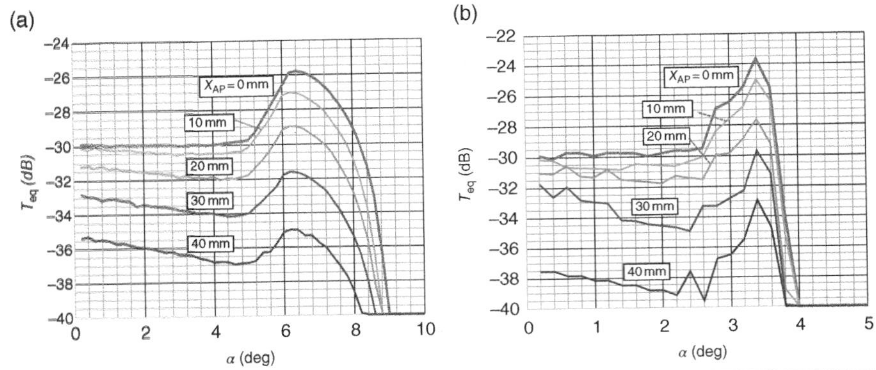

x_{AP} (mm)	T_{eq} (dB)	FoV α_{max} (deg)
0	>–30	8.2
<20	>–32	7.9
<30	>–34	7.6
<40	>–37	7.5

x_{AP} (mm)	T_{eq} (dB)	FoV α_{max} (deg)
0	>–30	3.7
<20	>–32	3.65
<30	>–34	3.63
<40	>–37	3.8

Fig. 7-25 Impact on the beam-to-PD power coupling factor T_{eq} versus beam incidence angle α when shifting the receiver across the beam's footprint (over a distance x_{AP} from 0 to 40 mm), if the defocusing factor p is not re-optimized, but fixed: for (a) the 4 × 4 PD matrix receiver with a fixed $p = 13.5\%$, and (b) the single PD receiver with a fixed $p = 7.0\%$. Same assumptions as in Fig. 7-24.

If there are active neighboring beams, crosstalk power from these beams may also be captured by the receiver. Because the distances between the signal beam and these neighboring beams will be small with respect to the distant ceiling unit, the incidence angles of the neighboring beams are approximately equal to the incidence angle α of the signal beam. Hence, the impact of such crosstalk on the receiver's beam-to-photodetector power coupling efficiency T_{eq}-versus-α performance shown in Fig. 7-25 may be found by including the (typically small) power penalty due to the crosstalk (as analyzed in Section 7.3.4.1, cf. Fig. 7-22) into T_{eq}.

7.4 Implementing the Receiver

We have designed a 4 × 4 PD matrix composed of four (1 × 4) arrays of photodiodes which each have an active area diameter of Ø150 μm and are spaced within an array at a pitch of 250 μm. These arrays can support 2.5 Gbit/s OOK-modulated data and are available from Albis Optoelectronics [7-12], who manufactured the photodiode arrays, interconnected them by wire bondings, and packaged them in a low-profile TO-46 can, see Figs. 7-26a and b.

The photodiode matrix is equipped with the taps according to the concept of Fig. 7-3b for connecting external resistors and thus minimizing the bias voltage

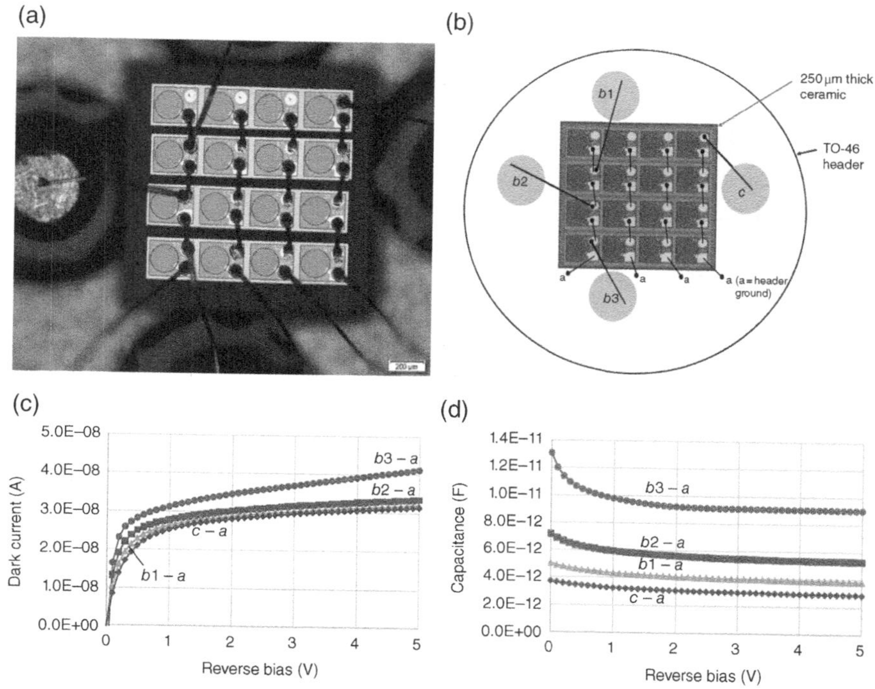

Fig. 7-26 4 × 4 PD matrix (assembled and characterized by Albis Optoelectronics). (a) 4 × 4 PD matrix, composed of four wire-bonded 1 × 4 arrays of Ø150 µm 2.5 Gbit/s photodiodes, (b) wiring diagram, (c) dark currents versus reverse bias voltage on the PD matrix, and (d) capacitances versus reverse bias voltage on the PD matrix (*a*: anode side of the whole matrix, *c*: cathode side, *b*1–*b*3: the three intermediate levels; see Fig. 7-3b)

unbalances when the photodiodes are unevenly illuminated (as discussed in Section 7.2.2). Measurements on the completed packaged PD matrix performed by Albis Optoelectronics confirmed the validity of our PD matrix concept. The results are shown in Figs. 7-26c and d. Fig. 7-26c shows that the dark current gradually declines when more rows of the matrix are cascaded; for the whole matrix, it is ~31 nA at 5 V reverse bias voltage. The measured cathode-to-anode capacitance of the matrix (*c*–*a* curve in Fig. 7-26d) was found to be ~3 pF for a matrix reverse bias voltage >1.5 V, which is indeed equal to the specified capacitance of ~2.3 pF of a single PD plus a parasitic capacitance of ~0.7 pF added by the TO-46 can. It may also be noted that the capacitance is only weakly dependent on the reverse bias voltage beyond 1.5 V. The other curves *b*1–*a* to *b*3–*a* illustrate how the capacitance indeed decreases by cascading more rows.

With the packaged 4 × 4 PD matrix, we built the OWC receiver circuit of which the circuit diagram is shown in Fig. 7-27a. A common-base transistor front-end

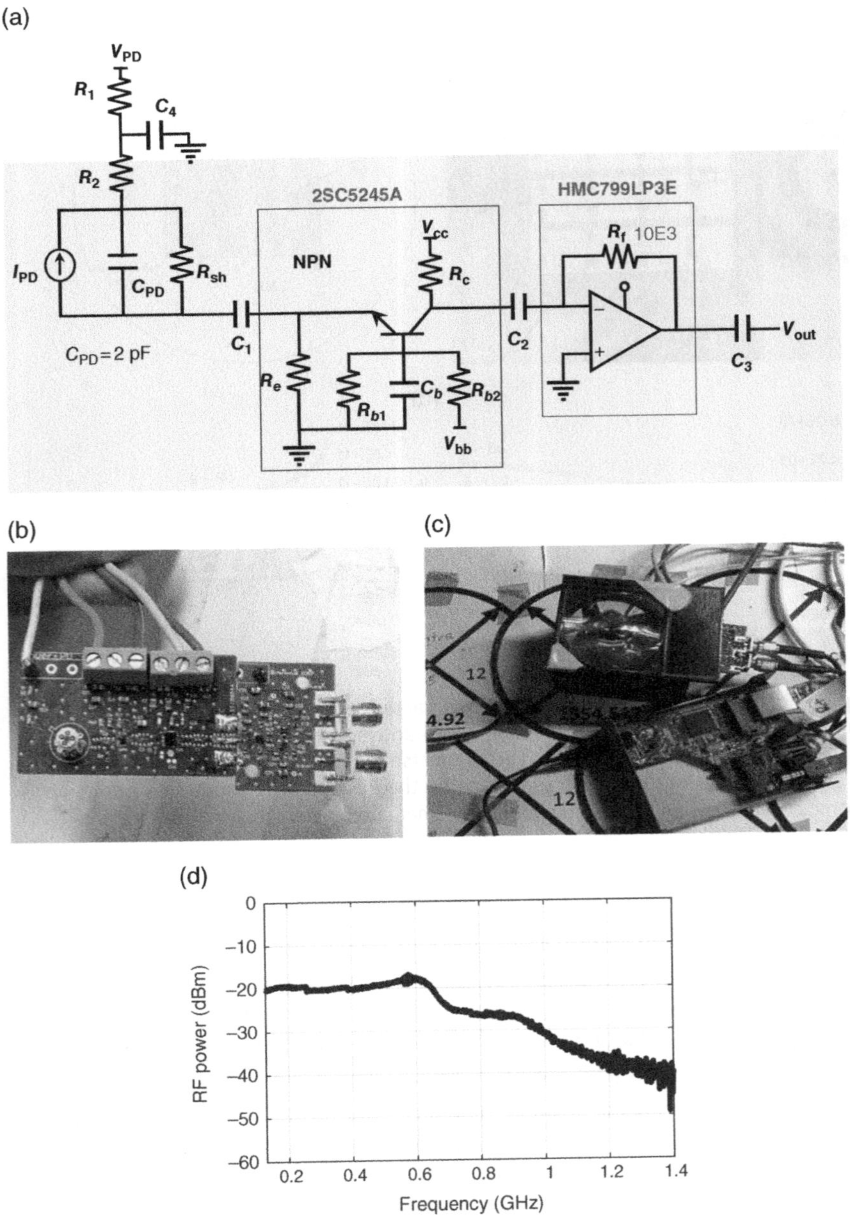

Fig. 7-27 Prototype of OWC receiver with 4 × 4 photodiode matrix. (a) OWC front-end receiver circuit. (b) OWC receiver, assembled on a printed circuit board including packaged 4 × 4 PD matrix module with front-end receiver and a differential output stage. (c) Measured frequency characteristics of the OWC receiver. (d) downstream OWC receiver module equipped with Ø50 mm Fresnel lens and combined with GbE media converter module providing an RJ45 outlet

amplifier stage was applied in order to minimize the impedance load to the PD matrix and thus maximize bandwidth, followed by a 700 MHz bandwidth commercially available 10 kΩ transimpedance amplifier (TIA). Fig. 7-27b shows the complete assembly, mounted on an epoxy printed circuit board (PCB). The measured frequency characteristics of this receiver are given in Fig. 7-27c; the measured −3 dB bandwidth is 670 MHz, so just slightly reduced with respect to the bandwidth of the TIA. The OWC receiver module was equipped with a Fresnel lens with a large D/f ratio ($f = 10$ mm, $D_{lens} = Ø50$ mm) in order to maximize its FoV, see Fig. 7-27d. A commercial bidirectional GbE SMF-to-RJ45 media converter was adapted and combined with our PD matrix front-end receiver circuit in order to create a complete OWC receiver module. This combination has been successfully deployed in our laboratory system experiments which demonstrate optical wireless high-definition GbE video streaming to a laptop computer.

The performance of the PD matrix is impacted by the circuit parasitic impedances inherent to the wire bonding. This performance may be improved by additional on-chip integration steps, particularly by adding predefined tracks to the chip's metallization mask which add only minimal parasitic effects and interconnect the photodiodes instead of the external wire bondings used in our experiments for the interconnecting.

7.5 Experimental Validation

The performance of prototypes of the OWC receiver has been evaluated in our laboratory testbed outlined in Fig. 7-28 (which we also used for the localization experiments, see Section 6.3.3).

We added a single-input dual-output converter stage after our novel wide-FoV receiver and measured the BER performance at each output as well as at the combined differential output. A receiver sensitivity (per PD element) of −30.5 dBm at 1 Gbit/s and of −29.0 dBm at 1.25 Gbit/s (GbE, Gigabit Ethernet speed) was measured (see Fig. 7-29).

Equipped with a large aperture Fresnel lens (Ø50 mm, $f = 10$ mm), we measured the FoV of the OWC receiver in our laboratory setup. As shown in Fig. 7-30, the beams launched by the PRA create slightly overlapping cells with a diameter of Ø10 cm. When moving the receiver a nearly error-free (BER $< 10^{-9}$) performance was achieved over a sizable area which is more than $\pm 3.5 \times \pm 3.5$ cells wide with a cell-to-cell center spacing of ~9 cm at a distance of ~180 cm from the beam-steerer (the pencil beam radiating antenna, PRA). This corresponds to a FoV angle $\alpha \approx \pm 10°$, which is in line with the simulation results reported in Fig. 7-15a.

Moreover, we adapted a commercially available GbE fiber-to-RJ45 media converter in order to combine it with our OWC receiver. As shown in Fig. 7-31, we

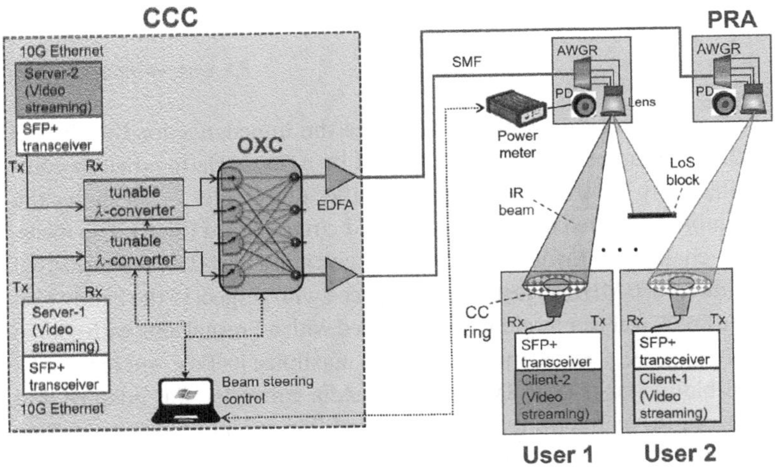

Fig. 7-28 Indoor OWC laboratory testbed

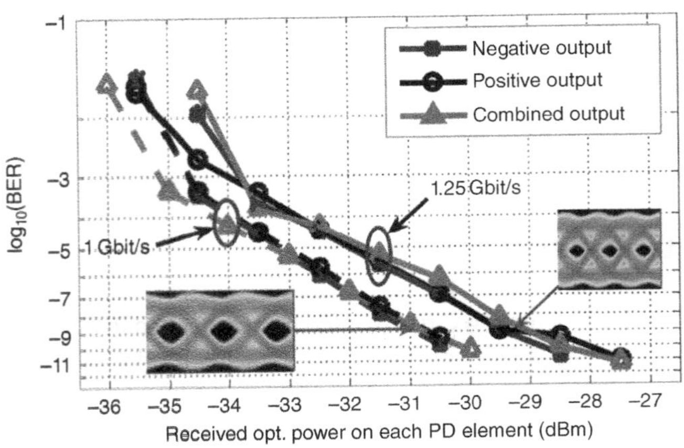

Fig. 7-29 BER performance for various OOK bitrates

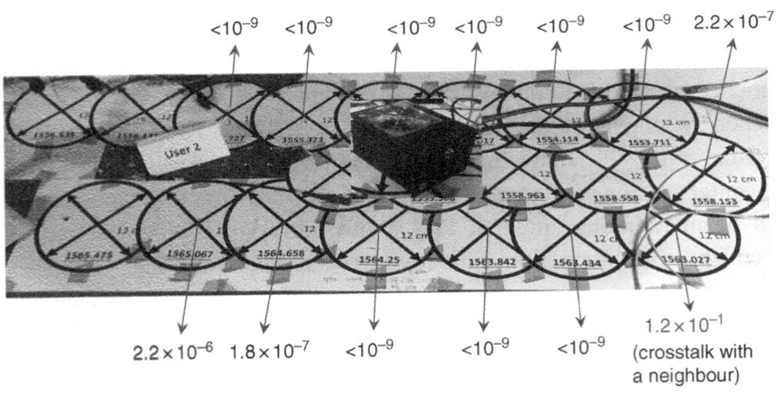

Fig. 7-30 Assessment of the OWC receiver's FoV at 1 Gbit/s; error-free transmission found within ±10° from the center cell (cell diameter Ø10 cm)

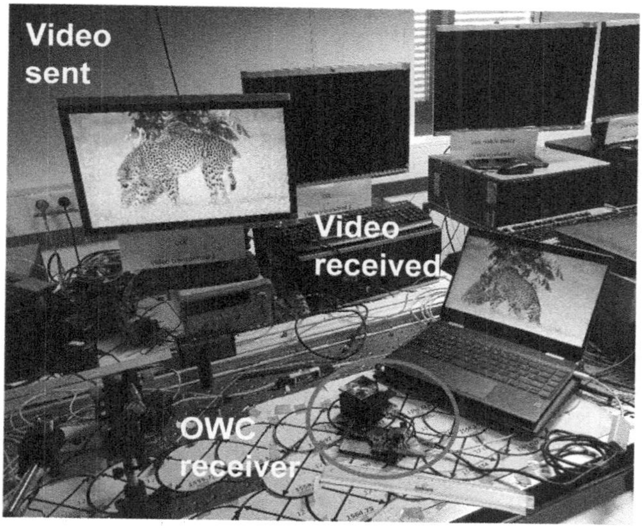

Fig. 7-31 GbE video streaming by OWC to laptop

successfully demonstrated real-time high-definition GbE video streaming at a laptop computer via its RJ45 input.

7.6 Scaling the PD Matrix

The impact on the receiver's FoV of scaling the PD matrix to other $N \times N$ dimensions has been analyzed by numerical integration for the case of a Gaussian beam profile and an ideal aberration-free thin lens. For downstream (launching a Gaussian beam with a waist of Ø100 mm and power $P_0 = 10$ dBm, and using a thin lens with $D_{\text{lens}} =$ Ø50 mm and $f = 10$ mm), at a required $T_{\text{eq}} = -30$ dB according to Fig. 7-32a, the attainable FoV is ±7.5° for a single PD. With a PD matrix, it grows from ±11° to ±21° when the matrix is enlarged from 2×2 to 8×8 photodiodes. For upstream (with a Gaussian beam of +2 dBm power and waist Ø15 mm, and a thin lens with $D_{\text{lens}} =$ Ø25 mm and $f = 5$ mm), at a required $T_{\text{eq}} = -22$ dB an FoV ≈ ±11° is attainable for a single PD. When the PD matrix is applied and its size is increased from 2×2 to 8×8, an FoV growing from ±11° to ±26° can be achieved.

Upscaling the PD matrix thus increases the FoV performance, but this FoV improvement becomes less significant at larger matrix sizes. The scalability of our PD matrix concept has also been validated at high data rates; 25 Gbit/s NRZ modulated transmission over 20 meters has been demonstrated with a 20×20 matrix of Ø30 μm PD-s [7-15].

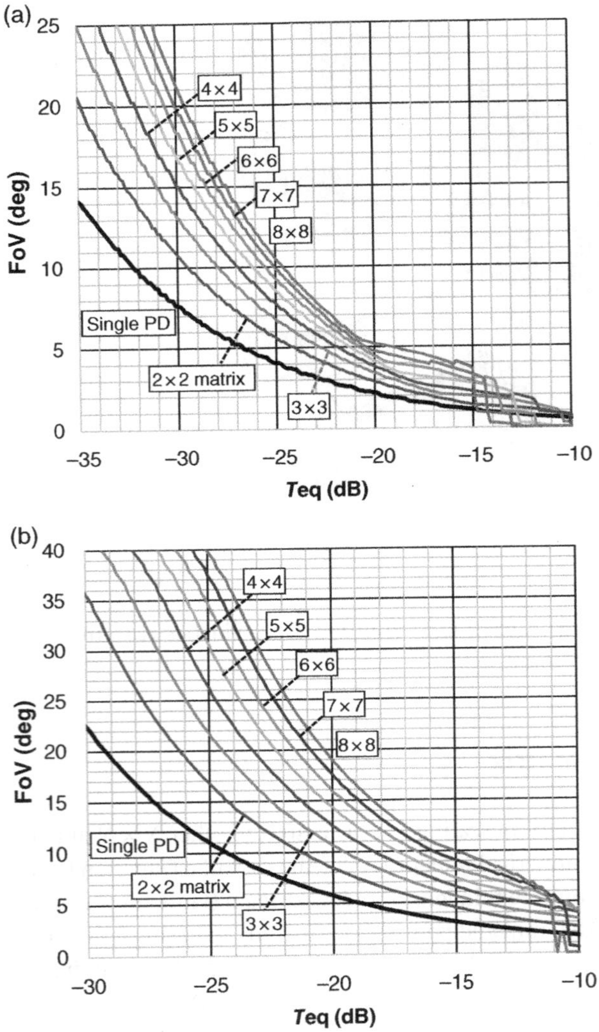

Fig. 7-32 Impact of scaling the PD matrix on the FoV. (a) *Downstream* (Gaussian beam Ø100 mm, thin lens Ø50 mm, $f = 10$ mm). (b) *Upstream* (Gaussian beam Ø15 mm, thin lens Ø25 mm, $f = 5$ mm)

7.7 Concluding Remarks

The photodiode matrix concept offers powerful solutions for realizing high-bandwidth receivers for optical wireless communication with an enlarged FoV and aperture. Thus, it eases optical beam alignment issues and increases the area in which the OWC receiver can be operated.

Numerical ray tracing shows that with a 4 × 4 PD matrix a clearly better FoV versus beam-to-photodetector coupling T_{eq} receiver performance can be achieved than with a single PD. It is found that the fill factor of the PD matrix has little impact on this performance when the PD matrix is over-filled (which is the typical condition when optimizing the FoV of the matrix-based receiver).

Moreover, with respect to alternative solutions such as angular diversity receivers (ADRs) and single-PD receivers with adaptable optics, the PD matrix receiver can be less complex, consume less power, and achieve a better SNR performance. Shifting the receiver in the user plane across the cells created by the discrete downstream beams will impact its performance; it brings a power penalty and its FoV is reduced as well. According to numerical analysis, a shift within 70% of the beam cell radius does not seriously affect the receiver's performance.

Experimental validation of a front-end receiver built with a 4 × 4 PD matrix (assembled by wire-bonding four commercial 1 × 4 linear arrays of Ø150 μm photodiodes) and a commercially available TIA (750 MHz bandwidth, 10 kΩ transimpedance) has shown good receiver performance at 1.25 Gbit/s (GbE speed).

Scaling the PD matrix to larger dimensions can further improve the receiver's FoV performance, albeit the improvement slows down at high dimensions.

References

[7-1] Optics law of conservation of etendue [on-line]. Available: https://en.wikipedia.org/wiki/Etendue

[7-2] R. Winston, "Dielectric compound parabolic concentrators," *Appl. Opt.*, vol. 15, no. 2, pp. 291–292, Feb. 1976.

[7-3] K. Wang, A. Nirmalathas, C. Lim, et al., "Full duplex gigabit indoor optical wireless communication system with CAP modulation," *Photon. Technol. Lett.*, vol. 28, no. 7, pp. 790–793, Apr. 2016.

[7-4] S. Collins, D.C. O'Brien, and A. Watt, "High gain, wide field of view concentrator for optical communications," *Opt. Lett.*, vol. 39, no. 7, pp. 1756–1759, Apr. 2014.

[7-5] A. Riaz, and S. Collins, "A wide field of view VLC receiver for smartphones," *Proc. ECOC 2020*, Brussels, Dec. 6–10, 2020, Paper Tu2G.4.

[7-6] Z. Cao, L. Shen, Y. Jiao, et al., "200 Gbps OOK transmission over an indoor optical wireless link enabled by an integrated cascaded aperture optical receiver," *Proc. OFC 2017*, Mar. 19–23, 2017, Los Angeles, Post-deadline paper Th5A.6.

[7-7] J. Zeng, V. Joyner, J. Liao, et al., "A 5Gb/s 7-channel current-mode imaging receiver front-end for free-space optical MIMO," *Proc. IEEE MWSCAS 2009*, Cancun, Aug. 2–5, 2009, pp. 148–151, https://doi.org/10.1109/MWSCAS.2009.5236130.

[7-8] Z. Zeng, M. Dehghani Soltani, M. Safari, and H. Haas, "Angle diversity receiver in LiFi cellular networks," *Proc. ICC 2019*, Shanghai, May 20–24, 2019.

[7-9] A.M.J. Koonen, K.A. Mekonnen, F.M. Huijskens, et al., "Novel broadband OWC receiver with large aperture and wide field of view", *Proc. ECOC 2020*, Brussels, Dec. 6–10, 2020, Paper Tu2G.6.

[7-10] A.M.J. Koonen, K.A. Mekonnen, F.M. Huijskens, et al., "Optical wireless GbE receiver with large field-of-view," *Proc. ECOC 2021*, Bordeaux, Sep. 13–16, 2021, Paper Th1B.6.

[7-11] A.M.J. Koonen, K.A. Mekonnen, Z. Cao, et al., "Beam-steered optical wireless communication for industry 4.0", *IEEE J. Sel. Top. Quantum Electron.*, vol. 27, no. 6, art. 6000510 (10 pp.), Nov./Dec. 2021, https://doi.org/10.1109/JSTQE.2021.3092837.

[7-12] Albis Optoelectronics, multi-channel top-illuminated monitor photodiode array [on-line]. Available: https://www.albisopto.com/albis_product/pd00jx-multi-channel-top-illuminated-monitor-photodiode-array/

[7-13] Fresnel lens [online]. Available: https://en.wikipedia.org/wiki/Fresnel_lens

[7-14] R. Ramaswami, K.N. Sivarajan, and G.H. Sasaki Ramaswami, "Ch. 5: Transmission system engineering," *Optical Networks – A Practical Perspective*, 3rd ed., 2010, Elsevier Inc., https://doi.org/10.1016/C2009-0-17339-7.

[7-15] T. Umewaza, A.Q. Matsumoto, K. Akahane, et al., "400-pixel high-speed photodetector for high optical alignment robustness FSO receiver," *Proc. OFC 2022*, San Diego, Mar. 6–10, 2022, Paper M4I.3.

8

Bidirectional All-Optical OWC Indoor System[1]

8.1 Introduction

Optical wireless communication (OWC) by means of two-dimensional (2D)-steered narrow infrared beams (BS-OWC) is a very promising technique for surmounting the bottlenecks of today's radio wireless systems, such as the heavy congestion that is often incurred in Wi-Fi networks. The small footprint of each infrared beam enables users to be connected individually at high user densities while they do not need to share capacity, whereas in wide-coverage LiFi and Wi-Fi systems, users have to share capacity, by means of MAC (medium access control) protocols. Most beam-steered OWC systems reported so far have focused on downstream connectivity only. In [8-1], we showed GbE video streaming by downstream BS-OWC links covering a sizable user area without a need to carefully align the orientation of the user's device, thanks to the wide field-of-view (FoV) of its receiver.

Bidirectional beam-steered systems which were reported so far mostly employed hybrid links (i.e., optical beam downstream (DS), mm-wave radio beam upstream (US); to be discussed in Section 8.2). Hybrid links, however, do not preserve some of the key advantages of OWC, such as security against eavesdropping and immunity against electromagnetic interference (EMI).

Alternatively, many all-optical bidirectional beam-steered systems reported were point-to-point (P2P) ones and simply duplicated the DS link into an US one (see Section 8.3). It should be noted that in an indoor network, however, the DS OWC links are typically emerging from a common point-to-multipoint

1 This chapter is largely based on T. Koonen, K. Addis Mekonnen, F. Huijskens, E. Tangdiongga, "Bi-directional all-optical wireless Gigabit Ethernet communication system using automatic self-aligned beam steering," *J. Lightwave Technol.*, vol. 41, no. 11, pp. 3446–3454, June 2023, https://doi.org/10.1109/JLT.2022.3231438.

Short-Reach Optical Wireless Communication: By Directed Narrow Beams, First Edition. Ton Koonen.

(P2MP) multicasting unit, which is mounted at the room's ceiling, whereas the US links are MP2P links from each user device to the US receiver (US Rx) at the ceiling. Such asymmetry in the connectivity patterns is not optimally served by duplicating the DS link into an US one.

Accurate steering of beams in bidirectional BS-OWC systems is required for both the DS and the US links, for which the user devices need to be localized and tracked. Various solutions to localize and track the user devices have been reported which require additional active infrastructure elements and a pre-established return channel to complete the control loop, as discussed in Chapter 6. Alternatively, we proposed in Chapter 6 to deploy passive retro-reflector elements that enable self-calibrated autonomous beam alignment without needing a return channel and drawing only minimal power from the (typically battery-operated) user device.

In this chapter, after reviewing concisely hybrid bidirectional systems in Section 8.2 and point-to-point all-optical bidirectional systems in Section 8.3, we describe in detail and demonstrate an all-optical bidirectional BS-OWC system with a P2MP architecture which we have introduced in [8-2]. It expands our previous BS-OWC GbE DS-only video streaming system [8-1] with a beam-steered GbE US link per user, established by automatic self-alignment of the US link next to the self-alignment of the DS link. We deploy the passive optical retro-reflecting structures at the receiver introduced in Chapter 6 which enable automation of the beam steering control without needing an active pre-existing feedback channel. In Section 8.4, the proposed architecture of the bidirectional OWC system is explained. The DS path design has been introduced previously in [8-1, 8-3, 8-4]. Section 8.5 introduces the US path design; the US beam steering principle has been discussed in Section 5.6. The US beam alignment to the US Rx's aperture has been automated, by using the hole-seeking center-of-gravity algorithm as described in Section 6.4, wherein its performance has been analyzed by simulations while adopting the system's design choices and has been experimentally validated in our OWC testbed. Section 8.6 describes the full experimental bidirectional OWC system which implements these choices and reports its performance for bidirectional TCP tests and GbE video transmission. Section 8.7 summarizes the results and makes concluding remarks on the application potential of our proposed system.

8.2 Hybrid OWC Downstream/RF Upstream Systems

In a hybrid optical/RF wireless bidirectional communication system, for the US path from the mobile device (MD) to the central communication controller (CCC), RF links (e.g., in the 57–64 GHz band) can be foreseen, to be picked up by an antenna at the pencil-radiating antenna (PRA) and carried over the fiber network to the CCC using radio-over-fiber techniques. As illustrated in Fig. 8-1, at the PRA

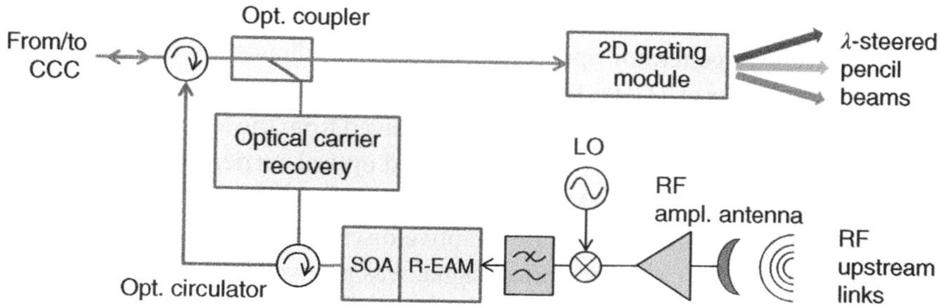

Fig. 8-1 Radio-over-fiber upstream path at PRA site, with optical carrier recovery. *Source:* [8-11] © [2016] IEEE

site, a nearly clean optical carrier can be recovered from part of the DS intensity-modulated optical signal (the pencil beam to be steered). The carrier recovery can be done by erasing the modulation by a semiconductor optical amplifier (SOA) which is operated in the gain-saturated regime [8-5]. Alternatively, it can be done by an injection-locked Fabry–Pérot laser diode [8-6]. Subsequently, a reflective modulator, composed of an SOA plus a reflective electro-absorption modulator (R-EAM) [8-7, 8-8], is used to modulate the frequency-downshifted US RF signal onto this clean carrier.

By means of a phased-array antenna at the PRA providing distinctive tunable nulls in the antenna pattern, also localization and tracking of the active MDs can be supported for aiding the optical pencil beam steering. Alternatively, localization and tracking may be done by means of a wide-view camera at the ceiling observing near-IR LED tags at the MD [8-9, 8-10]; more details have been discussed in Chapter 6.

A preliminary system experiment for assessing the feasibility of US transmission employing radio-over-fiber (RoF) techniques was carried out using the setup shown in Fig. 8-2 [8-11]. Optical carrier recovery was implemented by using an

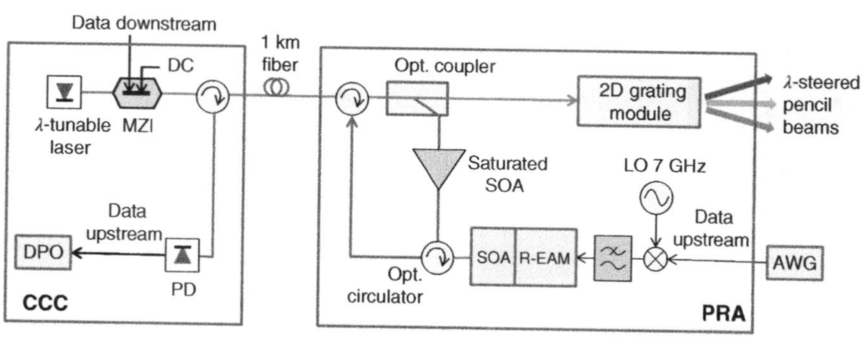

Fig. 8-2 Radio-over-fiber upstream system experiment. *Source:* [8-11] © [2016] IEEE

SOA working in its saturation region. The radio US signal was emulated by an arbitrary waveform generator operating in baseband followed by up-conversion using a local oscillator at 7 GHz. The resulting signal emulates the down-converted 60 GHz radio signal (with 7 GHz bandwidth) received from the MD. This radio signal then modulates the intensity of the recovered optical carrier (at $\lambda = 1550$ nm) using a 10 GHz bandwidth REAM-SOA. The US RoF signal is launched into the 1 km fiber link with 0 dBm power. Rate-adaptive discrete multitone (DMT) modulation was used. At the CCC, the optical signal was received by a 10 GHz receiver and captured by a digital phosphor oscilloscope for signal processing.

Fig. 8-3 depicts the waveforms at the input and output of the SOA at a bias current of 350 mA and optical input power of −1 dBm. In this operating condition, the SOA was in deep saturation and erased the modulation from 300 mV (peak-to-peak) to less than 20 mV. The influence of the SOA bias current and input power on the US performance is illustrated in Fig. 8-3. It was possible to achieve >10 Gbit/s transmission when the input optical power was > −5 dBm. Increasing the SOA bias current and input optical power further improved the performance as this enables the SOA to erase the downlink modulation even better.

Fig. 8-4 shows the adaptive DMT bit loading and BER performance for the radio signal after 1 km fiber transmission when the SOA was biased at 350 mA and the input optical power was −2 dBm. A maximum bit rate of 10 Gbit/s was achieved at an average BER < 10^{-3}. Up to 32-QAM was utilized for modulation of the DMT subcarriers.

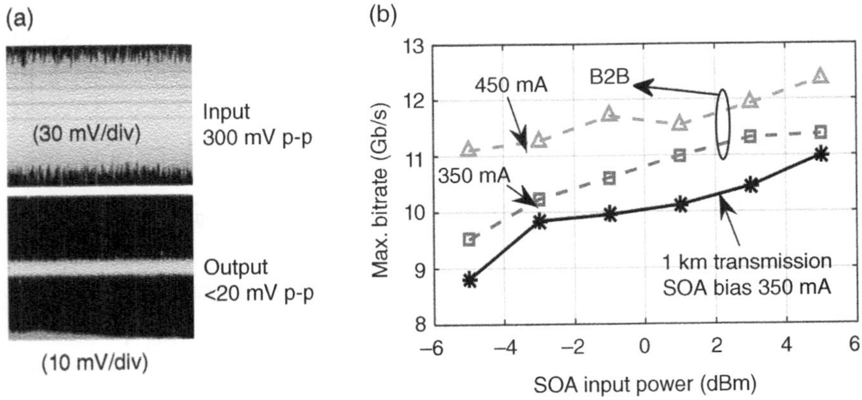

Fig. 8-3 Optical carrier recovery by the saturated SOA. *Source:* [8-11] © [2016] IEEE; (a) waveforms at the input and output of the SOA at bias current 350 mA and input optical power −1 dBm; (b) maximum upstream bitrate achievable

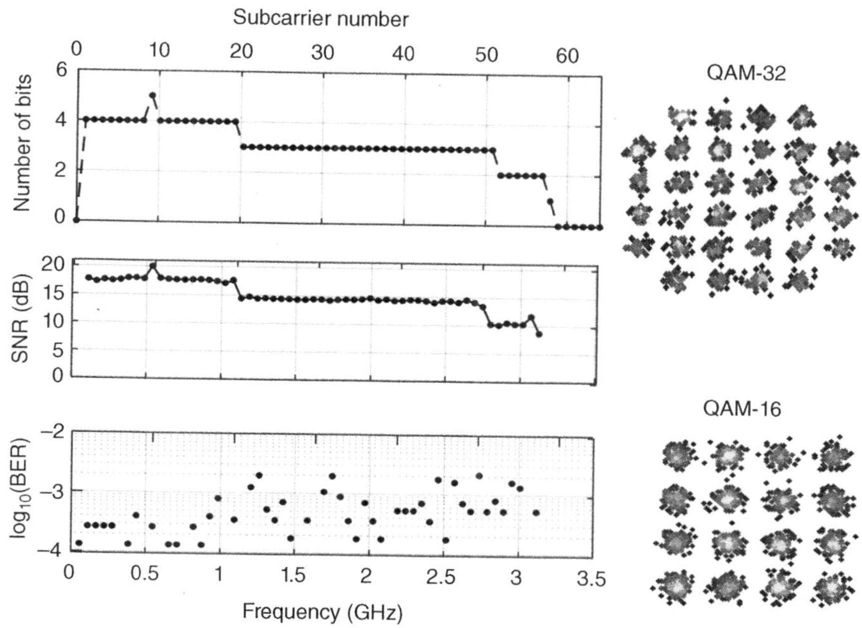

Fig. 8-4 DMT bit-loading profile and BER per subcarrier for the received upstream radio signal. *Source:* [8-11] © [2016] IEEE

8.3 All-Optical Bidirectional OWC Systems

Most all-optical bidirectional systems reported have simply duplicated the DS link into an US one, for example, using MEMS mirrors [8-12] or SLMs [8-13], as discussed in Section 5.2, or they covered very short distances only such as for docking systems [8-14].

A symmetric high-capacity full-duplex OWC system has been reported as shown in Fig. 8-5 [8-15]. Based on wavelength-controlled 2D beam steering with crossed

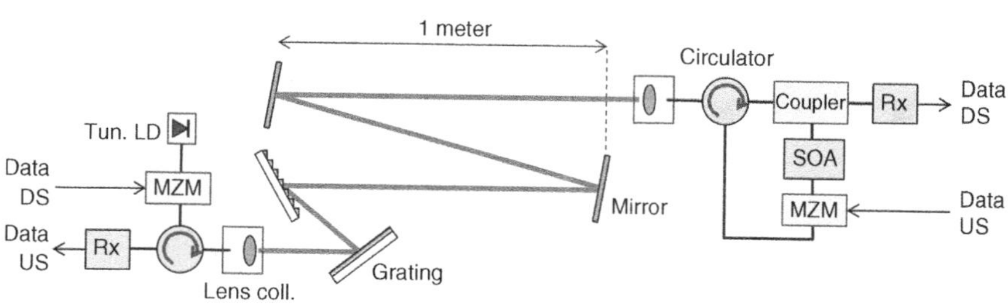

Fig. 8-5 Bidirectional all-optical wireless communication. *Source:* [8-15]/Ton Koonen

gratings (see Section 5.4), it employs the principle of reversibility of an optical path when operated at the same wavelength. Fiber-pigtailed triplet lens collimators were used to send/receive beams with 3.3 mm diameter, and optical circulators at the central site and at the user's device site separated the US and DS signals. The DS beam carries 10 Gbit/s in OOK modulation at a moderate optical extinction ratio of 3.6 dB. This enables the receiver at the device side to remove the modulation by means of an SOA operating in the gain-saturated regime and to regain an almost clean optical carrier. This carrier is subsequently re-modulated by a Mach Zehnder modulator (MZM) with 10 Gbit/s OOK US data. As the US signal thus has the same wavelength as the DS signal, it takes the same path backward through the crossed-gratings module. The user's device does not need an optical source but has a quite limited FoV and receiving aperture due to its SMF-based architecture.

8.4 Bidirectional OWC System Architecture

Fig. 8-6 shows the architecture of our proposed all-optical bidirectional OWC system. The ceiling central unit hosts the passive pencil beam-radiating antenna (PRA) unit which directs narrow downstream (DS) optical beams in two dimensions according to their wavelength; the DS data are fed from λ-tunable laser

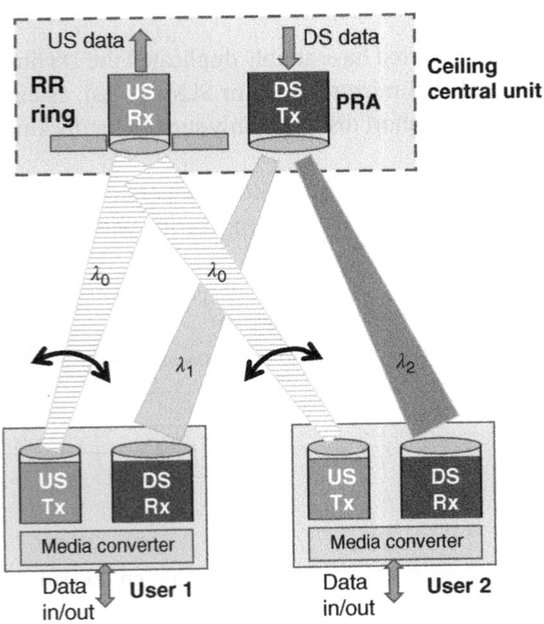

Fig. 8-6 Bidirectional BS-OWC system with automatic upstream beam self-alignment

transmitters remotely located in the central communication controller site (see Chapter 4) [8-3]. As described in Section 5.5, the PRA contains an arrayed waveguide grating router module acting as wavelength-demultiplexer, of which the output wavelength ports are arranged in a 2D fiber array. The position of each output fiber with respect to the axis of the subsequent lens determines the direction of the beam emitted at that wavelength. Thus, each user is DS-connected by his private λ-beam (λ_1, λ_2, ...), and many users can be connected simultaneously.

For US, a beam with arbitrary wavelength (λ_0) is preferred, in order to avoid costly λ-tunable sources and their control circuitry at the user's site. Therefore, as described in Section 5.6, the US beam steering in two dimensions from the US transmitter (US Tx) toward the US receiver (US Rx) at the ceiling central unit is done by mechanical means. This typically requires a pre-existing optical feedback loop from the ceiling unit to each user to control the pointing of the US beam and to assist in establishing the US path; hence, it requires setting up the DS path first. To circumvent this bootstrapping issue, as described in Chapter 6, we propose to use a ring of retroreflecting miniature corner cubes (RR ring) that surrounds the aperture of the US Rx at the ceiling. Such RR rings can be cut from commercially available RR foils commonly used for road signage, among others; we have demonstrated passive user localization with such RR foil technology before (see Chapters 4 and 6) [8-4]. At the user site, the US power reflected from the RR ring at the ceiling's US receiver (US Rx) is monitored and enables automatically aligning the US beam to the US Rx, by employing a dedicated RR hole-seeking center-of-gravity algorithm, as described in Section 6.4. Multiple users can deploy the RR ring simultaneously for their US beam alignment, as the RR ring reflects a US beam to its originating user only. The US Rx at the ceiling will receive US beams from multiple users. Therefore, for establishing each US link a US medium access control (MAC) protocol is needed, for example, a time division multiple access (TDMA) protocol similar to the ones deployed in commercial TDMA PON fiber-to-the-home systems [8-16].

It may be noted that our system uses two optical apertures in the ceiling central unit. Other BS-OWC system architectures, such as in [8-9], used bidirectional single aperture configurations for fiber-to-fiber coupling via the free-space link. Although in such solutions the ceiling unit for a single user may be more compact, for multiple users multiple ceiling units are needed. In our proposed architecture the two-aperture ceiling unit is shared among multiple users, requires feeding by only a single fiber, and therefore may then be more compact and easier to install.

8.5 Upstream Optical Path Design

Fig. 8-7 shows the design of the US optical path, starting at the user site and covering a distance d which ends at the PRA site at the ceiling.

Similar to the DS path design (see Section 5.4.1 and [8-3]), we choose some defocusing of the fiber with respect to the user's lens 1 (with focal length f_1) in order to

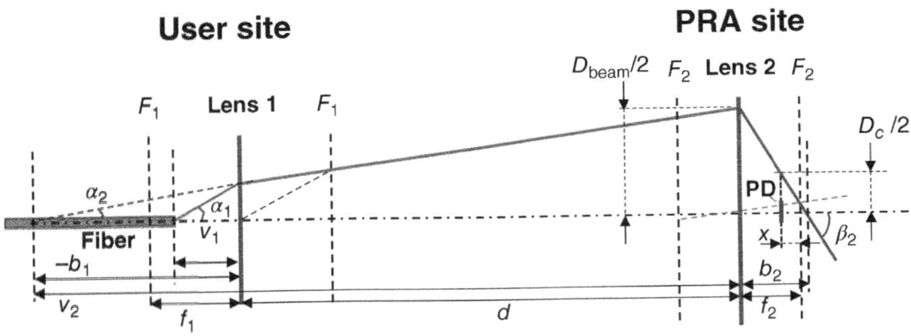

Fig. 8-7 Design of upstream optical path

obtain a slightly diverging US beam that eases US alignment. Using thin lens analysis, the US beam diameter D_{beam} at the PRA site with defocusing $p_1 = 1 - v_1/f_1$ is

$$D_{\text{beam}} = 2p_1 \tan\alpha_1 \left[d + f_1\left(\frac{1}{p_1} - 1\right)\right] \tag{8-1}$$

where $\tan\alpha_1 = \lambda/(\pi w_0)$ represents the divergence of the Gaussian beam with waist radius w_0 launched from the SMF; typically, for $\lambda = 1.5$ μm and $w_0 = 4.5$ μm, tan $\alpha_1 = 0.1$. The photodiode (PD) after lens 2 (with focal length f_2) at the US receiver in the PRA site is put slightly out of focus by some defocusing $p_2 = x/f_2 > 0$; this enlarges the receiving FoV α (see Section 7.3). The spot diameter D_c at the PD and the FoV α are given by

$$D_c = 2p_1 \tan\alpha_1 \left[f_2 + p_2\left\{d + f_1\left(\frac{1}{p_1} - 1\right) - f_2\right\}\right] \tag{8-2}$$

$$\tan\alpha = \frac{|D_c - D_{\text{PD}}|}{2f_2\,(1 - p_2)} \tag{8-3}$$

where D_{PD} is the photodiode's diameter. For achieving a large FoV, the focal length f_2 should be as small as possible, while the aperture of lens 2 should still be large enough to capture the full beam diameter. Assuming a uniform power profile in the beam, the beam-to-PD power coupling factor T at normal incidence at the PRA is

$$\begin{aligned} T &= 1 \quad \text{if } D_c \le D_{\text{PD}} \\ T &= \left(\frac{D_{\text{PD}}}{D_c}\right)^2 \quad \text{if } D_c > D_{\text{PD}} \end{aligned} \tag{8-4}$$

For lens 2 at the PRA, a Fresnel lens is preferred, as such lens type is particularly suited for light collection and can combine a large aperture with a short focal length in order to realize a large FoV (see Section 7.3, and [8-17]). The imaging performance of a Fresnel lens is less good than that of conventional plano-convex or biconvex lenses, but for just collecting light onto the PD, this is not an issue. For lens 1 at the user, a triplet lens is preferred. A triplet lens is the least complicated lens which is able to largely eliminate a range of lens aberrations [8-18]; as has been discussed in Section 6.4.2, this is important to attain adequate performance in the automated US beam steering alignment process without needing a costly more complex lens at the user's device. In our system demonstrator, to be discussed in Section 8.6, a Fresnel lens is used with a 25 mm aperture and focal length $f_2 = 5$ mm [8-19], and a Hastings triplet lens with $f_1 = 20$ mm and aperture 11.5 mm [8-20].

Neglecting lens aberrations, for $f_1 = 20$ mm and $f_2 = 5$ mm a US beam diameter $D_{\text{beam}} = 15$ mm needs a defocusing $p_1 = 2.78\%$ at a user-ceiling distance $d = 200$ cm, and $p_1 = 30.6\%$ at $d = 20$ cm. Fig. 8-8 shows how the beam-to-PD power coupling factor T and the half-angle FoV α then depend on the defocusing p_2. Note that T excludes the fill factor loss of the PD, which is about −7.2 dB for the composite 4 × 4 photodiode matrix with active area diameter $D_{\text{PD}} = 1.32$ mm adopted in our system (see Section 7.4 and [8-1]). Accepting $T > -10$ dB at $d = 200$ cm implies a defocusing at the US receiver $p_2 < 28\%$ and thus yields an FoV α up to 22° half-angle which amply exceeds the FoV of the DS receiver (which is about 10° half-angle, see Section 7.5), as required in the bidirectional system. Note that the FoV of the bidirectional system determines the user area that can be covered; this FoV is given by the minimum of the FoV of the DS link and the FoV of the US link (minus a possible tilt angle of the user device).

The analog 2D US beam steering by means of stepper motors moving the optical source (i.e., the fiber pigtail of a laser diode) at the user's device has been introduced in Section 5.6 and enables accurate alignment of the US beam to the

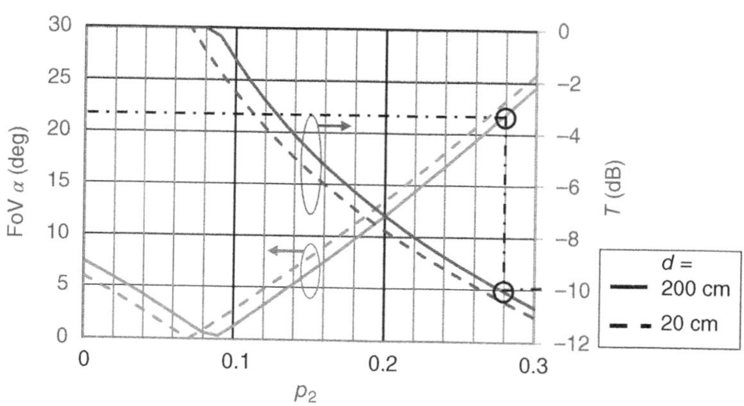

Fig. 8-8 Upstream beam coupling to US receiver at PRA site (US beam spot $D_{\text{beam}} = 15$ mm; $f_1 = 20$ mm, $f_2 = 5$ mm, $D_{\text{PD}} = 1.32$ mm)

aperture of the US Rx mounted at the PRA site at the ceiling, which is achieved by means of the center-of-gravity (CoG) algorithm, as introduced in Section 6.4.

8.6 System Demonstrator

Fig. 8-9 shows the configuration of the bidirectional OWC laboratory system. It adds the US part to our previous DS GbE video streaming setup having a reach of 2 meters (see Section 7.5 and [8-1]). By λ-tuning, the +10 dBm DS beams with a footprint diameter of Ø10 cm are 2D steered by the AWGR-based diffractive unit. Using a 4×4 photodiodes matrix with a diameter of Ø1.32 mm and a Ø50 mm $f = 10$ mm Fresnel lens, the DS-Rx has a FoV $\approx \pm 10°$ (see Section 7.5 and [8-1]). The automatic alignment of the DS beam to the DS-Rx has been achieved by monitoring the reflected power from an RR ring around the aperture of the DS-Rx at the user dongle, where this reflected power is monitored by a photodetector (PD) at the PRA site close to its lens which launches the DS beam. The monitored power is used for the λ-tuning control which drives the 2D DS beam steering as described in detail in Section 6.3.3 and [8-4]. Thus, maximization of this power yields the accurate alignment of the DS beam to the DS-Rx. Note that the DS beam is much wider than the aperture of the DS-Rx, in order to achieve wide area coverage within a limited λ-tuning range.

The demonstrator setup which has been realized in our laboratory is shown in Fig. 8-10, with a close-up view of the ceiling site in Fig. 8-11, and of the user site in Fig. 8-12.

To establish a DS link from the ceiling site two PRA-s are foreseen, with the possibility to switch between them in case there is a line-of-sight blockage for one of them, see Figs. 8-10 and 8-13. PRA1 is composed of a C-band AWGR from Kaiam with 96 flat-top channels spaced at $\Delta\nu = 50$ GHz having a bandwidth $BW = 35$ GHz, adjacent channel isolation $a_{ac} > 24$ dB, and polarization-dependent loss $PDL < 0.5$ dB, in combination with an L-band AWGR from Kylia with 48 channels at $\Delta\nu =$ 50 GHz having a $BW = 24$ GHz, $a_{ac} > 30$ dB, and $PDL < 0.4$ dB; this combination provides 144 output ports of which 128 fit in the entry aperture of the $f/0.95$ $f =$ 50 mm camera lens. Hence, PRA1 can support 128 beams. PRA2 is equipped only with a C-Band AWGR from USource with 96 channels at $\Delta\nu = 50$ GHz, $BW = 35$ GHz, $a_{nc} > 21$ dB, and $PDL < 0.65$ dB. Hence, it can support 96 beams.

In the US link, a 2 dBm beam at $\lambda \approx 1.5$ µm with Ø15 mm diameter footprint at the US-Rx is launched using a triplet lens with $f = 20$ mm. Note that the US beam power is well below the eye safety threshold of 10 dBm. Although the beam is slightly diverging, this safety aspect is particularly important as the user may be very close to his US unit. When well aligned, the US beam fits fully within the Ø25 mm aperture of the US-Rx at the ceiling; thus, sufficient link budget for good

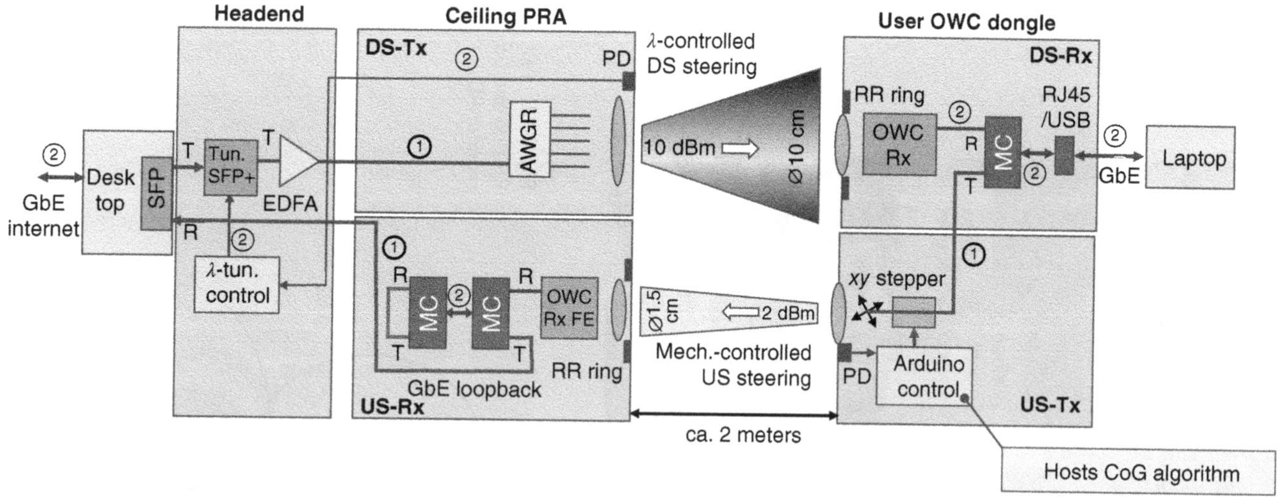

Fig. 8-9 Bidirectional OWC laboratory system, including both the downstream beam alignment and upstream automated beam alignment (①: fiber; ②: copper lines)

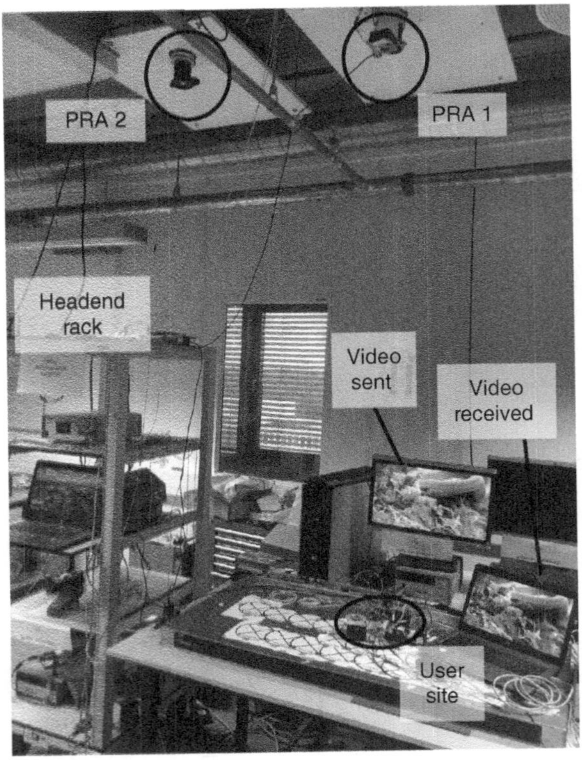

Fig. 8-10 Laboratory demonstration of GbE video streaming via bidirectional OWC link to a laptop computer

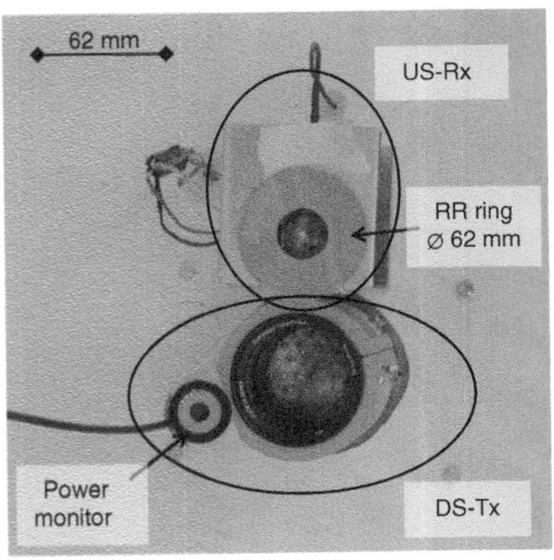

Fig. 8-11 PRA site: US-Rx with RR ring and DS-Tx with power monitor PD

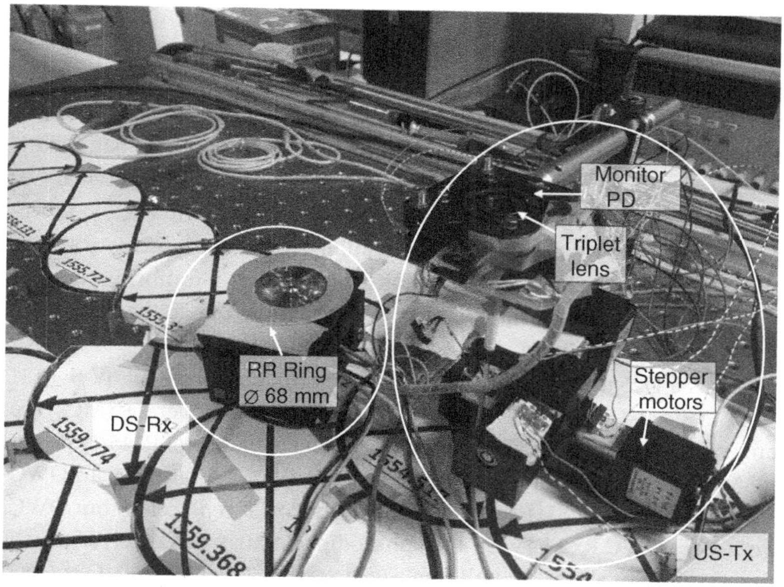

Fig. 8-12 User site: US-Tx with stepper motors, triplet lens, monitor PD, and DS-Rx with RR ring

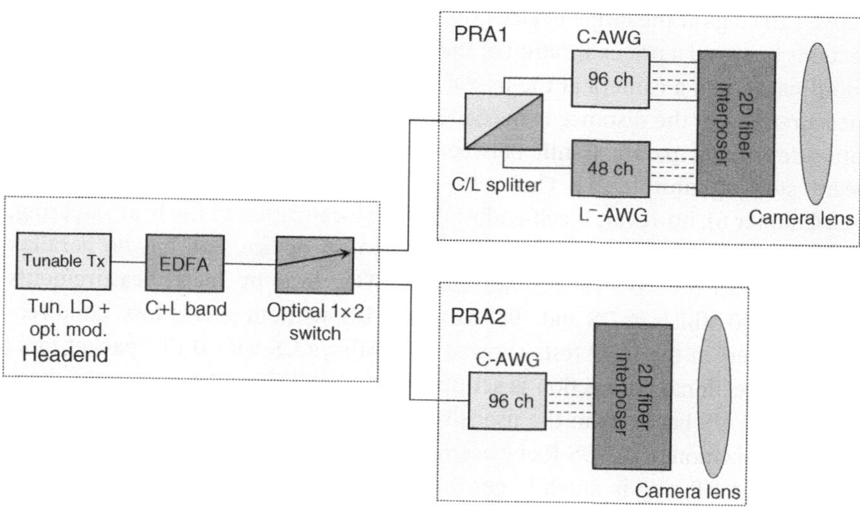

Fig. 8-13 Headend transmitter site with PRA selection switch and the two PRA-s

GbE US connectivity is achieved, although the US beam power is only 2 dBm. The beam is mechanically 2D steered by a NEMA11 *xy*-stepper motor stage, which is controlled by an Arduino board. This board performs the CoG automatic beam alignment algorithm according to Eq. (6-8) and gets its input from monitoring the power reflected by the RR ring at the PRA by means of three low-bandwidth Ø1 mm diameter photodiodes at the user site. As presented in Section 6.4.3, Fig. 6-29 shows the experimental results of monitoring this power. The 3D representation of the experimental data shows a similar shape as the analytical one shown in Fig. 6-23 in Section 6.4.1 which was calculated for a uniform scanning beam. The 3D shape of the experimental data is less smooth, due to the noise fluctuations in the monitored power.

The US-Rx has a Ø25 mm Fresnel lens with $f = 5$ mm, achieving a FoV $\approx \pm 12°$ (cf. Section 7.3.2 Fig. 7-15 and [8-1]). Commercial GbE media converters (MCs) are used to convert the bidirectional GbE data from optical to electrical RJ45/USB signals. These MCs have SMF optical inputs and outputs. For operating as an OWC receiver, the optical receiver SMF front-end of the first MC is replaced by our OWC receiver front-end (OWC Rx FE) with 4×4 photodiode matrix TIA receiver which has an enhanced FoV of about 10° half-angle (see Section 7.5 and [8-1]). Our CoG US beam alignment algorithm executed by the Arduino board successfully achieved US connectivity within 10 seconds for a step size of 6 mm, and even within 4 seconds for 7.5 mm. This time is mainly due to the relatively slow stepper motors. Other localization techniques (such as using camera observation and active LED tags at the target device [8-9, 8-10]) may need only a few tens of milliseconds but need a pre-calibration of the localization results to the beam steering coordinates, and a camera at the user device for US beam alignment. Moreover, measurement of the distance is needed in order to enable correction for the distance-dependent parallax shift between the camera observation unit and the beam steering unit [8-21]. Our retro-reflection-based localization technique (see Chapter 6), however, is self-calibrating the localization to the beam steering, requires only limited signal processing at the user device, and has no parallax issues. Within the FoV range, bidirectional TCP tests by Iperf measurements showed ~940 Mbit/s in DS and ~939 Mbit/s in US, without packet loss. Unidirectionally, Iperf in the UDP tests showed ~958 Mbit/s US with 0.18% packet loss.

The bidirectional connection is set up in two steps. First, the automatic alignment of the US beam upon the user's request is to be established. It starts with coarse localization of the US-Rx by beam scanning from the user device with large steps. As the RR ring is much larger than the aperture of the US-Rx, the coarse search time is reduced remarkably with respect to using no RR ring. Next, fine localization is done with small steps of about 6 mm using the CoG algorithm as proposed in Section 6.4.1, which processes the monitored power reflected by the RR ring (as represented in Fig. 6-29). This US-Rx localization takes about

10 seconds, mainly limited by the stepper motor speed. When the scan step size is enlarged, the alignment time will decrease inversely proportional to the square of the step size, but the accuracy will remarkably degrade (cf. Fig. 6-24).

Second, the DS link is aligned automatically by monitoring the retroreflected power from the passive RR ring at the DS-Rx of the user's laptop and using the wavelength-tuning control procedure we have reported before in [8-4] and have discussed in Section 6.3. The DS alignment time is less than 15 seconds and is mainly limited by the LabVIEW control program.

8.7 Conclusions

GbE bidirectional OWC transmission for high-density user connections using narrow infrared beams with automatic self-calibrated alignment of the US beam by means of a novel steering control protocol using the center-of-gravity (CoG) algorithm as well as automatic self-calibrated alignment of the DS beam has been designed, realized, and implemented in a laboratory testbed. Broadband receivers with a wide FoV $\approx \pm 10°$ were employed, which eased the required beam alignment accuracy. With low-aberration triplet lenses, a good accuracy of the CoG algorithm and hence adequate US beam steering has been achieved within 200 μm accuracy and 10 seconds, mainly limited by the speed of the stepper motors. The automatic DS beam alignment takes less than 15 seconds, mainly limited by the wavelength-tuning control of the DS beam steering. The time consumed by the US and DS beam steering limits the system's suitability for mobile users. Its major benefits are for the delivery of dense high-capacity services to nomadic or static users. TCP measurements show 940 Mbit/s transfer speeds per user within the FoV $\approx \pm 10°$. Real-time high-definition GbE video streaming to laptop computers via the bidirectional OWC link has been demonstrated.

References

[8-1] A.M.J. Koonen, K.A. Mekonnen, Z. Cao, et al., "Beam-steered optical wireless communication for industry 4.0," *IEEE J. Sel. Top. Quantum Electron.*, vol. 27, no. 6, art. no. 6000510, Nov./Dec. 2021, https://doi.org/10.1109/JSTQE.2021.3092837.

[8-2] A.M.J. Koonen, K. A. Mekonnen, F.M. Huijskens, and E. Tangdiongga, "Bi-directional all-optical wireless communication system with optical beam steering and automatic self-alignment," in *Proc. ECOC 2022*, Basel, Sep. 18–22, 2022, Paper Tu4F.1.

[8-3] A.M.J. Koonen, F. Gomez-Agis, F.M. Huijskens, and K.A. Mekonnen, "High-capacity optical wireless communication using two-dimensional IR beam steering," *J. Lightwave Technol.*, vol. 36, no. 19, pp. 4486–4493, Oct. 2018, https://doi.org/10.1109/JLT.2018.2834374.

[8-4] A.M.J. Koonen, K.A. Mekonnen, F.M. Huijskens, et al., "Fully passive user localization for beam-steered high-capacity optical wireless communication system," *J. Lightwave Technol.*, vol. 38, no. 10, pp. 2842–2848, May 2020, https://doi.org/10.1109/JLT.2020.2980428.

[8-5] H. Takesue, and T. Sugie, "Wavelength channel data rewrite using saturated SOA modulator for WDM networks with centralized light sources," *IEEE/OSA J. Lightwave Technol.*, vol. 21, no. 11, pp. 2546–2556, Nov. 2003.

[8-6] H. D. Kim, S.-G. Kang, and C.-H. Lee, "A low-cost WDM source with an ASE injected Fabry–Pérot semiconductor laser," *IEEE Photon. Technol. Lett.*, vol. 12, no. 8, pp. 1067–1069, Aug. 2000.

[8-7] N. Dupuis, J. Décobert, C. Jany, et al., "10-Gb/s AlGaInAs colorless remote amplified modulator by selective area growth for wavelength agnostic networks," *IEEE Photon. Technol. Lett.*, vol. 20, no. 21, pp. 1808–1810, Nov. 2008.

[8-8] K. Ławniczuk, O. Patard, R. Guillamet, et al., "40-Gb/s colorless reflective amplified modulator," *IEEE Photon. Technol. Lett.*, vol. 25, no. 4, pp. 341–343, Feb. 2013.

[8-9] A. Gomez, K. Shi, C. Quintana, et al., "A 50 Gb/s transparent indoor optical wireless communications link with an integrated localization and tracking system," *IEEE/OSA J. Lightwave Technol.*, vol. 34, no. 10, pp. 2510–2517, May 2016.

[8-10] N.Q. Pham, K. Mekonnen, E. Tangdiongga, et al., "User localization and upstream signaling for beam-steered infrared light communication system," *IEEE Photon. Technol. Lett.*, vol. 33, no. 11, pp. 545–548, 2021.

[8-11] A.M.J. Koonen, C.W. Oh, K. Mekonnen, et al., "Ultra-high capacity indoor optical wireless communication using 2D-steered pencil beams," *J. Lightwave Technol.*, vol. 34, no. 20, pp. 4802–4809, Oct. 2016, https://doi.org/10.1109/JLT.2016.2574855.

[8-12] K. Wang, A. Nirmalathas, C. Lim, et al., "Full-duplex gigabit indoor optical wireless communication system with CAP modulation," *IEEE Photon. Technol. Lett.*, vol. 28, no. 7, pp. 790–793, Apr. 2016.

[8-13] A. Gomez, K. Shi, C. Quintana, et al., "Beyond 100-Gb/s indoor wide field-of-view optical wireless communications," *IEEE Photon. Technol. Lett.*, vol. 27, no. 4, pp. 367–370, Feb. 2015.

[8-14] M. Faulwaßer, F. Deicke, and T. Schneider, "10 Gbit/s bidirectional optical wireless communication module for docking devices," *Globecom Workshop on OWC*, Austin, Dec. 2014.

[8-15] C.W. Oh, Z. Cao, E. Tangdiongga, and A.M.J. Koonen, "10 Gbps all-optical full-duplex indoor optical wireless communication with wavelength reuse", *Proc. OFC 2016*, Anaheim, Mar. 20–22, 2016, paper Th4A.6.

[8-16] Time division multiple access – passive optical network (TDMA-PON) [online]. Available: https://www.itu.int/en/ITU-T/studygroups/2017-2020/15/Documents/OFC2018-2-Q2_v5.pdf

[8-17] Fresnel lens concept [online]. Available: https://en.wikipedia.org/wiki/Fresnel_lens

[8-18] Triplet lens concept [online]. Available: https://en.wikipedia.org/wiki/Triplet_lens

[8-19] Fresnel lens [online]. Available: https://en.wikipedia.org/wiki/Fresnel_lenshttps://www.edmundoptics.com/p/098-x-098-020-focal-length-fresnel-lens/42289/

[8-20] Triplet lens [online]. Available: https://www.edmundoptics.com/f/hastings-triplet-achromatic-lenses/11968/

[8-21] Parallax errors in optical systems [online]. Available: https://en.m.wikipedia.org/wiki/Parallax

9

Conclusions and Outlook on Further Research

9.1 Conclusions

Optical wireless communication (OWC) techniques offer powerful solutions to resolve the imminent congestion in radio-based communication networks, particularly in local area wireless networks. By deploying steerable narrow infrared optical beams which address each user device individually, it can provide links with ultrahigh capacity where and when needed, while also minimizing the power consumption needed and offering high privacy. It can provide the ultimate capabilities for delivering broadband services wirelessly in human-to-machine and machine-to-machine communications. The small footprint of a beam enables the creation of a pico-cell per user device, and thus can create a high degree of spatial multiplexing which boosts the total wireless system's throughput. As the beams can deliver communication capacity on demand of individual users, the system achieves a high efficiency in bringing the data signals to the users and thus can reduce remarkably the power consumption needed with respect to a radio-based mm-wave communication system. Moreover, as nothing travels faster than light in vacuum (and air), a minimum in latency is achieved which makes it also eminently suited for time-critical machine-to-machine interactions. Radio beams travel equally fast, but due to the lower bandwidths accessible in a radio channel, extra processing time is needed for signal conditioning, which adds to the radio link's latency.

OWC by two-dimensional (2D)-steered narrow infrared beams is not foreseen to fully replace radio-based wireless communication, among others because it needs line-of-sight and the beam steering processes do not support high user mobility as well as radio-based communication in the ISM band (frequency band for industrial, medical and scientific purposes) and wide-beam VLC/LiFi communication

Short-Reach Optical Wireless Communication: By Directed Narrow Beams, First Edition. Ton Koonen.

can do. However, it can off-load high-speed traffic loads from radio-based networks and wide-beam OWC networks and thus gives these the room necessary to handle the booming amounts of low-speed intermittent traffic as, for example, generated by the emerging myriad of small Internet-of-Things devices. Its main benefits arise when broadband services need to be delivered wirelessly to/from nomadic or static users with no or limited in-service mobility demands.

In this book, the focus has been on the definition, design, implementation, and experimental validation of a short-reach indoor OWC system that deploys narrow, individually 2D-steerable infrared beams to deliver broadband services to multiple closely spaced users. The key techniques required for this system include: emitting and accurately steering multiple narrow beams, localization of the user devices, and receiving the beams for establishing high-capacity links without requiring tedious alignment. The main results are:

- *Definition of an indoor physically point-to-multipoint OWC system architecture*, along with its design rules based on wavelength-tuned individual beam steering using passive diffractive modules. The architecture is readily scalable to serve multiple users individually and wirelessly with broadband services. It is also maintenance-lean and effectively integrates with a Fiber-in-the-Home network architecture by consolidating the OWC beam generation and steering functions in a common central unit. *This is reported in Chapter 4.*
- *2D (semi-)discrete beam steering by wavelength tuning* in two implementation options (intended for downstream beam steering):
  - *A passive diffractive module deploying a pair of crossed diffraction gratings*. Its design rules have been established, and semi-discrete 2D steering over an angular area of 6° × 12° has been demonstrated by wavelength tuning over 125 nm in the 1.5 μm window, as well as delivery of 42.8 Gbit/s data capacity per beam using PAM-4 modulation over a 3 meters link.
  - *A passive diffractive module deploying an arrayed waveguide grating router (AWGR)* with its output fiber ports arranged in a 2D matrix put in front of a lens. The design rules have been set, and discrete 2D steering up to 128 beams has been demonstrated over an angular area of 17° × 17° by wavelength tuning using a C+L⁻-band composite AWGR with 128 output ports. A transport capacity of 112 Gbit/s per beam using PAM-4 modulation over a 3.4 meters link has been achieved with an 80-ports C-band AWGR, implying the feasibility of a total optical wireless system throughput beyond 8.9 Tbit/s.

The steering time with these diffractive modules is mainly limited by the time required to control the laser wavelength tuning. This time is within the range of milliseconds. *This is reported in Chapter 5.*

- *2D continuous beam steering by an active mechanically 2D translating module* deploying a stage of two linear stepper motor translators which moves the output fiber of a laser diode in front of a lens. This steering method is intended for upstream steering of a narrow beam from the user's module which is cost-sensitive and (typically) battery operated. The mechanical steering allows the laser diode to operate at an arbitrary wavelength, which reduces costs. The use of stepper motors which do not need powering as soon as beam alignment has been achieved saves energy from the user's battery. 2D angular tuning within a range of $\pm 10.8°$ by $\pm 10.8°$ has been realized, which fits within the FoV of the upstream receiver, as per the requirement. The steering time is mainly limited by the speed of the 2D stepper motor translators and their control, which may take a few seconds to cover the entire user area (128 cells). *This is reported in Chapter 5.*

- *Localization of the users that is needed for accurately directing the downstream and upstream beams.* This is done using beam scanning and a ring-shaped matrix of passive miniature retroreflectors placed around the aperture of the targeted receiver. Localization processing is required only at one side without any pre-existing return link; it enables setting up a connection from scratch, without position calibration. Alignment of the downstream beams is achieved through discrete scanning by wavelength tuning the oversized Ø10 cm beams by means of the diffractive beam steerer and processing the monitored reflected power. Experimentally, an adequate establishment of a downstream link has been achieved in less than 15 seconds, with the main limitation being the LabVIEW program's control of the scanning process and the user area of 128 cells (each Ø10 cm). Alignment of the narrow Ø15 mm upstream beams is achieved by nearly continuous scanning using the 2D mechanical stepper motor stage in conjunction with a center-of-gravity algorithm that processes the monitored reflected power. Adequate upstream beam alignment was experimentally achieved within 10 seconds, mainly limited by the low speed of the stepper motors. *This is reported in Chapter 6.*
- *Broadband receiver with large aperture and wide Field-of-View*: a novel receiver concept has been proposed which deploys a 2D matrix of interconnected photodiodes. While enlarging the light-detecting aperture, the $M \times M$ photodiode matrix (PD matrix) provides a bandwidth that matches with a single photodiode and an output current that is M times larger. By theoretical analysis, the compromise between the beam power-to-photodiode coupling and the FoV has been modeled and assessed for realistic beam profiles and lens designs by extensive (home-built, as described in the Appendix) ray tracing. The SNR performance of the PD matrix-based receiver has been analyzed and found to be better than that of an angular diversity receiver with a similar amount of photodiodes. A receiver with a 4×4 interconnected matrix of Ø150 μm photodiodes has been built in conjunction with a commercially available 750 MHz transimpedance amplifier and a large-NA Ø50 mm Fresnel lens with $f = 10$ mm. In our OWC laboratory testbed, experimentally, a -3 dB bandwidth of the receiver of 670 MHz has been achieved, and an FoV $\approx \pm 10°$ when using Gaussian beams with a waist

of Ø10 cm, which aligns well with the ray tracing analyses. Crosstalk from neighboring cells may occur when shifting the receiver away from the signal cell's center across the user plane; analysis has shown that with Gaussian Ø10 cm beams and a Ø50 mm receiver lens aperture, a shift up to 32 mm is allowed for a power penalty of <0.5 dB. Regarding the FoV performance, the 4 × 4 PD matrix receiver is clearly less affected by the shifting than the single-PD receiver. By scaling the photodiode matrix to larger $M \times M$ dimensions, the FoV versus beam-to-PD matrix coupling performance can be improved further, although the improvement steps shrink at large M. *This is reported in Chapter 7.*

- *Bidirectional all-optical OWC indoor system*: in order to adhere to the advantages of the physical point-to-multipoint architecture, and to implement the user's terminal cost-effectively, the upstream paths should not just be a copy of the downstream path. A dedicated design with an arbitrary-wavelength laser and mechanical 2D upstream beam steering with a 2D stepper motor stage as reported in Section 6.4 was made. By integrating
  - two ceiling units (containing the AWGR-based diffractive 2D steerer for the Ø10 cm downstream beams as described in Section 5.5, and the upstream receiver equipped with 4 × 4 photodiode matrix and a Ø25 mm Fresnel lens with $f = 5$ mm, see Chapter 7, surrounded by a retro-reflecting ring for upstream localization, see Section 6.4),
  - the user unit (with the mechanical 2D steerer for the Ø15 mm upstream beam, see Section 5.6, and the downstream receiver with 4 × 4 photodiode matrix, a Ø50 mm Fresnel lens with $f = 10$ mm, surrounded by a retro-reflecting ring for downstream localization, see Section 6.3, and a media converter providing the I/O functions to connect a laptop), and
  - the remote headend rack (containing the tunable wavelength transmitters, optical cross-connect switch for selecting a ceiling unit, and the control by means of a laptop computer),

the bidirectional OWC system demonstrator has been realized. Its performance for bidirectional GbE data communication has been verified; TCP measurements show 940 Mbit/s transfer speeds within the FoV of ±10° in the user area at link lengths of about 2.5 meter (ceiling-to-user plane). Live high-definition video streaming to laptop computers has been demonstrated successfully. *This is reported in Chapter 8.*

9.2 Outlook and Suggestions for Further Research

The research efforts reported in this book were directed toward the design, analysis, implementation, and experimental validation of an OWC system for broadband short-reach indoor communication, including its key functions, to nomadic or stationary users. As the beam-steered OWC research is still in its early phase, it is a fertile domain for the next research steps. Some of them are discussed in the following sections.

9.2.1 Applying the OWC Key Functions in Other OWC Domains

Further research steps may be taken in translating these efforts toward other OWC application domains in which medium-reach and long-reach communication is needed. Such domains will often involve outdoor settings in which atmospheric conditions may lead to fluctuations in the losses and the phase of the optical path (scintillation). Additional signal conditioning methods (such as space–time block coding [9-1]) may be investigated to reduce the impact of these fluctuations.

The key functions as investigated in this book for short-reach OWC may mutatis-mutandis be applicable for OWC in other domains, such as

- for medium reach: underwater settings (e.g., for controlling underwater unmanned maintenance/repair robots; typically blue-to-green light beams are used) [9-2], road traffic (car-to-car, car-to-roadside), street level (shopping areas), sport stadiums, backhauling (supporting mobile network antenna stations), drone communication (for inter-drone communication and coordination), fixed wireless drop links to connect remote buildings, etc.
- for long/ultra-long reach: ground-to-satellite, inter-satellite, ground-to-plane, and inter-plane communications.

In general, at longer path lengths, the angular accuracy of the beam steering and localization becomes more delicate, which requires making their discretization steps in the steering much smaller (or even allow only continuous steering). Moreover, path losses will grow, and beam power and/or receiver sensitivity need to be improved. On the other hand, the angular steering range and the FoV of the receiver can be reduced.

9.2.2 Photonic Integration

To bring the OWC modules to a higher technology readiness level (TRL), photonic integrated circuit (PIC) technologies are needed, particularly where mass applications are sought, such as for indoor short-range OWC. PIC technologies may be beneficial to reduce the physical volume of these modules and their assembly efforts, to reduce their power consumption, and to improve their robustness and reproducibility. For example, the AWGR-based diffractive 2D beam steerer concept as proposed in Section 5.5 may be realized in a passive PIC built in a silicon-on-oxide technology, where the AWGR with a very high port count is co-integrated with a 2D array of surface-emitting slanted waveguides, see Fig. 9-1. Thus, when co-assembling this PIC with a lens, a compact and robust passive diffractive 2D beam steerer can be built.

9.2.3 Next 2D Beam-Steering Concepts

The diffractive beam steerer concepts explored in this book map a one-dimensional variable, the wavelength λ, to 2 orthogonal angular dimensions φ

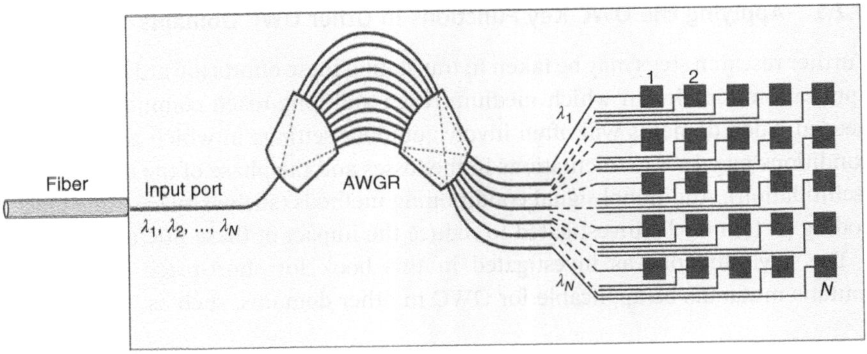

Fig. 9-1 Integrated passive AWGR-based beam steerer (AWGR with N output waveguides, terminated in vertically emitting slanted endfaces 1..N)

and ψ in order to provide coverage of the 2D user plane (cf. Fig. 4-3). Such mapping inevitably implies discontinuities in the steering (cf. Fig. 5-11 and Fig. 5-22). As the steering cannot be fully continuous, the beams can only make discrete steps and create discrete cells in addressing the user plane. A beam cannot stay centered on the user's receiver when the latter is moving in an arbitrary direction, so the performance of the downstream link may vary depending on the user's position. When 2D variables could be mapped to the orthogonal angular dimensions φ and ψ, continuous steering would be possible, the beam could stay centered on the user's receiver when it moves, and the issue of performance variations when moving could be eliminated. Part of the attractiveness of the single variable λ for the 2D tuning is that it can be done with a remote optical source (a tunable laser diode), that this parameter is not prone to random changes when it is brought (by fiber) to the beam steerer, and the steering unit (typically at the ceiling) is passive, which is attractive for operation, maintenance, and installation. Next to the one-dimensional λ, one can think of a second-dimensional optical variable, for example, the polarization angle θ of a linearly polarized output signal of the λ-tunable laser. In previous work, doubling the number of output beams of an AWGR-based beam steerer without increasing crosstalk levels has been proposed by using two orthogonal polarization states for the even and odd wavelength channels [9-3]. By introducing a continuously variable linear polarization angle θ, both the wavelength λ and the linear polarization state θ can be mapped to the two orthogonal angular beam steering dimensions φ and ψ, and 2D continuous steering would be possible. This (λ, θ) tuning can be done independently in each of the optical sources in the remote headend station (in the central communication controller, CCC). Polarization-maintaining fiber will have to be used for bringing the

(λ, θ) tuned signals to the 2D beam steerer. A new passive 2D steering module needs to be devised that can steer each beam under the joint control of the beam signal's wavelength λ and its linear polarization state θ. Thus, such a passive module can handle multiple beams independently, and the number of beams to be steered can readily be scaled.

9.2.4 Next Localization Schemes

Localizing the user by means of beam scanning takes less time if the beam diameter becomes increased, but then the localization becomes also less accurate. To reduce scanning time but maintain accuracy, it will be beneficial to adapt the beam diameter from relatively large in the initial coarse scanning phase, to small in the finalizing fine scanning phase. Such adaptive control of the beam diameter may be achieved by a lens with a variable focal length, for example, a tunable liquid lens [9-4], and may be driven by the localization processor. This concept may not be suited for the downstream beams in the diffractive beam steerers reported in Sections 5.4 and 5.5, as they use a lens that is jointly used by all downstream beams. For the upstream links, where only a single beam is launched, this concept may be interesting to explore in depth.

9.2.5 Next OWC Receiver Concepts

Higher data rates and/or longer reaches will require a higher OWC receiver sensitivity. Coherent detection techniques in optical fiber transmission have been key for introducing more comprehensive (two-dimensionally modulated, such as quadrature amplitude modulation, QAM) signals with larger information content per symbol, and thus higher data rates within the same bandwidth. Moreover, by means of mixing with a local oscillator, signal amplification is achieved at the receiver's input and thus the receiver sensitivity is enhanced. In line with these coherent technologies for fiber-based transmission, an OWC receiver equipped with coherent detection techniques would be attractive for higher data rates and sensitivity enhancement. The performance of the OWC coherent receiver should not depend on the polarization state of the incoming signal (which may change due to random scintillation). In analogy with coherent receivers in optical fiber networks, the OWC coherent receiver may be designed to include a polarization-diversity scheme. Besides, to reduce the impact of phase noise of the local oscillator laser, the design may include a phase diversity scheme as well. A research challenge will be to devise a coherent OWC receiver which is based on the FoV-enlarging 2D photodiode matrix concept we reported in Chapter 7 of this book and which is extended with the coherent polarization- and phase-diversity detection schemes.

9.2.6 New OWC Networking Aspects[1]

As already mentioned in Section 8.4, in the present bidirectional indoor OWC system demonstrator the upstream receiver in the ceiling unit may need to handle the upstream traffic from multiple users. Therefore, for each upstream link, an upstream medium access protocol is needed for which a TDMA protocol – such as what is commonly used in TDMA-PON – may be investigated [9-5].

If required, higher data rates per optical wireless link may be created by putting multiple wavelength channels into a beam. When deploying the AWGR-based beam steerer with a single input fiber (see Section 5.5, Fig. 5-17), multiple wavelength channels can be multiplexed into the same beam if their wavelengths are separated by (a multiple of) the FSR of the AWGR, causing them to emerge from the same output port of the AWGR. This requires a large wavelength tuning range of the beam-generating laser diodes which can cover multiple FSRs of the AWGR. Alternatively, an AWGR with multiple input fiber ports may be employed (e.g., see [9-6]). By making use of the wavelength-cyclic routing characteristics of an AWGR [9-7], additional input fibers can carry wavelength channels that are cyclically multiplexed into the beam emerging from the same output port of the AWGR. For instance, Fig. 9-2 illustrates this concept for two input fibers of the AWGR, effectively doubling the capacity of a beam. Basically, the number of input fibers of the AWGR may be as large as the number N of output fibers, while preserving the cyclic routing properties of the AWGR. Thus, the capacity of a beam can be N-fold the capacity of a single wavelength channel (see Fig. 9-3).

At the receiver side, the wavelength channels in the beam subsequently need to be separated (that is, demultiplexed) by free-space optical filters which will feed them to separate OWC receivers.

While this multi-input AWGR solution includes multiple independent wavelength channels within a single optical beam and multiple such beams are steered independently in two dimensions, the end-to-end solution of an OWC system transporting such beams also requires a receiver which can capture the beam and separate the wavelength channels within it, that is, a wavelength-demultiplexing free-space optical receiver. This wavelength demultiplexing cannot be achieved straightforwardly by, for example, putting predefined wavelength filters in the receiver, as the wavelengths will change depending on the 2D steering direction established by the AWGR-based beam steerer. Using the cyclic wavelength routing property of the AWGR deployed in the beam steerer, the multiple wavelengths in a beam will

1 This section is partly based on Dutch patent application N2037871, "An optical wireless communication system comprising a transmitter and a receiver as well as a corresponding method of operating the optical wireless communication system and a corresponding transmitter and receiver for operating in the optical wireless communication system," inventor A.M.J. Koonen, filed June 6, 2024.

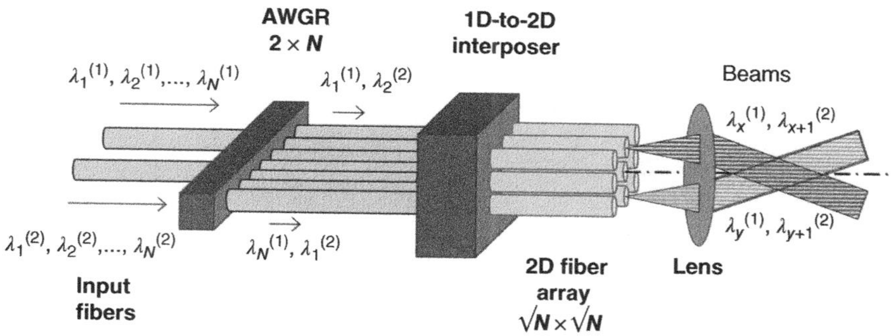

Fig. 9-2 Wavelength multiplexing of signals into the same 2D-steered beam by adding input ports to the AWGR

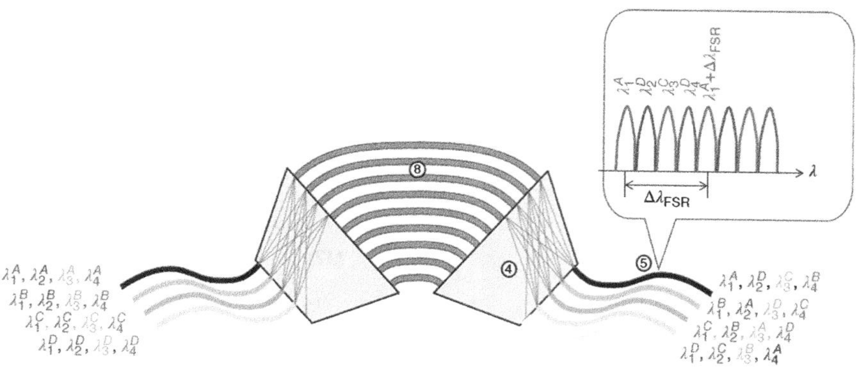

Fig. 9-3 Cyclic wavelength routing properties of an AWGR with multiple input ports

be adjacent to each other and spaced by the AWGR's wavelength grid, but their absolute wavelengths vary when the beam is steered. Hence, the wavelength demultiplexing at the OWC receiver may be done with a wavelength filter having periodic characteristics. In integrated optics technology, such a filter can be realized readily with a Mach-Zehnder Interferometer (MZI) configuration (e.g., see [9-8]). A basic MZI stage and its periodic filtering characteristics are shown in Fig. 9-4.a and .b, respectively. It separates the incoming wavelengths into two groups, where the grid spacing $\Delta\lambda$ of its periodic filtering is (with optical path length difference $n\,\Delta L$ between the MZI arms)

$$\Delta\lambda = \frac{\lambda^2}{\lambda + n\,\Delta L} \approx \frac{\lambda^2}{n\,\Delta L} \tag{9-1}$$

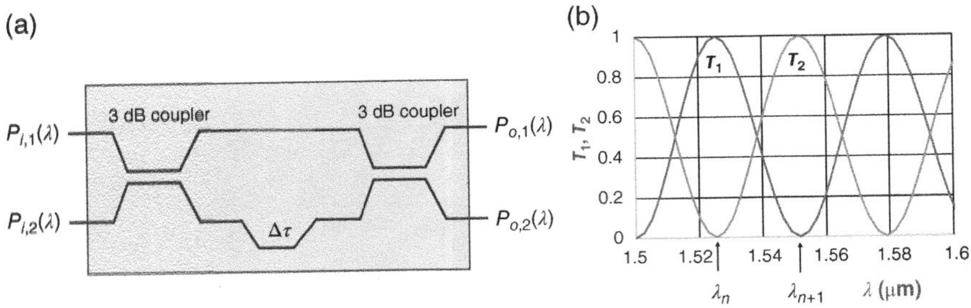

Fig. 9-4 Basic Mach-Zehnder interferometer in integrated optics technology. (a) Configuration in integrated optics waveguide technology, with delay $\Delta\tau = n\,\Delta L/c_0$ in one arm and c_0 light velocity in vacuum. (b) Periodic wavelength filtering characteristics

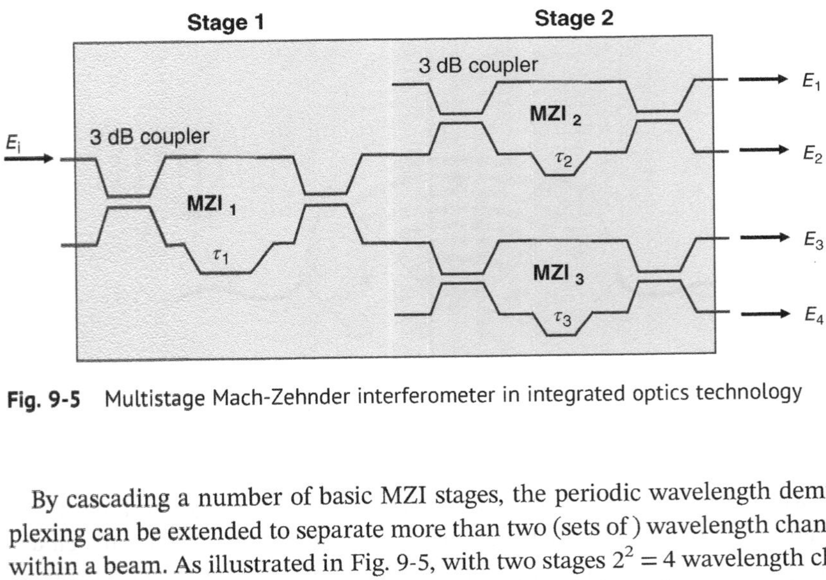

Fig. 9-5 Multistage Mach-Zehnder interferometer in integrated optics technology

By cascading a number of basic MZI stages, the periodic wavelength demultiplexing can be extended to separate more than two (sets of) wavelength channels within a beam. As illustrated in Fig. 9-5, with two stages $2^2 = 4$ wavelength channels per beam can be demultiplexed. The wavelength-dependent power transfer characteristics of a four-channel two-stage MZI demultiplexer based on the model shown in Fig. 9-5 can be derived as

$$T_1 = \left|\frac{E_1}{E_i}\right|^2 = \sin^2\left(\frac{\pi n \Delta L_1}{\lambda}\right)\cos^2\left(\frac{\pi n \Delta L_2}{\lambda}\right)$$

$$T_2 = \left|\frac{E_1}{E_i}\right|^2 = \sin^2\left(\frac{\pi n \Delta L_1}{\lambda}\right)\sin^2\left(\frac{\pi n \Delta L_2}{\lambda}\right)$$

$$T_3 = \left|\frac{E_1}{E_i}\right|^2 = \cos^2\left(\frac{\pi n \Delta L_1}{\lambda}\right)\sin^2\left(\frac{\pi n \Delta L_3}{\lambda}\right)$$

$$T_4 = \left|\frac{E_1}{E_i}\right|^2 = \cos^2\left(\frac{\pi n \Delta L_1}{\lambda}\right)\cos^2\left(\frac{\pi n \Delta L_3}{\lambda}\right) \tag{9-2}$$

where $\Delta L_i = \frac{c_0\,\Delta_{\tau_i}}{n}$ with $i = 1$–3. For proper separation of four wavelength channels, it can be derived that the delay lines ΔL_2 and ΔL_3 need to differ slightly: if $\Delta L_3 = \Delta L_1/2$ then

$$\Delta L_2 = \frac{1}{2}\,\Delta L_1 - \frac{3\lambda_n}{4n} \tag{9-3}$$

which yields the power transfer characteristics of the two-stage MZI depicted in Fig. 9-6, where an MZI_1 frequency grid spacing of $\Delta\nu_1 = 100$ GHz (and therefore a wavelength channel spacing of 50 GHz, as this spacing is half of $\Delta\nu_1$) is chosen which implies with the speed of light $c_0 = 300{,}000$ km/s and waveguide refractive index $n = 1.5$ for the delay line ΔL_1

$$\Delta L_1 = \frac{c_0}{n\Delta\upsilon} = 2\,\text{mm} \tag{9-4}$$

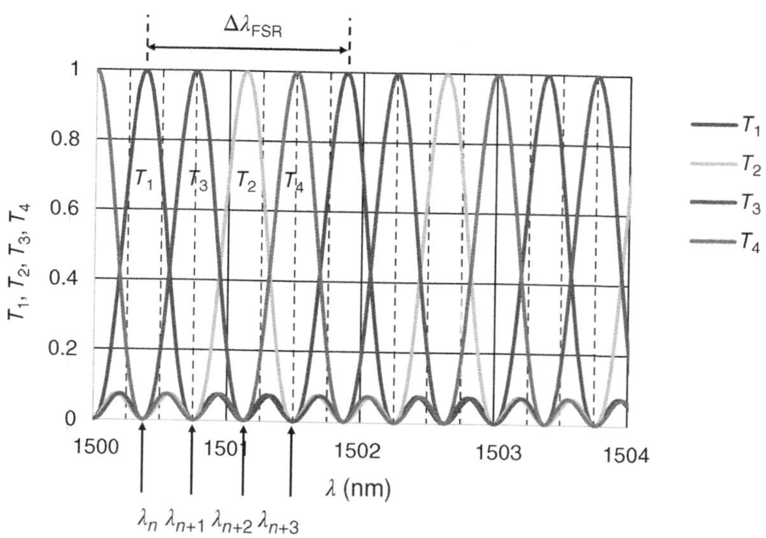

Fig. 9-6 Periodic filtering characteristics of a two-stage Mach-Zehnder interferometer in integrated optics technology

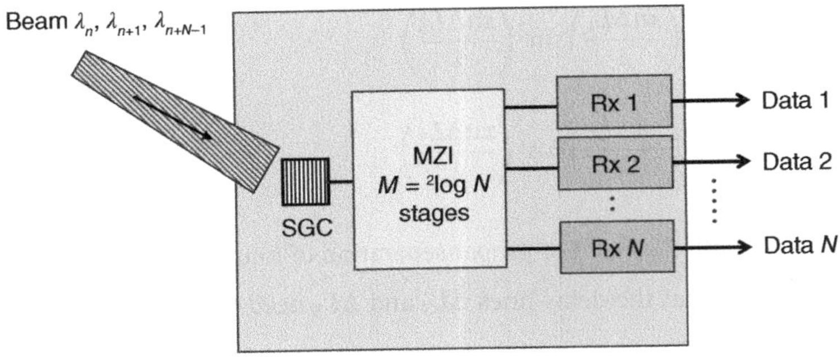

Fig. 9-7 OWC receiver for demultiplexing $N = 2^M$ wavelengths within a steered beam (SGC: surface grating coupler)

The transfer characteristics T_1 through T_4 show how a set of four closely spaced wavelengths λ_n through λ_{n+3} can be separated. They also show that this wavelength separation function is maintained when the set is shifted over a wavelength $\delta\lambda = \Delta\lambda_{FSR}/4$ which matches the spacing between the wavelength channels λ_n through λ_{n+3}. Thus, as strived for, the wavelength channel separation function is not dependent on the absolute wavelength positions of the channels, but just on their relative spacing with respect to each other as defined by the cyclic property of the AWGR in the 2D beam steerer.

Based on such an M-stage MZI structure, a receiver can be built that can separate $N = 2^M$ wavelengths in a beam and feed these to N separate receiver units. As shown in Fig. 9-7, the beam may be captured from free space into an optical waveguide in a photonic-integrated circuit (PIC) by means of a surface grating coupler (SGC; cf. Fig. 3-14), and subsequently be fed to the multistage MZI which is realized in the same PIC. The MZI outputs are feeding the separated wavelength channels to the respective photonic receivers (Rx_1, .., Rx_N) which deliver the demultiplexed data streams (data 1, .., data N).

In narrow-beam-steered OWC communication systems, as studied for indoor application in this book, the beams may be regarded as “virtual fibers”. Building on this analogy, further studies on OWC modules could investigate the equivalents of fiber network functions including optical cross-connecting, add-drop multiplexing, label-controlled wavelength routing, optical repeating with optical amplification, and others.

References

[9-1] Space–time block code [on-line]. Available: https://en.wikipedia.org/wiki/Space_time_block_code

[9-2] K. Wang, T. Song, Y. Wang, et al., “Evolution of short-range optical wireless communications,” *J. Lightwave Technol.*, vol. 41, no. 4, pp. 1019–1040, Feb. 2023, https://doi.org/10.1109/JLT.2022.3215590.

[9-3] X. Zhang, C. Li, Y. Jiao, et al., "Crosstalk-mitigated AWGR-based two-dimensional IR beam-steered indoor optical wireless communication system with a high spatial resolution," *J. Lightwave Technol.*, vol. 37, no. 15, pp. 3713–3720, Aug. 2019, https://doi.org/10.1109/JLT.2019.2917835.

[9-4] Edmund Optics liquid lenses [on-line]. Available: https://www.edmundoptics.com/knowledge-center/application-notes/imaging/introduction-to-liquid-lenses/

[9-5] Time division multiple access – passive optical network (TDMA-PON) [on-line]. Available: https://www.itu.int/en/ITU-T/studygroups/2017-2020/15/Documents/OFC2018-2-Q2_v5.pdf

[9-6] Z. Fan, J. Zhao, and J.-J. He, "High performance demonstration of a 16×16 silica-based cyclic arrayed-waveguide grating router," *Proc. 2021 19th Int. Conf. Optical Communications and Networks (ICOCN)*, Qufu, Aug. 23–27, 2021.

[9-7] Arrayed waveguide grating [on-line]. Available: https://en.wikipedia.org/wiki/Arrayed_waveguide_grating

[9-8] G. Keiser, Ch. 10: WDM concepts and components, *Optical Fiber Communications*, 3rd ed., 2000, Singapore: McGraw-Hill.

Appendix A: 3D Ray Tracing

A.1 Elementary 3D Ray Tracing Algorithms

Algorithms for 3D ray tracing in an optical system by means of vector calculus have been developed and implemented in MATLAB. Optical systems consisting of segmented homogeneous media are considered (e.g., lenses consisting of several homogeneous elements), in which the rays follow piecewise linear paths. This piecewise linear ray tracing reduces calculation times and makes the tracing of high numbers of rays feasible, which enables a comprehensive analysis of lens aberrations and their impact on optical imaging characteristics.

A.1.1 3D Ray Propagation Modeling

An optical ray $\overline{s(t)}$ emerging from a point $\overline{S_0}$ into a direction $\overline{e_s}$ maybe modeled by

$$\overline{s(t)} = \overline{s_0} + t\,\overline{e_s} \tag{A-1}$$

where the unity vector $\overline{e_s}$ in the 3D space is given by its azimuth angle φ and elevation angle θ according to (see Fig. A-1)

$$\overline{e_s} = (\sin\theta\cos\varphi,\ \sin\theta\sin\varphi,\ \cos\theta) \text{ where } |\overline{e_s}| = 1 \tag{A-2}$$

and where t is the parameter equal to the physical path length traveled (in arbitrary units).

A point $\overline{x}$ is located in a plane A which is defined by a point $\overline{P_A}$ located in A and the normal unity vector $\overline{n_A}$ of A with $|\overline{n_A} = 1|$ if (see Fig. A-2)

$$\overline{n_A} \circ \left(\overline{x} - \overline{P_A}\right) = 0 \quad \textit{where } \circ \textit{ denotes the vector inner product operator} \tag{A-3}$$

Short-Reach Optical Wireless Communication: By Directed Narrow Beams, First Edition. Ton Koonen.

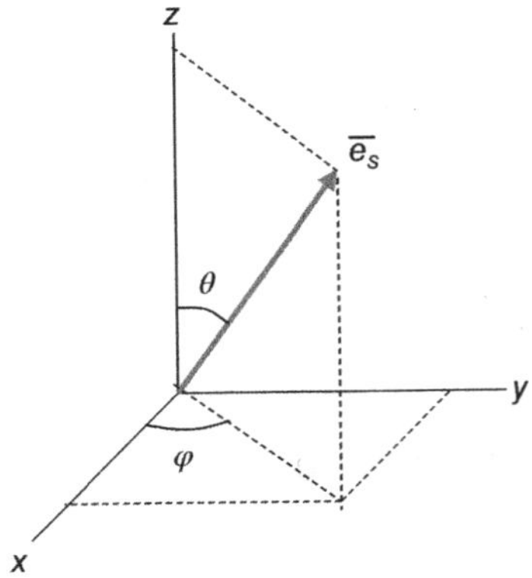

Fig. A-1 Unity vector $\overline{e_s}$

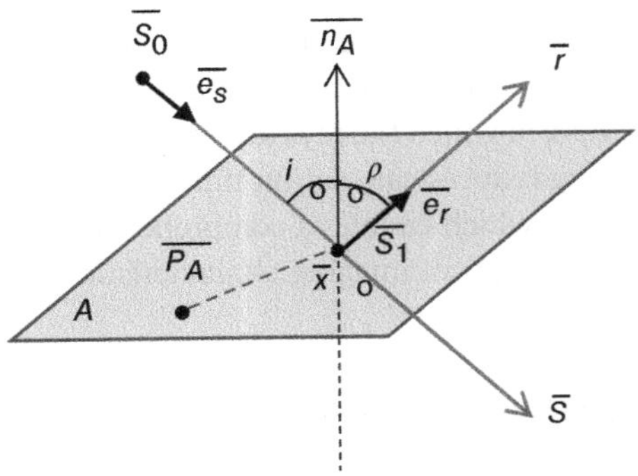

Fig. A-2 Transmission and reflection at plane A

The intersection point $\overline{S_1}$ of the ray $\overline{s}$ with the plane A can then be found by

$$\overline{S_1} = \overline{S_0} + t\,\overline{e_s} \quad \text{where } t = \frac{\overline{n_A} \circ \left(\overline{P_A} - \overline{S_0}\right)}{\overline{n_A} \circ \overline{e_s}} \tag{A-4}$$

The distance d of a point $\overline{Q}$ to the ray $\overline{s(t)}$ can be found from (see Fig. A-3)

$$\left(\overline{Q} - \overline{s(t)}\right) \perp \overline{e_s} \quad \text{hence} \quad t = \left(\overline{Q} - \overline{S_0}\right) \circ \overline{e_s} \tag{A-5}$$

where $\overline{a} \circ \overline{b}$ denotes the inner-product of vectors $\overline{a}$ and $\overline{b}$, which yields

$$d^2 = \left|\overline{Q} - \overline{s(t)}\right|^2 = \left|\overline{Q} - \overline{S_0}\right|^2 - \left[\left(\overline{Q} - \overline{S_0}\right) \circ \overline{e_s}\right]^2 \tag{A-6}$$

Fig. A-3 Distance from point $\overline{Q}$ to ray $\overline{s(t)}$

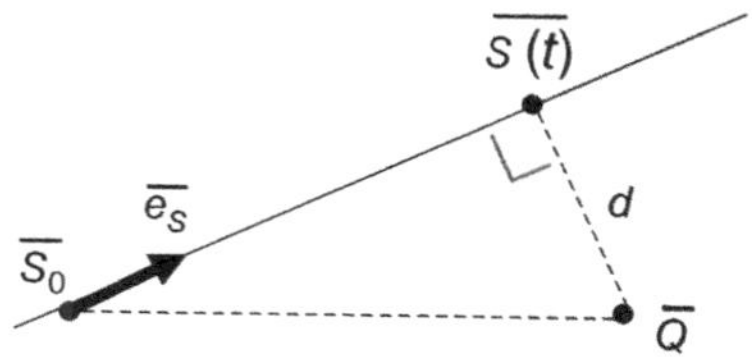

Fig. A-4 Impact of a limiting aperture

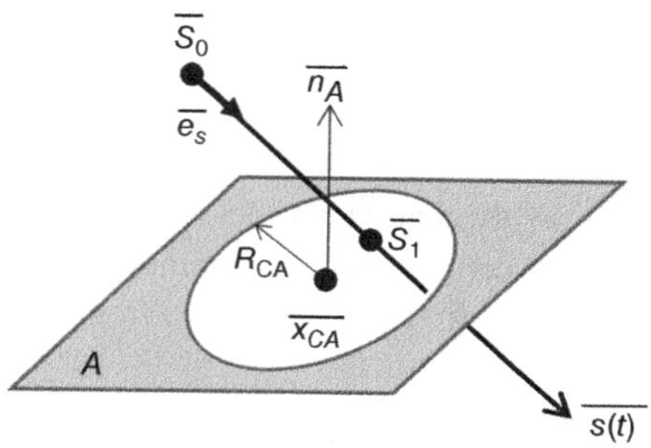

Light rays may be blocked at a limiting aperture; see Fig. A-4. Assume a circular aperture with radius R_{CA} at position $\overline{x_{\mathrm{CA}}}$ in a plane A. The intersection point $\overline{S_1}$ with A is at

$$\overline{S_1} = \overline{S_0} + t\ \overline{e_s} \quad \text{for} \quad t = \frac{\overline{n_A} \circ (\overline{x_{\mathrm{CA}}} - \overline{S_0})}{\overline{n_A} \circ \overline{e_s}} \tag{A-7}$$

where $\overline{n_A}$ is the normal unity vector ($|\overline{n_A} = 1|$)

The ray passes through the aperture if $|\overline{S_1} - \overline{x_{\mathrm{CA}}}| < R_{\mathrm{CA}}$.

A.1.2 Reflection

The ray $\overline{r(t)}$ which is the reflection of $\overline{s(t)}$ in point $\overline{S_1}$ follows from the law of reflection $i = \rho$. Its unity vector $\overline{e_r}$ is found by mirroring the $\overline{n}$ – component in $\overline{s}$ (see Fig. A-2)

$$\overline{e_r} = \overline{e_s} - 2\,(\overline{e_s} \circ \overline{n})\,\overline{n} \quad (\text{where } |\overline{e_r}| = 1) \tag{A-8}$$

and

$$\overline{r(t)} = \overline{S_1} + t\,\overline{e_r} \tag{A-9}$$

A.1.3 Ideal Thin Lens

A positive lens brings rays together by refracting them toward its axis, while a negative lens steers them away from its axis. A ray is refracted by a positive ideal thin lens according to the model shown in Fig. A-5, where the refracted ray is defined by the intersection point $\overline{S_1}$ of the incident ray with the lens

$$\overline{S_1} = \overline{P_0} + t_1\,\overline{e_{s0}} \quad \text{with} \quad t_1 = \frac{\overline{n_L} \circ \left(\overline{O} - \overline{P_0}\right)}{\overline{n_L} \circ \overline{e_{s0}}} \tag{A-10}$$

where $\overline{n_L}$ denotes the orientation of the lens and $\overline{O}$ denotes its center. The direction of the refracted ray $\overline{e_{s1}}$ is found by determining the intersection point $\overline{P_1}$ of a ray through $\overline{O}$ parallel to the incident ray with the lens' focal plane. The focal plane contains the focal point $\overline{F}$ and is located at a distance equal to the focal length f. The direction $\overline{e_{s1}}$ follows from

$$\overline{P_1} = \overline{O} + t_2\,\overline{e_{s0}}$$

$$\overline{n_L} \circ \left(\overline{P_1} - \overline{F}\right) = 0$$

$$t_2 = \frac{\overline{n_L} \circ \left(\overline{F} - \overline{O}\right)}{\overline{n_L} \circ \overline{e_{s0}}} = \frac{f}{\overline{n_L} \circ \overline{e_{s0}}}$$

$$\overline{e_{s1}} = \frac{\overline{P_1} - \overline{S_1}}{\left|\overline{P_1} - \overline{S_1}\right|} \tag{A-11}$$

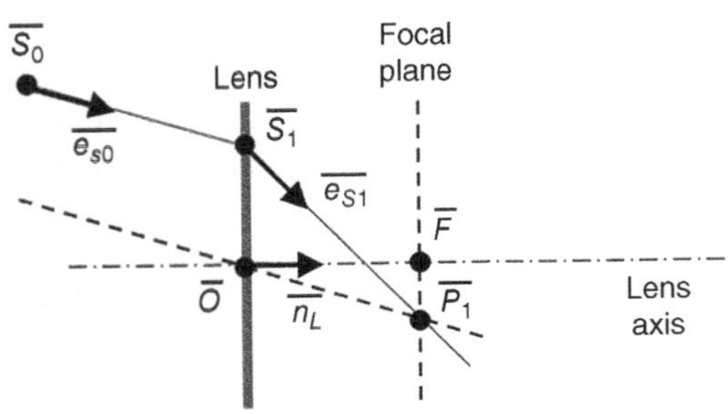

Fig. A-5 Refraction by a positive ideal thin lens

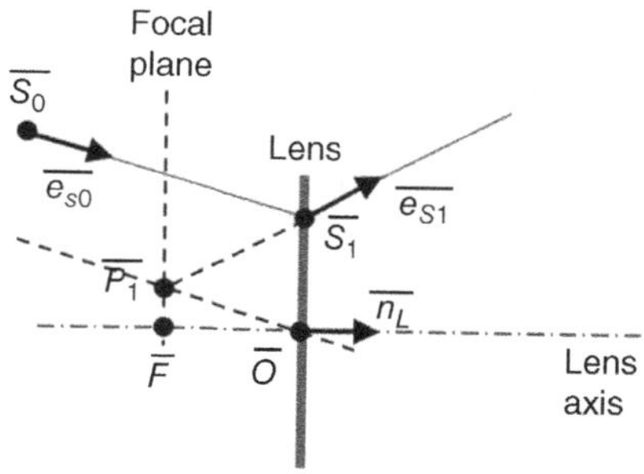

Fig. A-6 Refraction by a negative ideal thin lens

In a similar way, a negative ideal thin lens can be modeled according to Fig. A-6 by

$$\overline{S_1} = \overline{S_0} + t_1 \cdot \overline{e_{s0}}$$

$$t_1 = \frac{\overline{n_L} \circ \left(\overline{O} - \overline{S_0}\right)}{\overline{n_L} \circ \overline{e_{s0}}}$$

$$\overline{P_1} = \overline{O} - t_2 \cdot \overline{e_{s0}}$$

$$\overline{n_L} \circ \left(\overline{P_1} - \overline{F}\right) = 0$$

$$t_2 = \frac{\overline{n_L} \circ \left(\overline{F} - \overline{O}\right)}{\overline{n_L} \circ \overline{e_{s0}}} = \frac{f}{\overline{n_L} \circ \overline{e_{s0}}}$$

$$\overline{e_{s1}} = \frac{\overline{S_1} - \overline{P_1}}{\left|\overline{S_1} - \overline{P_1}\right|} \tag{A-12}$$

A.1.4 Refraction at a Plane Surface Boundary

The refraction of a ray by a boundary plane between a medium with refractive index n_1 and a medium with n_2 is governed by Snellius law (see Fig. A-7)

$$n_1 \sin i = n_2 \sin r \tag{A-13}$$

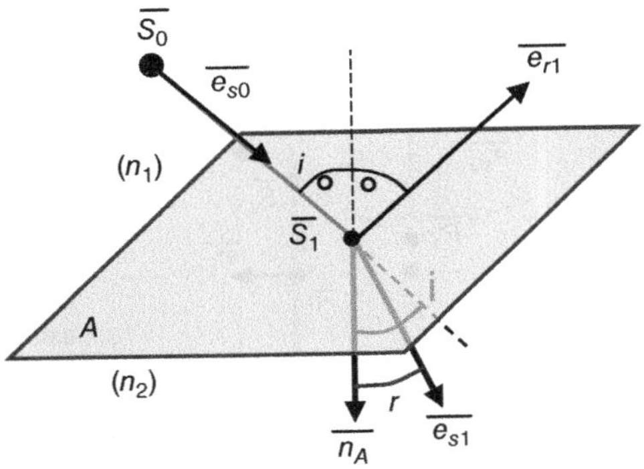

Fig. A-7 Refraction and reflection of a ray at a surface boundary between media with different refractive indices

The direction of the refracted ray $\overline{e_{s1}}$ follows from

$$\overline{e_{s1}} = a\,\overline{e_{s0}} + b\,\overline{n_A} \tag{A-14}$$

in which a and b are to be solved :

$$\overline{e_{s0}} \circ \overline{n_A} = |\overline{e_{s0}}|\,|\overline{n_A}| \cos i = \cos i$$

$$\overline{e_{s1}} \circ \overline{n_A} = |\overline{e_{s1}}|\,|\overline{n_A}| \cos r = \cos r$$

$$\overline{e_{s0}} \circ \overline{e_{s1}} = |\overline{e_{s0}}|\,|\overline{e_{s1}}| \cos(i - r) = \cos i \cos r + \sin i \sin r$$

$$\rightarrow a = \frac{n_1}{n_2} \text{ and } b = \cos r - \frac{n_1}{n_2} \cos i \tag{A-15}$$

where $\cos r = \sqrt{1 - \left(\frac{n_1}{n_2}\right)^2 \left(1 - (\cos i)^2\right)}$ and $\cos i = \overline{e_{s0}} \circ \overline{n_A}$

Total reflection will occur at surface A when $n_1 > n_2$ if incidence angle i exceeds the critical angle

$$i > \arcsin\left(\frac{n_2}{n_1}\right) \quad \text{so if} \quad \cos i < \sqrt{1 - \left(\frac{n_2}{n_1}\right)^2} \tag{A-16}$$

A.1.5 Refraction at a Spherical Surface Boundary

By deploying a spherical surface boundary, ray refraction by lens operation occurs. Fig. A-8 shows the entry of a ray at the convex side of such spherical boundary between media with different refractive indices n_1 and n_2. This boundary is part of a sphere with a center $\overline{x_M}$ and radius R.

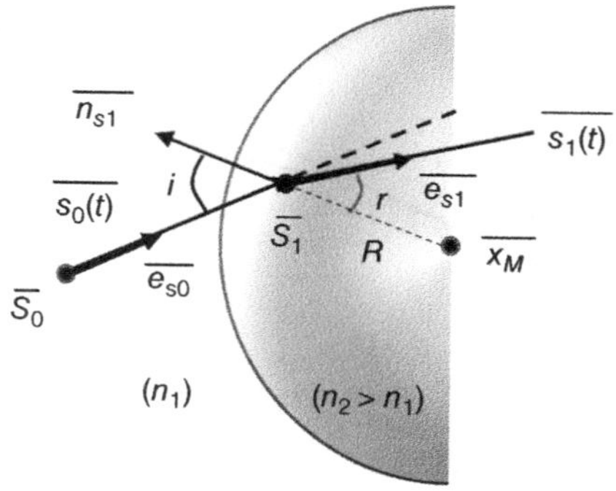

Fig. A-8 Ray entry at convex side of spherical surface boundary between media with refractive indices n_1 and $n_2 > n_1$

For the intersection point $\overline{S_1}$ of the incident ray $\overline{s_0(t)} = \overline{S_0} + t\,\overline{e_{s0}}$ holds $\left|\overline{S_1} - \overline{x_M}\right| = R$, hence

$$t = -\overline{e_{s0}} \circ \left(\overline{S_0} - \overline{x_M}\right) - \sqrt{\left(\overline{e_{s0}} \circ \left(\overline{S_0} - \overline{x_M}\right)\right)^2 + R^2 - \left|\overline{S_0} - \overline{x_M}\right|^2} \qquad \text{(A-17)}$$

The unity normal vector $\overline{n_{S1}}$ in $\overline{S_1}$ of the spherical surface is $\overline{n_{S1}} = \dfrac{\overline{S_1} - \overline{x_M}}{R}$ and the incidence angle of ray $\overline{e_{s0}}$ at $\overline{S_1}$ is given by $\cos i = -\overline{e_{s0}} \circ \overline{n_{S1}} \geq 0$, from which the direction $\overline{e_{s1}}$ of the refracted ray follows by

$$\overline{e_{s1}} = a \cdot \overline{e_{s0}} + b \cdot \overline{n_{S1}} \quad \text{with } a = \frac{n_1}{n_2} \text{ and } b = -\cos r + \frac{n_1}{n_2}\cos i \qquad \text{(A-18)}$$

where $\cos r = \sqrt{1 - \left(\dfrac{n_1}{n_2}\right)^2 \left(1 - (\cos i)^2\right)}$

Similarly, Fig. A-9 shows the entry of a ray at the concave side of a spherical surface boundary. The intersection point $\overline{S_1} = \overline{S_0} + t\,\overline{e_{s0}}$ is on the spherical surface, hence, $\left|\overline{S_1} - \overline{x_M}\right| = R$ and now

$$t = -\overline{e_{s0}} \circ \left(\overline{S_0} - \overline{x_M}\right) + \sqrt{\left(\overline{e_{s0}} \circ \left(\overline{S_0} - \overline{x_M}\right)\right)^2 + R^2 - \left|\overline{S_0} - \overline{x_M}\right|^2} \qquad \text{(A-19)}$$

With the unity normal vector $\overline{n_{S1}} = \dfrac{\overline{S_1} - \overline{x_M}}{R}$ and incidence angle i of ray $\overline{e_{s0}}$ at $\overline{S_1}$ given by $\cos i = \overline{e_{s0}} \circ \overline{n_{S1}} \geq 0$ follows

$$\overline{e_{s1}} = a\,\overline{e_{s1}} + b\,\overline{n_{S1}} \quad \text{with } a = \frac{n_1}{n_2} \text{ and } b = \cos r - \frac{n_1}{n_2}\cos i \qquad \text{(A-20)}$$

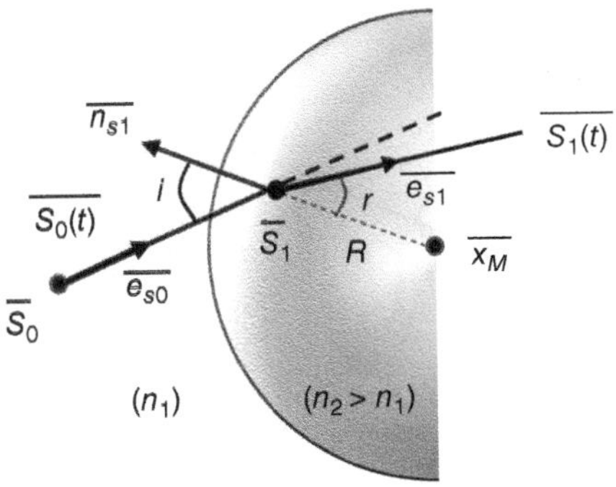

Fig. A-9 Ray entry at concave side of spherical surface boundary between media with refractive indices n_1 and $n_2 < n_1$

where $\cos r = \sqrt{1 - \left(\frac{n_1}{n_2}\right)^2 (1 - (\cos i)^2)}$

Total reflection at the spherical surface will occur if $n_1 > n_2$ and $\frac{n_1}{n_2} \sin i = \sin r > 1$ so if

$$\cos i < \sqrt{1 - \left(\frac{n_2}{n_1}\right)^2} \qquad \text{(A-21)}$$

A.2 Homogeneous Ray Launching

In order to perform a ray tracing analysis of the imaging by an optical system for different intensity profiles of the incident beam, adequate modeling of an extended optical source is needed. This source should have a specified diameter and emit a specified number of rays which are launched according to a homogeneously distributed spatial density and follow a specified optical intensity profile.

As shown in Fig. A-10, a circular extended source may be assumed with an aperture radius R_a, of which the ray emitters are grouped in $N + 1$ adjacent concentric rings, each having a width $\Delta r = r_n - r_{n-1}$ where r_n and r_{n-1} are the outer and inner radius of ring n, respectively. The mth ray emitter in ring n ($n = 0, 1, \ldots, N$) is located at polar coordinates $(r_s(n), \varphi_{mn})$, where $m = 1, 2, \ldots, M(n)$, $r_s(n) = (r_n + r_{n-1})/2$ with $r_s(0) = 0$, and $\varphi_{mn} = 2\pi(m - 1)/M(n)$. For a homogeneous spatial distribution, one ray is emitted per elementary area A_E. The area $A(n)$ of ring n is

$$A(n) = \pi\left(r_n^2 - r_{n-1}^2\right) \quad \text{for } n \geq 1 \text{ where } A(0) = \pi r_0^2 \qquad \text{(A-22)}$$

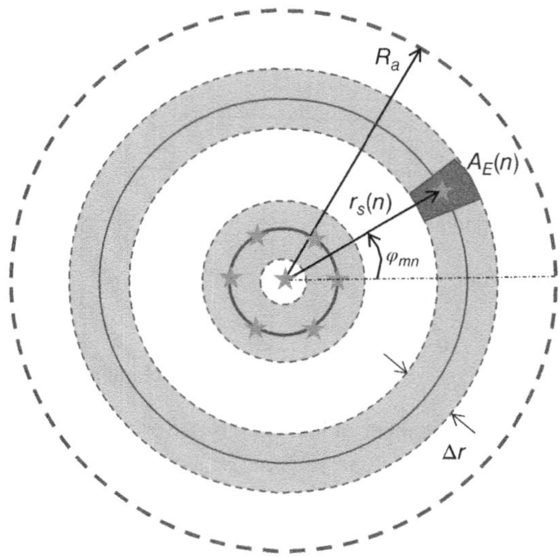

Fig. A-10 Extended optical source launching homogeneously distributed rays

For equidistant launching, the distance in the nth ring between the emitters $d \approx 2\pi n\ \Delta r/M(n)$ should be equal to the ring's width Δr, hence $M(n) \approx 2\pi n$. As $M(n)$ has to be an integer, we take $M(n) = 6n$ for $n \geq 1$, with $M(0) = 1$.

All the elementary areas should jointly fill the whole aperture, so when in total K rays are to be launched

$$A_E = \frac{A(n)}{M(n)} = \frac{\pi\left(r_n^2 - r_{n-1}^2\right)}{6n} = A(0) = \pi r_0^2 = \frac{\pi R_a^2}{K} \tag{A-23}$$

yielding $r_0 = \frac{R_a}{\sqrt{K}}$ and thus

$$r_n^2 = r_0^2\left(1 + \sum_{l=1}^{n} M(l)\right) = r_0^2\left[1 + 3\,n\,(1 + n)\right] \text{ so}$$

$$r_n = R_a\sqrt{\frac{1 + 3n(1 + n)}{K}} \tag{A-24}$$

The total number of rays K which are launched when N emitter rings are deployed is

$$K = \sum_{n=0}^{N} M(n) = 1 + \sum_{n=1}^{N} 6n = 1 + 3N\,(1 + N) \tag{A-25}$$

Fig. A-11 illustrates the fast growth of the number of launched rays K when the number of rings N is increased.

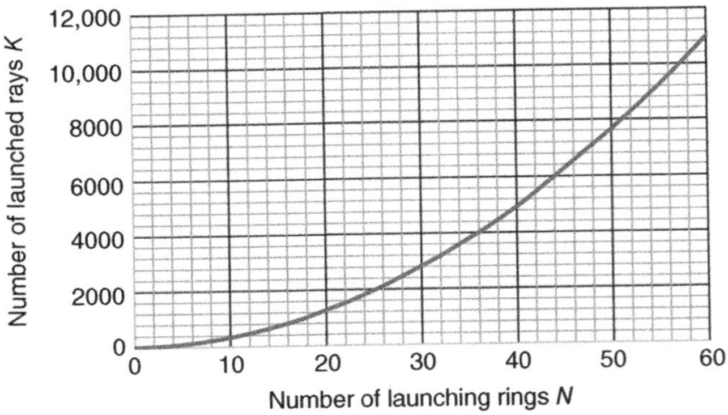

Fig. A-11 Number of launched rays K versus number of launching rings N

The optical intensity profile of a Gaussian beam propagating in the +z direction is given by [A-1]

$$I(r,z) = I_0 \left(\frac{w_0}{w(z)}\right)^2 \exp\left(\frac{-2r^2}{w(z)^2}\right) \tag{A-26}$$

where I_0 is the intensity at the center of the beam at its waist, w_0 is the waist radius at $z = 0$, and the beam waist $w(z)$ is

$$w(z) = w_0 \sqrt{1 + \left(\frac{z}{z_R}\right)^2} \tag{A-27}$$

with Rayleigh range $z_R = \pi {w_0}^2 n_{medium}/\lambda$, where n_{medium} is the refractive index of the medium. The power P_{aper} of the beam with total power P_{beam} passing through an aperture with radius R_{aper} in the transversal plane at position z is then

$$P_{aper}(R_{aper}, z) = \int_{\varphi=0}^{2\pi} \int_{r=0}^{R_{aper}} I(r,z) r \, dr d\varphi = P_{beam} \left\{1 - \exp\left(\frac{-2{R_{aper}}^2}{w^2(z)}\right)\right\} \tag{A-28}$$

For creating a Gaussian beam with total power P_{beam}, waist radius w, and launching aperture R_a, the power launched by each of the $M(n)$ emitters in ring n for $n \geq 1$ (with $P(n)$ the power launched by ring n) is

$$P_s(n) = \frac{P(n)}{M(n)} = \frac{P_{beam}}{6n} \left\{\exp\left(-\frac{2r_{n-1}^2}{w^2}\right) - \exp\left(-\frac{2r_n^2}{w^2}\right)\right\} \quad \text{for } n \geq 1$$

Table A-1 Fraction of Gaussian beam power passing through an aperture

R_{aper}/R_{beam}	P_{aper}/P_{beam}	$1 - P_{aper}/P_{beam}$
1	0.864664717	0.1353353
1.53	0.990737673	0.0092623
2	0.999664537	0.0003355
3	0.999999985	1.523E−08

$$P_s(0) = P(0) = P_{beam}\left\{1 - \exp\left(-\frac{2r_0^2}{w^2}\right)\right\} \quad \text{for } n = 0 \tag{A-29}$$

As the Gaussian beam profile theoretically extends infinitely, a finite launching aperture implies omitting part of the beam's power. The fraction of the beam power that is passing the aperture versus the aperture/beam diameter ratio is shown in Table A-1.

For creating a beam with a uniform intensity profile and with total power P_{beam} and radius R_{beam}, the power P_{aper} passing through an aperture with radius R_{aper} is

$$\text{if } R_{beam} < R_{aper}\text{: launching aperture } R_a = R_{beam} \text{ and } P_{aper} = P_{beam}$$

$$\text{if } R_{beam} \geq R_{aper} : R_a = R_{aper} \quad \text{and} \quad P_{aper} = \left(R_{aper}/R_{beam}\right)^2 P_{beam} \tag{A-30}$$

The power launched by each of the $M(n)$ emitters in ring n for $n \geq 1$ (with $P(n)$ the power launched by ring n) is

$$P_s(n) = \frac{P(n)}{M(n)} = \frac{1}{3}\left(\frac{\Delta r}{R_a}\right)^2 P_{aper} \quad \text{for} \quad n \geq 1$$

$$P_s(0) = P(0) = \left(\frac{\Delta r}{2R_a}\right)^2 P_{aper} \quad \text{for} \quad n = 0 \tag{A-31}$$

Examples of homogeneously launching $K = 1027$ rays for a Gaussian beam with a waist diameter of Ø15 mm and of Ø100 mm are given in Fig. A-12a and c, respectively. For a uniform beam with a diameter of Ø15 mm and of Ø100 mm these examples are given in Fig. A-12b and d, respectively. Fig. A-11 shows the number of rings N needed for the homogeneous launching of K rays, and Table A-2 the launching data for ring n.

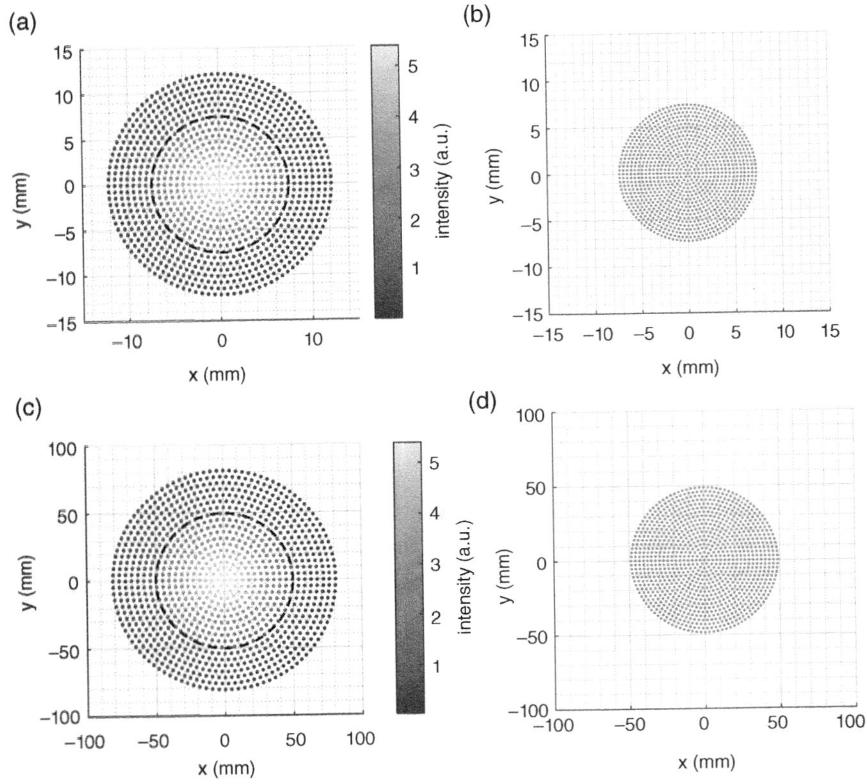

Fig. A-12 Homogeneous ray launching (for $K = 1027$ rays launched). (a) Gaussian beam with D_{beam} = Ø15 mm waist diameter (*dashed circle*). (b) Uniform beam with D_{beam} = Ø15 mm. (c) Gaussian beam with D_{beam} = Ø100 mm waist diameter (dashed circle). (d) Uniform beam with D_{beam} = Ø100 mm

Table A-2 Homogeneous ray launching from ring n, when launching with N rings (where $N = 10$ rings cover a beam diameter of Ø25 mm).

Ring	**Elements**	***K***		**Ring area**	**Ray area**	**Ring width**
n	$M(n)$	Cum # rays	r_n	A_n	AE_n	w_n
0	1	1	0.687062	1.483003	1.483003	
1	6	7	1.817796	8.898015	1.483003	1.130733
2	12	19	2.994834	17.79603	1.483003	1.177039
3	18	37	4.179236	26.69405	1.483003	1.184401
4	24	61	5.366127	35.59206	1.483003	1.186891
5	30	91	6.554155	44.49008	1.483003	1.188028

Table A-2 (Continued)

Ring	Elements	K		Ring area	Ray area	Ring width
n	$M(n)$	Cum # rays	r_n	A_n	AE_n	w_n
6	36	127	7.742797	53.38809	1.483003	1.188642
7	42	169	8.931808	62.28611	1.483003	1.189011
8	48	217	10.12106	71.18412	1.483003	1.18925
9	54	271	11.31047	80.08214	1.483003	1.189413
10	60	331	12.5	88.98015	1.483003	1.18953
11	66	397	13.68962	97.87817	1.483003	1.189616
12	72	469	14.8793	106.7762	1.483003	1.189682
13	78	547	16.06903	115.6742	1.483003	1.189733
14	84	631	17.2588	124.5722	1.483003	1.189773
15	90	721	18.44861	133.4702	1.483003	1.189806
16	96	817	19.63844	142.3682	1.483003	1.189833
17	102	919	20.8283	151.2663	1.483003	1.189855
18	108	1027	22.01817	160.1643	1.483003	1.189873
22	132	1519	26.7778	195.7563	1.483003	1.189924
26	156	2107	31.53757	231.3484	1.483003	1.189953
32	192	3169	38.67739	284.7365	1.483003	1.189978
37	222	4219	44.62732	329.2266	1.483003	1.18999
41	246	5167	49.3873	364.8186	1.483003	1.189997
50	300	7651	60.09732	444.9008	1.483003	1.190007
58	348	10267	69.6174	516.0849	1.483003	1.190012
71	426	15337	85.08759	631.7591	1.483003	1.190017
82	492	20419	98.17779	729.6373	1.483003	1.190019
91	546	25117	108.888	809.7194	1.483003	1.190021
100	600	30301	119.5982	889.8015	1.483003	1.190022
115	690	40021	137.4485	1023.272	1.483003	1.190023
129	774	50311	154.1088	1147.844	1.483003	1.190024
141	846	60067	168.3891	1254.62	1.483003	1.190024
163	978	80197	194.5696	1450.377	1.483003	1.190025
173	1038	90307	206.4699	1539.357	1.483003	1.190025
183	1098	101017	218.3701	1628.337	1.483003	1.190025

Data: $M(n)$: number of ray emitters in ring n; K: total number of rays launched; r_n: radius of ring n (in mm); A_n: area of ring n (in mm^2), AE_n: area of an emitter in ring n (in mm^2); w_n: width of ring n (in mm).

A.3 Aspheric Fresnel Lens Modeling

A Fresnel lens is composed of many concentric circular grooves with carefully designed facets. Notifying that the light refraction is taking place at the surfaces, such a lens may perform the same light-focusing function as a plano-convex lens (see Fig. A-13). However, a Fresnel lens can be much thinner and thus have less volume and weight. By optimizing the shape of each concentric groove, a lens with an aspherical profile can be emulated. Thus, a Fresnel lens may combine a large aperture D with a small focal length f, and achieve a large numerical aperture NA according to

$$\mathrm{NA} = n \sin\theta = n\ \sin\left(\mathrm{atan}\left(\frac{D}{2f}\right)\right) = \frac{n}{\sqrt{1+(2f/D)^2}} \tag{A-32}$$

A Fresnel lens is typically used for light collecting, such as in solar cells, and thus may also be attractive for application in the optical beam receivers in OWC systems. Fresnel lenses can be manufactured readily by injection-molding with optical-quality plastics. Their imaging properties may be less good than those of a conventional lens due to the discrete profile of the Fresnel lens with its distinct grooves. By applying a higher groove density and lower groove profile, the imaging can be improved.

We devised a ray modeling approach for an ideal aspherical Fresnel lens profile which has an infinitesimally small profile height, see Fig. A-14. The lens is

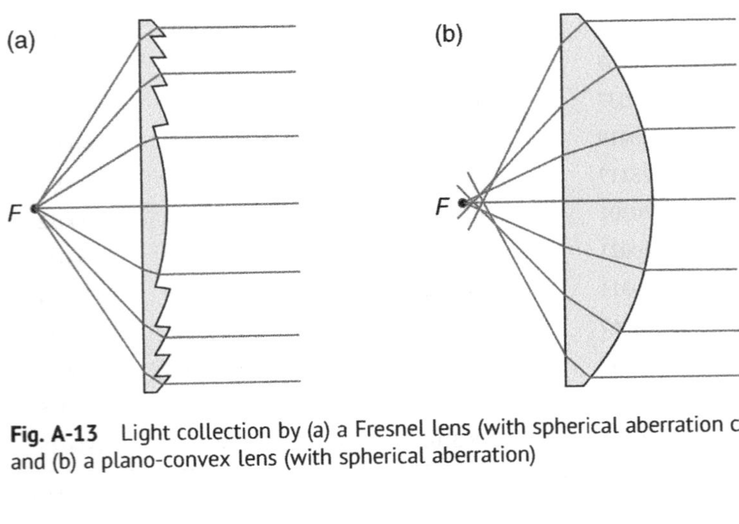

Fig. A-13 Light collection by (a) a Fresnel lens (with spherical aberration compensated) and (b) a plano-convex lens (with spherical aberration)

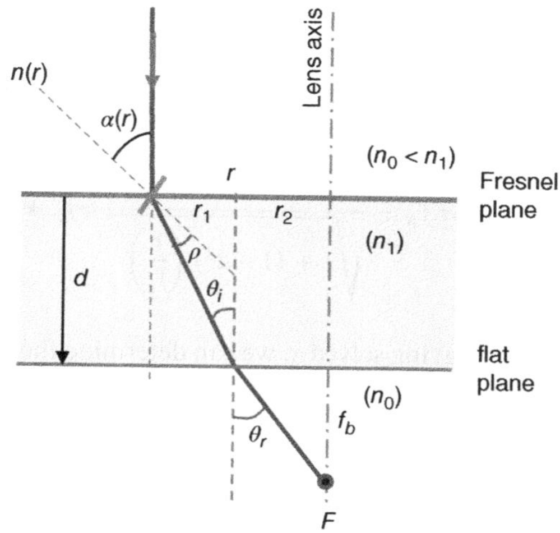

Fig. A-14 Modeling an ideal aspheric lens, fully compensated for spherical aberration

optimized by tailoring the inclination angle α (r) of each concentric groove such that spherical aberration [A-2] is fully compensated: that is, independent of the radial position r at which the ray is incident, each ray passes through the focal point F which is positioned at a back focal length (BFL) f_b from the lens' planar facet.

For this, three design equations need to be fulfilled:

$$1)\ r = r_1 + r_2 = d \tan\theta_i + f_b \tan\theta_r$$

$$2)\ n_1 \sin\theta_i = n_0 \sin\theta_r \quad \text{(Snellius law)}$$

$$3)\ n_0 \sin\alpha(r) = n_1 \sin\rho = n_1 \sin(\alpha(r) - \theta_i) \qquad \text{(A-33)}$$

which implies solving x from

$$d\,x + \frac{x\,f_b}{\sqrt{a^2 + (a^2 - 1)\,x^2}} = r \quad \text{where } x = \tan\theta_i \text{ and } a = \frac{n_0}{n_1} \qquad \text{(A-34)}$$

By Newton–Raphson's zero-seeking algorithm (see [A-3]), which is known to converge fast (roughly doubling the number of correct digits every iteration step), x can be solved from $g(x) = 0$ by the iteration

$$x_{n+1} = x_n - \frac{g(x_n)}{g'(x_n)} \qquad \text{(A-35)}$$

wherein $g(x) = d\,x + \frac{x\,f_b}{\sqrt{a^2 + (a^2 - 1)x^2}} - r$ and $g'(x) = d + \frac{a^2 f_b}{[a^2 + (a^2 - 1)x^2]^{3/2}}$

As the starting point for the iteration, we may take $x_0 = \tan\theta_i = \frac{r}{d}\left(1 - \frac{f_b}{f_e}\right)$ where the effective focal length (EFL) $f_e = \dfrac{r}{\tan\theta_r}$ follows from

$$f_e = f_b + \frac{a\,d}{\sqrt{1 + (1 - a^2)\left(\frac{r}{f_e}\right)^2}} \approx f_b + a\,d \tag{A-36}$$

After having solved x, we can determine the optimized groove inclination angle $\alpha\,(r)$ at radial position r by

$$\tan\alpha(r) = \frac{x}{1 - a\sqrt{1 + x^2}} \tag{A-37}$$

A.3.1 Some Fresnel Lens Design Results

The profiles for a Fresnel lens fully compensated for spherical aberration have been calculated when using 10,000 discretization steps in Newton–Raphson's zero-seeking algorithm. The results for a Fresnel lens with focal length $f = 10$ mm and aperture D_{lens} = 50 mm, as used in the ray tracing for the downstream receiver in Chapter 7 are shown in Fig. A-15a and b, when the maximum profile height of the lens (cf. Fig. A-13a) $h_{\max} = 0.2$ and 0.5 mm, respectively. For a Fresnel lens with focal length $f = 5$ mm and aperture $D_{\text{lens}} = 25$ mm, as used in the ray tracing for the upstream receiver in Chapter 7, the results are shown in Fig. A-16a and b, when $h_{\max}$ = 0.2 and 0.5 mm, respectively.

While the Fresnel lens has been corrected for spherical aberration, other off-axis aberrations (such as coma, and astigmatism) [A-2] will occur when the beam incidence angle $\alpha \neq 0$. These aberrations will affect the imaging by the Fresnel lens at larger incidence angles α. By ray tracing, the off-axis imaging quality for the Fresnel lens with $f = 10$ mm and D_{lens} = 50 mm has been assessed when a Gaussian beam with a waist diameter of Ø100 mm (the downstream case in Chapter 7) is received. The results are shown in Fig. A-17 for beam incidence angles α = 0 and α = 5° when no defocusing is applied (so p=0) and when a defocusing p = 20% is applied. These results clearly show the impact of the off-axis aberrations (notably coma) of the Fresnel lens on the imaging.

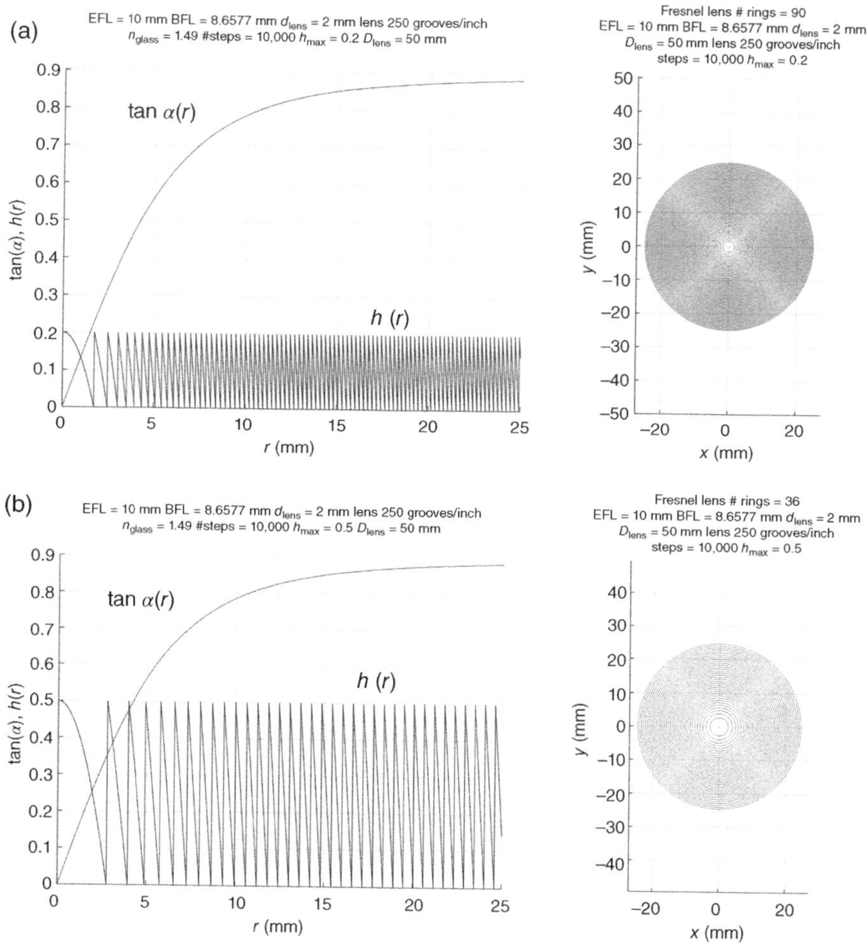

Fig. A-15 Fresnel lens with $f = 10$ mm and aperture $D_{\text{lens}} = 50$ mm, with maximum profile height (a) $h_{\max} = 0.2$ mm, and (b) $h_{\max} = 0.5$mm (curves: lens facet inclination angle $\alpha(r)$, facet profile height $h(r)$)

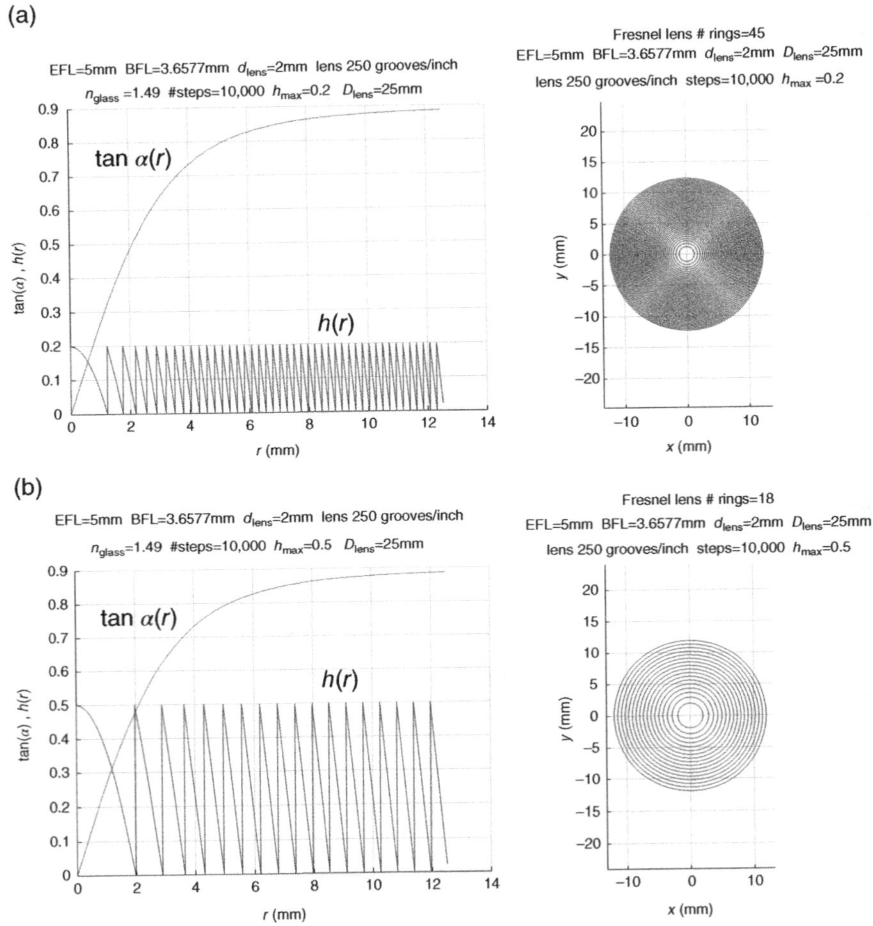

Fig. A-16 Fresnel lens with $f = 5$ mm and aperture $D_{\text{lens}} = 25$ mm, with maximum profile height (a) $h_{\max} = 0.2$ mm, and (b) $h_{\max} = 0.5$ mm (curves: lens facet inclination angle $\alpha(r)$, facet profile height $h(r)$)

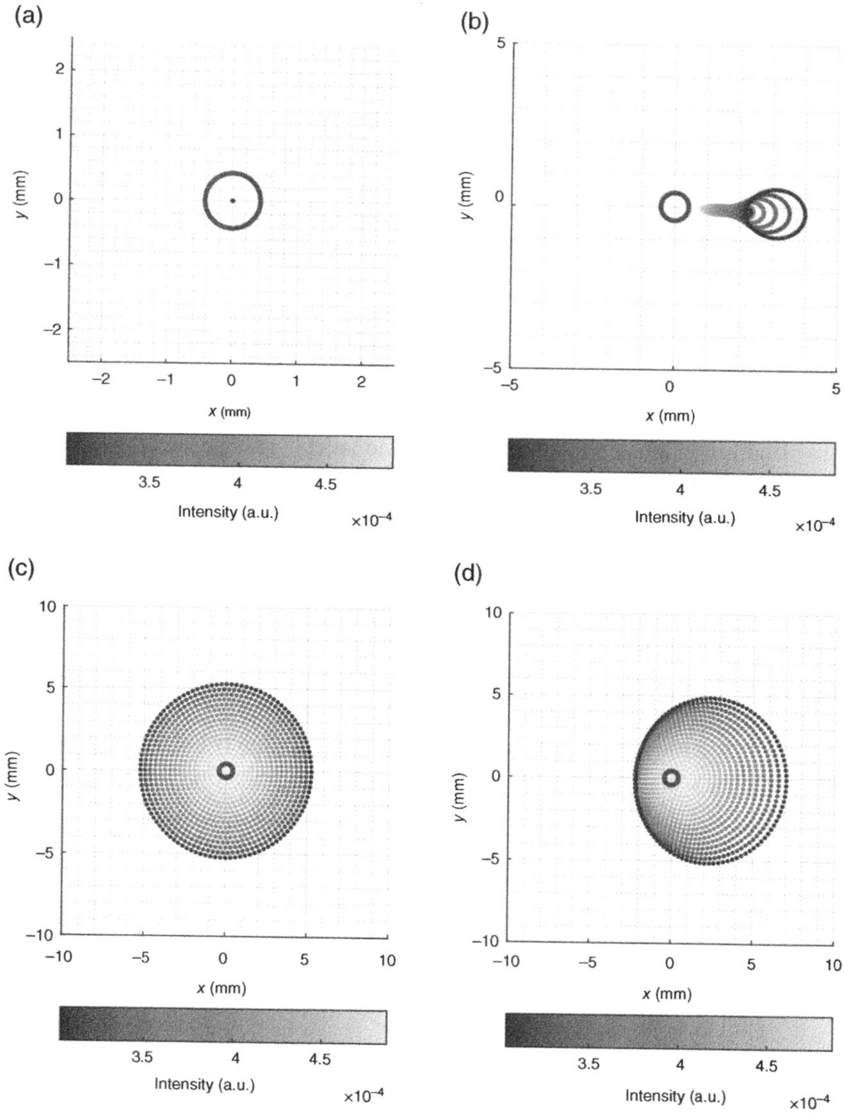

Fig. A-17 Imaging of Gaussian beam Ø100 mm by Fresnel lens $D_{\text{lens}} = Ø50$ mm, $f = 10$ mm for focused ($p = 0$) and defocused ($p = 0.2$) operation, at beam incidence angles $\alpha_{\text{beam}} = 0$ and $\alpha_{\text{beam}} = 5°$ (the thick circle represents the 4 × 4 photodiode matrix with diameter of Ø859 μm, see Section 7.3). (a) $p = 0$, $\alpha_{\mathbf{beam}} = 0$. (b) $p = 0$, $\alpha_{\mathbf{beam}} = 5°$. (c) $p = 0.2$, $\alpha_{\mathbf{beam}} = 0°$. (d) $p = 0.2$, $\alpha_{\mathbf{beam}} = 5°$

A.4 Modeling of Matrix of Corner Cube Retroreflectors

Corner cubes (CCs) are well-known passive optical elements that act as retroreflectors; they reflect an incoming ray in the same direction as where it came from but with an offset between the incoming and the reflected ray [A-4]. A 2D analysis has been presented in Section 6.3. With the 3D ray tracing tools presented in this Appendix, in particular the modeling of optical reflections in Section A.1.2, a 3D analysis can be made of the behavior of a matrix of CCs which has been introduced in Chapter 6 for enabling device localization.

The 3D analysis is built on the model for a single CC element shown in Fig. A-18. The CC element is a pyramid that has three facets – ABD, BDC, and CDA – which are mutually perpendicular flat reflective surfaces. The sides AD, BD, and CD are orthogonal to each other, and the ground facet ABC is an equilateral triangle.

A ray that is incident on the CC via its ground facet is subsequently three times reflected inside the CC and then emerges out of the CC. It can be traced by using Eqs. (A-8) and (A-9). Fig. A-19 shows how an incoming ray starting at point S is reflected inside the CC in points 1, 2, and 3. It subsequently emerges from the CC in order to end in point F located next to S in the launching plane. Thanks to the orthogonality of the three CC facets, the outgoing ray is exactly parallel to the incoming ray but does have an offset with respect to it given by the distance S–F.

A CC has a limited aperture, which is smaller than its footprint. As visualized in Fig. A-20a, when an incoming ray hits a first reflection point 1 too close to one of the CC's edges, it may reach a second reflection point 2 but may fail to reach a third point inside the CC so it cannot complete the full retro-reflecting process. Hence,

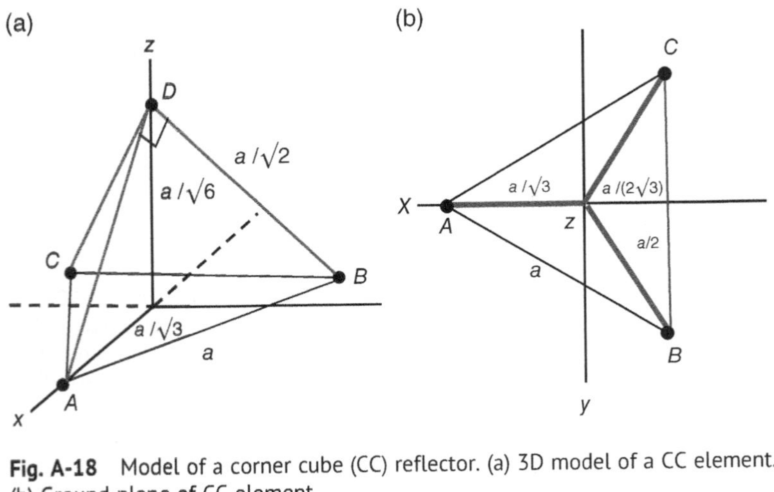

Fig. A-18 Model of a corner cube (CC) reflector. (a) 3D model of a CC element. (b) Ground plane of CC element

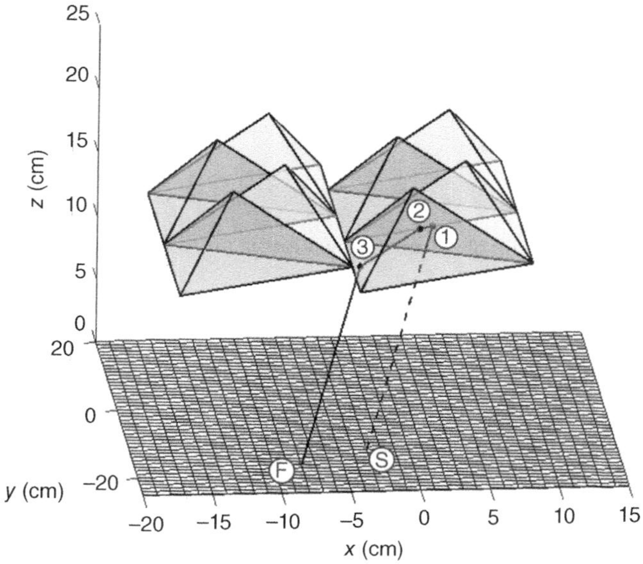

Fig. A-19 3D tracing of a ray retroreflected by a CC: a ray (*dashed*) is launched at point S, is reflected inside the CC subsequently at points 1, 2, and 3, and emerges from the CC to land at point F in the plane from where the ray was launched; launching angle, $\alpha = 15°$

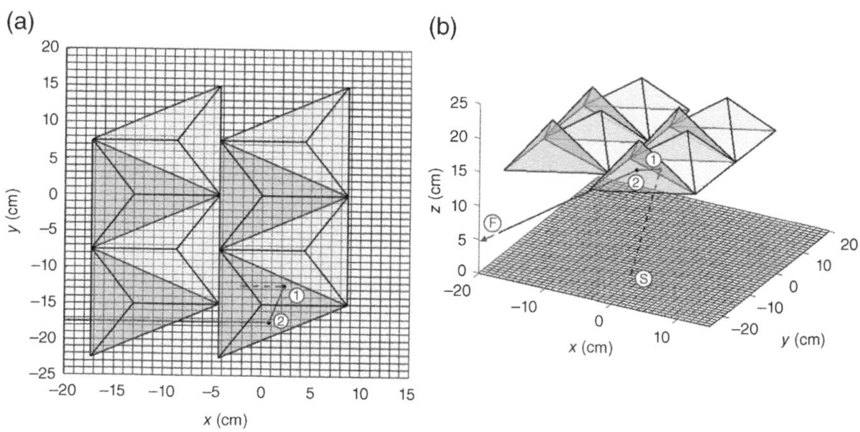

Fig. A-20 Limited aperture of a CC (ray launching angle $\alpha = 15°$). (a) Top view. (b) Side view (non-retroreflected ray toward point F diverges from the incoming ray (*dashed*) started at S)

the ray emerging from the CC is not parallel to the incoming ray; the end point F where this non-retroreflected ray lands in the launching plane may be quite far from the ray's starting point S (see Fig. A-20b).

References

[A-1] Gaussian beam [on-line]. Available: https://en.wikipedia.org/wiki/Gaussian_beam

[A-2] F.L. Pedrotti, L.M. Pedrotti, and L.S. Pedrotti, Aberration theory, *Introduction to Optics*, 3rd ed., 2018, Cambridge: Cambridge University Press.

[A-3] Newton–Raphson's method, a numerical root-finding algorithm [on-line]. Available: https://en.wikipedia.org/wiki/Newton%27s_method

[A-4] Corner reflector [on-line]. Available: https://en.wikipedia.org/wiki/Corner_reflector

Index

a

aberrations, lens 10, 111, 136, 141, 208, 261
  astigmatism 276
  Fresnel lens 205
  off-axis 128, 177, 276
  Petzval field curvature 129, 131
  spherical lens 10, 81, 205, 274–76
  triplet lens 241
ambient light 136
antenna
  directivity 81
  horn 76, 80, 140, 151
  radiation efficiency 80
application domains of OWC
  exhibition hall 65, 77, 173
  hospital 66, 173
  industry 4.0 54, 73, 77
  intra-data center 54, 66, 79, 173
  offices 7, 50, 54, 66, 69, 77
  residential settings 7, 54, 66, 77
area coverage 51–52
  increase the coverage 106, 127, 226
  by narrow beams 75, 121, 136, 141
  wide coverage 8
atmospheric effects 4, 7, 66, 82, 251

b

bandwidth
  of AWGR beam steerer 76, 136, 139
  of wavelength-steered OWC link 102, 104
battery power, save
  by reduced processing in user device 86, 189, 230, 249
  by reduced upstream beam power 9, 32, 142, 173
beam
  blocking. *See* line-of-sight
  Gaussian 236, 249, 270
  Gaussian at receiver 195, 205, 225, 276
  Gaussian from steerer 134, 181
  Gaussian retroreflected 157–63, 170
  spot deformation 177
  squint 39, 42
  steering time 249
  uniform at receiver 101, 202, 206, 208, 215, 236
  uniform from steerer 134, 174, 180, 242, 271
beam, narrow infrared
  high capacity 5, 8, 75, 141

Short-Reach Optical Wireless Communication: By Directed Narrow Beams, First Edition. Ton Koonen.

beam, narrow infrared (*cont'd*)
high directivity 82
per user 52, 84, 97, 243, 247
beam power coupling
to photodiode 187, 220, 225, 242
to photodiode matrix 103, 189, 193, 201, 250
to receiver 100–5
beam scanning 117. *See also* localization by beam scanning
beam steering, optical
1D steering with diffraction grating 113–20
2D steering with arrayed waveguide grating router (AWGR) 84, 98, 106, 113, 128–45, 151, 254
2D steering with crossed diffraction gratings 98, 113, 121–28
2D steering with multiple wavelengths per beam 254
2D steering with stepper motors 142–45, 152, 181–83, 248–50
by pencil-radiating antenna (PRA) 98, 100, 105
by PRA in bidirectional OWC system 233–35, 240
by PRA in hybrid OWC/RF system 76, 151, 230
by PRA in system demonstrator 140, 141, 168, 237, 238
by PRAs with rerouting switch 250
comparison of technologies **144**
covering user area by PRA 102, 134
user localization by PRA 152, 168
with MEMS 109–11
with SLM 109–11
beam steering, radio 7, 16, 40
by phased array antenna (PAA) 38
by PAA and wavelength tuning 40–43
by PAA with true time delay (TTD) lines 39–41, 44
bidirectional wireless communication
all-optical 10
all-optical beam-steered 173, 229, 233, 234
all-optical beam-steered-system concept 233, 234
all-optical beam-steered-system demonstrator 250
hybrid optical/radio 65, 76, 97, 139, 151, 230
LiFi 71
BROWSE-indoor OWC network architecture and system concept 8, 10, 75, 97, 106, 113

c

camera
localization by camera 73, 84, 151, 231, 242
optical camera communication (OCC) 73
capacity on demand 7, 54, 96, 247
CapEx (capital expenditure) 17, 20, 69, 89
Cat-5e cable 16, 19, 22, 55
cat's eye reflector. *See* retroreflector, optical
center of gravity, algorithm 10, 174, 230, 243, 249
accuracy 175–81
central communication controller 76, 97, 99, 100, 140, 153, 230
collimator
lensed fiber 111, 124, 125, 136, 139, 234
communication
human-to-machine 247
machine-to-machine 6, 54, 79, 247
compound parabolic concentrator (CPC) 74, 103, 111, 187

control channel 72, 96, 111
 embedded 72, 96, 98, 111
corner cube, 154. *See also* retro-reflector
 dead zone 157, 160, 161, 166
 effective aperture 154, 156, 165
 imperfections in geometry 163, 170
 matrix of miniature corner cubes 152, 165, 168
 matrix of miniature corner cubes in foil 9, 166, 235
 performance of matrix **160, 163**
crossconnect, optical 31, 97, 153, 169, 250, 258
crosstalk
 between beams 144, 189, 252
 between users 6, 97
 from neighbouring cells 211, 250
 from neighbouring rooms 4

d

defocusing with respect to a lens
 at beam steerer 81, 129–41, 142, 153, 160–68
 at OWC receiver 105, 198, 205–13, 221, 274
 for upstream path 239–41
digital signal processing (DSP) 137
docking system, optical wireless 233

e

eaves-dropping 229
economics of home networks 7, 17
electro-absorption modulator
 reflective 231
electro-magnetic interference (EMI) 4, 8, 65, 229
energy efficiency 8, 65
 of beam-steered OWC 6, 55, 82, 89, 172
etendue, law of conservation 74, 86, 103, 187
 breaking the law 188
extinction ratio, optical 214, 234
eye pattern 125, 127, 138
eye safety 9, 142, 152, 173
 for infrared beams 136, 138, 238
 standards IEC 60825, ANZI Z-136 48, 49, 67, 98

f

fiber. *See also* plastic optical fiber (POF)
 bending sensitivity 26
 graded-index multimode (GI-MMF) 24
 multimode (MMF) 16, 19, 47
 polarization-maintaining (PMF) 127, 252
 single-mode (SMF) 19, 24, 138
 splicing 24
fiber array
 2D array 82, 128, 144, 153, 166, 235
 2D array realization 139, 170
Fiber-to-the-Home (FttH) 7, 13, 27, 52, 153
Fiber-to-the-Room (FttR) 56
Field of View 74, 233
 of optical beam steerer 128, 143
 of optical wireless receiver. *See* receiver, optical wireless
file transfer, *See* Gigabit Ethernet link
fill factor of photodiode matrix 87, 104, 189, 202, 206, 208
 impact 208–10, 227, 237
fill factor of user area 133–35
Forward Error Correction (FEC) 28, 125
free-space optics (FSO)
 long reach 5, 66, 251, 253
 medium reach 5
Free Spectral Range (FSR) 41, 98, 115

g

Gigabit Ethernet
 link 28, 189, 223

Gigabit Ethernet (*cont'd*)
  video streaming 84, 141, 173, 223, 229
  video streaming demonstrator 10, 238, 250
grating, arrayed waveguide-router (AWGR) 9, 54, 84, 144, 240
  C band AWGR 76, 130, 139, 144, 170, 238, 248
  C+L band AWGR 139, 144, 170, 238
  wavelength-cyclic routing in AWGR 254
grating, diffraction 8
  angular dispersion 116, 118, 122
  angular tuning range **117**
  arrayed waveguide 117, 119, 122
  blaze angle 117
  echelle grating 117, 122, 124
  order of interference 114, 126, 128
  reflection grating 96, 98, 115, 116, 123, 127, 144
  transmission grating 109, 123, 127
  virtually imaged phased array (VIPA) 119, 126

i

IEEE standards. *See also* wireless communication standards
  IEEE 802.11bb for LiFi 69, 79
  IEEE 802.15.7 for VLC 70
indoor network
  converged optical fiber network 7, 14, 16, 27, 36, 53
  evolution roadmap 53
  Gigabit Home Networking (G.hn) standard 29, 78–79
  opaque network node (O/E/O converting) 20, 54
  optically transparent node 19, 54
  topology 16–18, 20–23, 54, 100
Internet of Things (IoT) 2, 3, 31, 63, 64
ITU grid optical frequencies 113, 139

l

laser diode
  arbitrary wavelength laser 9, 142, 145, 235, 249
  injection-locked laser 231
  tunable laser 8, 45, 145, 163, 252
  VCSEL, vertical cavity surface emitting laser 111
  VCSEL array 111
  wavelength switching 113
  wavelength tuning range 85, 100, 102, 116, 128, 254
  wavelength tuning time 96, 113, 144
latency 8, 54, 66, 79, 247
  in OWC *versus* in fiber communication 96, 172
  in signal processing 6, 126
  in VLC systems 71
LED
  blue LED plus phosphor coating 69
  quad-color 69
  tri-color 69
lens
  aspherical. *See* Fresnel lens
  back focal length 275
  camera 128, 131, 139, 140, 170, 238
  doublet 136
  effective focal length 142, 205, 276
  f/D number 128, 131, 139
  Fresnel lens-in ceiling receiver 236, 249
  Fresnel lens in user receiver 181, 205, 206, 223, 238, 250
  Fresnel lens modeling 10, 274–82
  ideal thin 103, 165, 195, 204–9, 210, 264
  plano-convex 177–80, 274
  spherical 10, 205, 266
  triplet 124, 142, 177–81, 234, 237, 243
  tunable liquid 253
license. *See* spectrum license

LiFi system 80, 83, 89
LiFi 2.0 83, 111
line of sight
for 60 GHz radio 32
for beam-steered OWC 65, 97, 102, 172, 247, 248
for LiFi 71
for optical wireless communication 4, 5, 7, 67, 89
for VLC 50
localization
accuracy 151
by beam scanning of ceiling unit 77, 173–80, 181, 250
by beam scanning of user devices 74, 102, 163–68, 170, 250, 253
by camera of user devices 151–53
by triangular algorithms 73, 149
by visible light communication (VLC) 149
of user 102, 139, 238
self-calibrated 9, 74, 152, 164, 183, 230, 243

m

Mach Zehnder
interferometer (MZI) 34, 255
modulator (MZM) 34, 136, 137, 234
media converter 21, 223, 242, 250
MEMS, micro-electro-mechanical system
mirror 52, 96, 109, 143, 233
switch 40, 169
mode group division multiplexing. *See* spatial multiplexing
modulation
color shift keying (CSK) 71
DC-biased optical orthogonal frequency division multiplexing (DCO-OFDM) 71, 84
discrete multitone (DMT) 27, 125
on-off keying (OOK) 26, 28, 29, 96, 188
orthogonal frequency division multiplexing (OFDM) 27–31, 34, 36, 69, 79
pulse amplitude modulation with 4 levels (PAM-4) 29, 71, 76, 79, 106, 125, 137, 248
quadrature amplitude modulation (QAM) 27, 29, 34, 47, 48, 51, 84, 111, 232, 253
multicasting, point-to multipoint 77, 230
multiple-input multiple-output (MIMO)
in optical wireless communication 5, 50, 71, 73, 84, 86, 111
in radio communication 32, 37, 77

n

non-imaging optics. *See* compound parabolic concentrator

o

OpEx (operational expenditure) 17, 19, 69, 89
optical carrier recovery 231
optical circulator 165, 234
optical phased array (OPA). *See* grating, diffraction, arrayed waveguide
optical wireless communication
by infrared beams 7, 65, 66
by visible light (VLC) 50–51, 65, 69–72, 86, 95–97
by visible light (VLC) / LiFi 7, 77, 80, 89
underwater 4, 251

p

parallax error 9, 152, 183, 245
pencil-radiating antenna (PRA). *See* beam steering, optical
phased array antenna (PAA) 39, 80
photodiode

photodiode (*cont'd*)
metal-semiconductor-metal (MSM) 188
responsivity 67, 189, 214
uni-travelling-carrier (UTC) 88, 138, 188
photodiode matrix 107, 189, 226, 253
bandwidth 9, 253
beam coupling 202–8
constellation 242
DC biasing 193–97
electrical modelling 189–92
frequency characteristics 192–93
implementation of receiver 220–23
packaged 220–21
scaling 9, 87, 225–26
SNR performance 198–99
uneven illumination 194–95
photonic integrated circuit
for optical beam steering 84, 96
for optical wireless communication 8, 251
for OWC receiver 138
for radio beam steering 42
plastic optical fiber (POF) 19, 24, 79
bending sensitivity 26
duplex POF 21
Gigabit Ethernet transceiver 55
graded-index (GI-POF) 26, 29
multicore (MC-POF) 26
multi-Gbit/s experiments **30**
multimodal dispersion 19, 26, 27, 48
splicing 19, 23
step-index (SI-POF) 26, 29
polarization dependence
of arrayed waveguide grating router (AWGR) 128, 145, 238
of diffraction grating 116–17, 128, 145
of surface grating coupler (SGC) 75
power consumption
for beam alignment 145
in indoor network 19, 21, 55, 248
in OWC receiver 75, 126, 188–89
in Wi-Fi access device 82
power line communication (PLC) 13
privacy 27
by beam-steered OWC 52, 83, 89, 173, 247
by OWC 4, 6, 48, 66, 76

r

radio
60 GHz mm-wave 3
mm-wave 4, 77, 79, 229, 247
sub-THz 3, 63
ultra-wideband (UWB) 34, 35, 139, 141
radio access point (RAP) 40
radio-over-fiber (RoF) 7, 16, 97
analog (ARoF) 31, 54
beam steering 37–38, 40
digital (DRoF) 31, 54
fiber dispersion impact 34
intermediate frequency over fiber (IFoF) 31
picocell network 3, 31, 55
return path in hybrid OWC/RF network 100, 106, 230
ultra-wideband (UWB) 34
wavelength routing in network 45
ray tracing
modelling corner cube 156, 160, 166, 280–82
modelling Fresnel lens 274–82
modelling ideal thin lens 264–65
modelling reflection 10, 154, 159, 161, 268, 278
modelling refraction 10, 266, 274
ray launching 158, 164, **272**, 281
receiver, aperture of optical wireless receiver 75, 86–88, 125, 187–89, 211, 254

ceiling receiver 142, 173, 181, 235, 250
lensed ceiling receiver 237, 243
lensed user receiver 139, 171, 174, 183
user receiver 100–2, 131, 151, 238, 243
receiver, Field of View of optical wireless receiver 5, 8, 9, 51, 74, 187–89
impact of beam deformation 211
impact on receiver alignment 86, 234
increasing the Field of View 65, 188, 252
with a lens 105, 202, 205–8, 212
receiver, optical wireless
multiwavelength 254–58
shifting the receiver 211–20
tilt of the receiver 100, 103, 181, 237
receiver, optical wireless receiver circuitry
angular diversity receiver (ADR) 51, 75, 86, 188, 199–200, 227, 253
coherent receiver 253
common base transistor and TIA after photodiode matrix 221, 227, 253
matrix of photodiodes 189–201
transimpedance amplifier (TIA) after photodiode matrix 190, 192, 199, 201–2
transimpedance amplifiers in angular diversity receiver 86
residential gateway 16, 17, 27
retroreflector
optical 84, 154, 280. *See also* corner cube

s

security 6, 27, 52, 229
spatial light modulator (SLM) 96, 109, 110, 143
spatial multiplexing
by optical wireless cells 5, 50, 65, 77, 83, 96, 247
by radio cells 3, 63
in optical fiber 48
in VLC MIMO systems 71
mode group division multiplexing 16, 47
spectrum license 4, 63, 65
spot deformation 211
stepper motor. *See* beam steering, optical 2D
surface grating coupler (SGC) 75, 88, 188, 258

t

technology readiness level (TRL) 10, 251
time division multiple access (TDMA) protocol 235, 254
transceiver, SFP+ 139, 169
transimpedance amplifier. *See* receiver, optical wireless receiver circuitry
transmission control protocol (TCP) 138, 183, 230, 242, 250
true time delay (TTD). *See* radio beam steering
turbulence. *See* atmospheric effects

u

underwater communication. *See* optical wireless communication, underwater
USB dongle 71, 239, 242
user
mobile 86, 106, 183, 251
mobile with OWC device 153, 173
mobile with RF device 40, 75, 151
mobile with RF/OWC device 230
user interface rates 15
user mobility
nomadic 6, 52, 74, 183, 243, 248, 250
static user 6, 52, 248
support by VLC 247

V

video streaming. *See* Gigabit Ethernet
VLC, visible light communication. *See* optical wireless communication by visible light

W

wavelength conversion
  by cross-gain modulation in SOA 47
Wi-Fi 13, 15, 53, 229
  frequency bands 4
  line of sight 4
  localization 151
  optical wireless communication
    compared with Wi-Fi 7, **68**, 69
  standards 4, 32, **33**, 45, 67, 78, 82
wireless communication standards
  indoor 32, **33**

Publications of the Author

A non-exhaustive list is given of publications relevant to the material reported in this book of which Ton Koonen is the first author.

Journal Publications (First Author)

[1] A.M.J. Koonen, K.A. Mekonnen, F.M. Huijskens, E. Tangdiongga, "Bi-directional all-optical wireless Gigabit Ethernet communication system using automatic self-aligned beam steering," *IEEE/OPTICA J. Lightwave Technol.*, vol. 41, no. 11, pp. 3446–3454, Jun. 2023, https://doi.org/10.1109/JLT.2022.3231438.

[2] A.M.J. Koonen, K.A. Mekonnen, Z. Cao, F.M. Huijskens, N.-Q. Pham, E. Tangdiongga, "Beam-steered optical wireless communication for industry 4.0," *IEEE J. Sel. Top. Quantum Electron.*, vol. 27, no. 6, art. 6000510 (10 pp.), Nov./Dec. 2021, https://doi.org/10.1109/JSTQE.2021.3092837.

[3] A.M.J. Koonen, K.A. Mekonnen, F.M. Huijskens, N.-Q. Pham, Z. Cao, E. Tangdiongga, "Fully passive user localization for beam-steered high-capacity optical wireless communication system," *IEEE/OSA J. Lightwave Technol.*, vol. 38, no. 10, pp. 2842–2848, May 2020, https://doi.org/10.1109/JLT.2020.2980428.

[4] A.M.J. Koonen, K. A. Mekonnen, Z. Cao, F. M. Huijskens, N.-Q. Pham, E. Tangdiongga, "Ultra-high capacity wireless communication by means of steered narrow optical beams," *Philos. Trans. R. Soc. A*, 19 pp, Mar. 2020, https://doi.org/10.1098/rsta.2019.0192.

[5] A.M.J. Koonen, F. Gomez-Agis, F.M. Huijskens, K.A. Mekonnen, Z. Cao, E. Tangdiongga. "High-capacity optical wireless communication using two-dimensional IR beam steering," *IEEE/OSA J. Lightwave Technol.*, vol. 36, no. 19, pp. 4486–4493, Oct. 2018, https://doi.org/10.1109/JLT.2018.2834374.

Short-Reach Optical Wireless Communication: By Directed Narrow Beams, First Edition. Ton Koonen.

[6] A.M.J. Koonen, "Indoor optical wireless systems: Technology, trends, and applications," *IEEE/OSA J. Lightwave Technol.*, vol. 36, no. 8, pp. 1459–1467, Apr. 2018, https://doi.org/10.1109/JLT.2017.2787614 .

[7] A.M.J. Koonen, C.W. Oh, K.A. Mekonnen, Z. Cao, E. Tangdiongga, Ultra-high capacity indoor optical wireless communication using 2d-steered pencil beams," *IEEE/OSA J. Lightwave Technol.*, vol. 34, no. 20, pp. 4802– 4809, Oct. 2016, https://doi.org/10.1109/JLT.2016.2574855.

[8] A.M.J. Koonen, E. Tangdiongga, "Photonic home area networks," *IEEE/OSA J. Lightwave Technology*, vol. 32, no. 4, pp. 591–604, Feb. 2014, https://doi.org/10.1109/JLT.2013.2283145.

[9] A.M.J. Koonen, H.P.A. van den Boom, E. Ortego Martinez, A. Pizzinat, Ph. Guignard, B. Lannoo, C.M. Okonkwo, and E. Tangdiongga, "Cost optimization of optical in-building networks," *OSA Opt. Exp.*, vol. 19, no. 26, pp. B399–B405, Nov. 2011, https://doi.org/10.1364/OE.19.00B399.

[10] A.M.J. Koonen, M. Garcia Larrode, "Radio-over-MMF techniques – part II: microwave to millimeter-wave systems," *IEEE/OSA J. Lightwave Technol.*, vol. 26, no. 15, pp. 2396–2408, Aug. 2008, https://doi.org/10.1109/JLT.2008.927182.

Conference Publications (First Author)

[1] Ton Koonen, "High-capacity optical wireless communication by directed narrow beams," *Proc. 28th Optoelectronics and Communications Conference (OECC2023)*, Shanghai, Jul. 2–6, 2023, Keynote paper OECC2023-0319-30, 6 pp.

[2] Ton Koonen, K. Mekonnen, "Recent progress in high-capacity optical wireless communication," *Proc. 2022 Asia Communications and Photonics Conference (ACP 2022)*, Shenzhen, Nov. 5–8, 2022, Keynote paper, pp. 1819–1822.

[3] A.M.J. Koonen, K. A. Mekonnen, F.M. Huijskens, E. Tangdiongga, "Bi-directional all-optical wireless communication system with optical beam steering and automatic self-alignment," *Proc. ECOC 2022*, Basel, Sep. 18–22, 2022, Paper Tu4F.1.

[4] Ton Koonen, Z. Cao, "Photonic integrated circuits for optical wireless communication," *Proc. 2021 Asia Communications and Photonics Conference (ACP 2021)*, Shanghai, Oct. 24–27, 2021, Tutorial paper M5D.1, 3 pp.

[5] A.M.J. Koonen, K.A. Mekonnen, F.M. Huijskens, N.-Q. Pham, Z. Cao, E. Tangdiongga, "Optical wireless GbE receiver with large field-of-view," *Proc. ECOC 2021*, Bordeaux, Sep. 13–16, 2021, Paper Th1B.6.

[6] Ton Koonen, "Optical technologies to disclose the spatial diversity dimension in systems and networks," *Proc. ACP 2020*, Beijing, Oct. 24–27, 2020, Plenary talk, Paper M1A1.

[7] A.M.J. Koonen, K.A. Mekonnen, F.M. Huijskens, Z. Cao, E. Tangdiongga, "Novel broadband OWC receiver with large aperture and wide field of view", *Proc. ECOC 2020*, Brussels, Dec. 6–10, 2020, Paper Tu2G.6.

[8] Ton Koonen, K. Mekonnen, F. Huijskens, N.-Q. Pham, Z. Cao, E. Tangdiongga, "Recent advances in ultra-broadband optical wireless communication," *Proc. 2019 Asia Communications and Photonics Conference (ACP 2019)*, Chengdu, Nov. 2–5, 2019, Invited paper M3C.4, 3 pp.

[9] A.M.J. Koonen, K.A. Mekonnen, F.M. Huijskens, E. Tangdiongga, "Fully passive user localisation for beam-steered high-capacity optical wireless communication system," in *Proc. ECOC 2019*, Dublin, Sep. 22–26, 2019, paper Tu.3.C.4.

[10] Ton Koonen, J.-Y. Sung, Z. Cao, K. Mekonnen, F. Huijskens, E. Tangdiongga, "Recent advances in optical technologies for creating ultra-high capacity wireless indoor networks," *Proc. 2018 Asia Communications and Photonics Conference (ACP 2018)*, Hangzhou, Oct. 26–29, 2018, Invited paper S4E.2, 3 pp.

[11] A.M.J. Koonen, A.M. Khalid, C.W. Oh, F. Gomez-Agis, and E. Tangdiongga, "High-capacity optical wireless communication using 2-dimensional IR beam steering," *Proc. OECC 2017*, Singapore, Jul. 31–Aug. 4, 2017, Paper 1-4K-3.

[12] A.M.J. Koonen, F. Gomez-Agis, Z. Cao, K.A. Mekonnen, F.M. Huijskens, E. Tangdiongga, "Indoor ultra-high capacity optical wireless communication using steerable infrared beams," *Proc. MWP 2017*, Beijing, Oct. 23–26, 2017, Keynote paper.

[13] A.M.J. Koonen et al., "High-capacity optical wireless communication using AWG router for 2-dimensional IR beam steering," in *Proc. OECC 2017*, Singapore, Jul. 31–Aug. 4, 2017, Paper 1-4K-3.

[14] A.M.J. Koonen, C.W. Oh, A.M. Khalid, K.A. Mekonnen, M. Torres Vega, Z. Cao, and Eduward Tangdiongga, "2D beam-steered high-capacity optical wireless communication," *Proc. IEEE Summ. Top. 2016*, Newport Beach, Jul. 11–13, 2016, Paper TuC2.2.

[15] A.M.J. Koonen, C.W. Oh, K. Mekonnen, E. Tangdiongga, "Ultra-high capacity indoor optical wireless communication using steered pencil beams," in *Proc. MWP 2015*, Paphos, Oct. 25–29, 2015, Paper WeC-5.

[16] A.M.J. Koonen, C.W. Oh, E. Tangdiongga, "Reconfigurable free-space optical indoor network using multiple pencil beam steering," *Proc. OECC/ACOFT 2014*, Melbourne, Jul. 6–10, 2014, Paper Tu3F-1.

[17] Ton Koonen, "Home area networks," *Proc. OFC/NFOEC 2013*, Los Angeles, Feb. 22–27, 2013, Tutorial OTh1G.1, 86 slides.

[18] Ton Koonen, Y. Shi, S. Zou, C. Okonkwo, H. van den Boom, E. Tangdiongga, "Recent progress in photonic in-building networks," *Proc. OECC 2012*, Busan, Jul 2–6, 2012, Invited paper 6A3-1, 2 pp.

[19] A.M.J. Koonen, Y. Shi, N.C. Tran, C.M. Okonkwo, H.P.A. van den Boom, E. Tangdiongga, "A perspective from Europe on in-home networking," *Proc. OFC 2012*, Los Angeles, Mar. 4–8, 2012, Invited paper NTh1D.1, 3 pp.

[20] A.M.J. Koonen, N.C. Tran, and E. Tangdiongga, "The merits of reconfigurability in WDM-TDM optical in-building networks," *Proc. OFC/NFOEC 2011*, Los Angeles, Mar. 6–10, 2011, Paper JWA63.

[21] A.M.J. Koonen, H.P.A. van den Boom, I. Tafur Monroy, G.-D. Khoe, "High capacity multi-service in-house networks using mode group diversity multiplexing", *Proc. OFC/NFOEC 2004*, Anaheim, Feb. 22–27, 2004, Paper FE-H.

[22] A.M.J. Koonen, H. van den Boom, G.-D. Khoe, "Broadband access and in-house networks – extending the capabilities of multimode fibre networks", *Proc. ECOC 2003*, Rimini, Sep. 21–25, 2003, Paper Mo3.1.5.

Patents

[1] A.M.J. Koonen, "An optical wireless communication system comprising a transmitter and a receiver as well as a corresponding method of operating the optical wireless communication system and a corresponding transmitter and receiver for operating in the optical wireless communication system," Dutch patent application N2037871, filed Jun. 6, 2024.

[2] A.M.J. Koonen, "Bi-directional all-optical wireless communication system with autonomous optical beam steering," priority date Jun. 17, 2022, Dutch patent NL2032214B1, granted Jan. 5, 2024, application PCT/NL2023/050339, filed Jun. 16, 2023, WO202344115A1, published Dec. 21, 2023.

[3] A.M.J. Koonen, "An optical wireless communication receiver with large photodetector surface area, large field of view and high bandwidth," priority date Oct. 31, 2019, Patent PCT/EP2020/080594, published May 6, 2021, EP4052390A1 published Sep. 7, 2022, EP 4052390B1 granted Oct. 23, 2024.

[4] A.M.J. Koonen, "Two-dimensional optical beam steering module," priority date Nov. 29, 2016, Patent PCT/EP2017/080716, published June 7, 2018, US patent appl. US2021/0109286A1, published Apr. 15, 2021.

[5] Antonius Marcellus Jozef Koonen, Peter G.M. Baltus, Antonio Liotta, "Two-dimensional optical beam steering module," priority date Jan. 20, 2012, US patent 9246589 B2, WO patent 2013107853, filing date Jan. 18, 2013.

[6] A.M.J. Koonen, "Werkwijze en stelsel voor het via een multi-modus optische vezel overdragen van microgolfsignalen" (Method and system for transport of microwave signals via a multimode optical fibre), Dutch patent NL1019047, issued Apr. 2, 2003.

[7] A.M.J. Koonen, "Re-configurable fibre wireless network", US Patent no. 6,674,966, filed Oct. 12, 1999, issued Jan. 6, 2004.

Miscellaneous

[1] Ton Koonen, "Light work(s)," Valedictory lecture, Eindhoven University of Technology, Sep. 24, 2021. ISBN 978-90-386-5365-5, NUR 959. Available: https://research.tue.nl/en/publications/light-works

[2] A.M.J. Koonen, "Optical wireless systems: technology, trends and applications," webinar IEEE Phot. Soc., Feb. 21, 2018 [online]. Available: https://ieeephotonics.org/webinars/optical-wireless-communication/

Printed and bound by CPI Group (UK) Ltd, Croydon, CR0 4YY

19/09/2026

14965106-0001